SOLUTIONS MANUAL

for

COLLEGE CHEMISTRY

An Introduction to General, Organic, and Biochemistry

Fourth Edition

Morris Hein
Leo R. Best
Scott Pattison

Brooks/Cole Publishing Company
Pacific Grove, California

Brooks/Cole Publishing Company
A Division of Wadsworth, Inc.

Printed in the United States of America

10 9 8 7 6 5 4 3 2 1

QD 31.2.H43 1988 87-24633
540--dc19 CIP

ISBN 0-534-08564-4

Sponsoring Editor: Harvey Pantzis
Editorial Assistant: Heidi Wieland
Production Coordinator: Dorothy Bell
Cover Design: Vernon T. Boes

TO THE STUDENT

This solutions manual contains answers to all the end-of-chapter questions and to all the problems.

This book will be valuable to you if you use it properly. This book is not intended to be a substitute for answering the questions by yourself. It is important that you go through the process of writing down the answers to questions and the solutions to problems before you see them in the answer book. Once you have seen the answer, much of the value of whether you have learned the material or not is lost. This does not mean that you cannot learn from the answers. If you absolutely cannot answer a question, study the answer carefully to see where you are having difficulty.

You should spend sufficient time answering each assigned question and problem. Only after you have made a serious attempt to answer a question or solve a problem should you resort to the solutions manual. Check the answers to compare them with your own. If your answers are correct you will have immediate gratification that you have reasonable mastery of the material. If an answer is incorrect you need to reread and restudy the material.

Problems are an integral part of this chemistry course. In order to solve problems in chemistry you must be willing to work at it. Set up the problem in proper form using correct units. Use the sample problems in your text as a guide. Don't use short cuts. Although there may be more than one solution to a problem, only one is given in this book. So don't be alarmed if your solution does not match exactly with the one given in this manual as long as the results are the same.

As you use this manual you will see hundreds of answered questions and solved problems using thousands of words and numbers. We have endeavored to prepare as error free a manual as possible. But please bear with us that no one is perfect and that an occasional error may show up. We would appreciate your letting us know of any errors that you find. Good luck!

TABLE OF CONTENTS

I. Answers to Questions and Problems in the Text

Chapter 1 Introduction

1. No. The geocentric concept developed from incorrect interpretation of the observed apparent movement of the sun relative to the earth.

2. The principal goals of the alchemists were to find a method for prolonging human life and to change the base metals such as iron and copper into gold.

3. The first use of the chemical balance in chemistry is analogous to the first use of the telescope in astronomy. These instruments made it possible to make measurements and observations with great precision and thus put both sciences on a truly quantitative basis.

4. (a) Law
   (b) Observation
   (c) Hypothesis
   (d) Theory
   (e) Observation
   (f) Theory
   (g) Hypothesis

5. Numerous examples might be listed - WD-40, Preparation H, Crazy Glue, etc.

6. Correct statements are: a, b, d, f, g, h, and i.

Chapter 2 Standards for Measurement

B. 1. Kilometer = 1000 meters; 1 meter = 10 decimeters

$$1\ \text{km} \times \frac{1000\ \text{m}}{\text{km}} \times \frac{10\ \text{dm}}{\text{m}} = 10000\ \text{dm}$$

2. 25°C = 77°F   100°C = 212°F   212°F - 77°F = 135°F

3. Specific heat: Al = 0.900 J/g°C; Fe = 0.473 J/g°C Aluminum will absorb 0.900 J/g°C while iron absorbs 0.473 J/g°C. Therefore, when 10 J are added to 1 g each of these metals, iron will become warmer than aluminum.

4. The volumetric flask is a more precise measuring instrument than the graduated cylinder. This is because the narrow opening at the point of measurement (calibration mark) in the neck of the flask means that a small error made in not filling exactly to the mark will make a smaller percentage error in total volume than will a similar error with the cylinder.

5. The three materials would sort out according to their densities with the most dense (mercury) at the bottom and the least dense (carbon tetrachloride) at the top. In the cylinder the solid magnesium would sink in the carbon tetrachloride and float on the liquid mercury.

6. Order of increasing density: ethyl alcohol, cottonseed oil, sulfur, salt, aluminum.

7. An argon-filled balloon would sink in a methane atmosphere because the density of argon is greater than that of methane.

8. The density of ice must be less than 0.926 g/mL and greater than 0.789 g/mL.

C. 1. The principal advantages are: The conversion factors within the metric system are all multiples of 10 rather than the irregular units of the American system; e.g., 3; 12; 5,280; 16; 4; etc. The prefixes used to represent larger and smaller amounts are the same for all kinds of unit measurements in the metric system.

2. Abbreviations:

(a) Gram = g	(b) Kilogram = kg	(c) Milligram = mg
(d) Microgram = μg	(e) Centimeter = cm	(f) Millimeter = mm
(g) Micrometer = μm	(h) Angstrom = Å	(i) Milliliter = mL
(j) Microliter = μL	(k) Liter = L	

3. (a) Significant
(b) 0.007 Not significant
(c) 4200 May or may not be significant
(d) 3.0030 Significant
(e) 100.00 Significant
(f) 8.00×10^2 Significant

4. Rule 1. When the first digit beyond those you want to retain is 4 or less, that digit and all others beyond it are dropped. The last digit retained is not changed.

   Rule 2. When the first digit beyond those you want to retain is 5 or greater, that digit and all others beyond it are dropped and the last digit retained is increased by one.

5. Heat is a form of energy and as such can be added to or removed from a system. Temperature is a measure of the intensity of heat or how hot a system is. Temperature determines the direction of heat flow (from hot to cold).

6. The ethyl alcohol will become hotter than the water because it has the lower specific heat.

   Ethyl alcohol: $\Delta t = 100\ \text{J} \times \frac{\text{g}^\circ\text{C}}{2.138\ \text{J}} \times \frac{1}{25\ \text{g}} = 1.9^\circ\text{C}$

   Water: $\Delta t = 100\ \text{J} \times \frac{\text{g}^\circ\text{C}}{4.184\ \text{J}} \times \frac{1}{25\ \text{g}} = 0.96^\circ\text{C}$

7. Density is the ratio of the mass of a substance to the volume occupied by that mass. Density has the units of mass over volume. Specific gravity is the ratio of the density of a substance to the density of a reference substance (usually water at a specified temperature for solids and liquids). Specific gravity has no units.

8. The following statements are correct: a,c,d,e,g,h,i,j,l,n,o,r, and s.

D. 1. Number of significant figures (the significant figures are underlined)

   (a) $0.0\underline{2}\,\underline{5}$ (2) (b) $\underline{2}.\underline{0}\,\underline{0}$ (3) (c) $\underline{2}\,\underline{2}.\underline{4}$ (3) (d) $0.0\,0\,\underline{8}\,\underline{1}$ (2)

   (e) $0.0\,\underline{4}\,\underline{0}\,\underline{4}$ (3) (f) $\underline{1}\,\underline{2}\,\underline{9}.\underline{0}\,\underline{4}\,\underline{2}$ (6) (g) $\underline{5}.\underline{5}\,\underline{0} \times 10^3$ (3)

   (h) $\underline{1}.\underline{9}\,\underline{9}\,\underline{5} \times 10^6$ (4)

2. Rounding off to three significant figures

   (a) 93.246 = 93.2 (b) 8.8726 = 8.87

   (c) 0.02854 = 0.0285 (d) 21.25 = 21.3

   (e) 4.644 = 4.64 (f) 129.509 = 130

   (g) 34.250 = 34.3 (h) $1.995 \times 10^6 = 2.00 \times 10^6$

3. Exponential notation

   (a) $2{,}900{,}000 = 2.9 \times 10^6$ (b) $0.0456 = 4.56 \times 10^{-2}$

   (c) $0.58 = 5.8 \times 10^{-1}$ (d) $4082.2 = 4.0822 \times 10^3$

   (e) $0.00840 = 8.40 \times 10^{-3}$ (f) $40.30 = 4.030 \times 10^1$

(g) $12{,}000{,}000 = 1.2 \times 10^{7}$ (h) $0.0000055 = 5.5 \times 10^{-6}$

4. Solving mathematical problems

(a)
$$\begin{array}{r} 12.62 \\ 1.5 \\ 0.25 \\ \hline 14.37 \end{array} = 14.4$$

(b) $4.68 \times 12.5 = 58.5$

(c) $2.25 \times 10^{3} \times 4.80 \times 10^{4} = 1.08 \times 10^{8}$

(d) $\frac{182.6}{4.6} = 40.$ (e) $\frac{452 \times 6.2}{14.3} = 2.0 \times 10^{2}$

(f)
$$\begin{array}{r} 1986 \\ 23.48 \\ 0.012 \\ \hline 2009.492 \end{array} = 2009$$

(g) $0.0394 \times 12.8 = 0.504$

(h) $2.92 \times 10^{-3} \times 6.14 \times 10^{5} = 1.79 \times 10^{3}$

(i) $\frac{0.4278}{59.6} = 7.18 \times 10^{-3}$ (j) $\frac{29.3}{284 \times 415} = 2.49 \times 10^{-4}$

5. Fractions to decimals (three significant figures)

(a) $\frac{5}{6} = 0.833$ (b) $\frac{3}{7} = 0.429$

(c) $\frac{12}{16} = 0.750$ (d) $\frac{9}{18} = 0.500$

6. Solving for X.

(a) $3.42X = 6.5$

$$\frac{3.42X}{3.42} = \frac{6.5}{3.42}$$

$$X = \frac{6.5}{3.42} = 1.9$$

(b) $\frac{X}{12.3} = 7.05$

$$X = 7.05 \times 12.3 = 86.7$$

(c) $\frac{0.525}{X} = 0.25$

$$0.525 = 0.25X$$

$$X = \frac{0.525}{0.25} = 2.1$$

(d) $0.298X = 15.3$

$$X = \frac{15.3}{0.298} = 51.3$$

(e) $\frac{X}{0.819} = 10.9$

$$X = 10.9 \times 0.819 = 8.93$$

(f) $\frac{8.4}{X} = 282$

$$8.4 = 282X$$

$$X = \frac{8.4}{282} = 0.030$$

7. Solving for an unknown.

(a) $°C = \frac{212 - 32}{1.8} = 100$

(b) $°F = 1.8(22) + 32 = 72$

(c) $K = 25 + 293 = 318$

(d) $\frac{8.9\ g}{mL} = \frac{40.90\ g}{V}$

$V = \frac{40.90\ g}{8.9\ g/mL} = 4.6\ mL$

8. Unit conversions.

(a) $28.0\ cm \times \frac{1\ m}{100\ cm} = 0.280\ m$

(b) $1000.\ m \times \frac{1\ km}{1000\ m} = 1.000\ km$

(c) $9.28\ cm \times \frac{10\ mm}{1\ cm} = 92.8\ mm$

(d) $150\ mm \times \frac{1\ m}{1000\ mm} \times \frac{1\ km}{1000\ m} = 1.50 \times 10^{-4}\ km$

(e) $0.606\ cm \times \frac{1\ m}{100\ cm} \times \frac{1\ km}{1000\ m} = 6.06 \times 10^{-6}\ km$

(f) $4.5\ cm \times \frac{1\ m}{100\ cm} \times \frac{1\ Å}{10^{-10}\ m} = 4.5 \times 10^{8}\ Å$

(g) $6.5 \times 10^{-7}\ m \times \frac{10^{10}\ Å}{1\ m} = 6.5 \times 10^{3}\ Å$

(h) $12.1\ m \times \frac{100\ cm}{1\ cm} = 1.21 \times 10^{3}\ cm$

(i) $8.0\ km \times \frac{1000\ m}{1\ km} = 8.0 \times 10^{3}\ m$

(j) $315\ mm \times \frac{1\ cm}{10\ mm} = 31.5\ cm$

(k) $25\ km \times \frac{1000\ m}{km} \times \frac{1000\ mm}{m} = 2.5 \times 10^{7}\ mm$

(l) $1\ nm = 10^{-9}\ m$; $1\ nm = 10^{-7}\ cm$; $12\ nm \times \frac{10^{-7}\ cm}{nm} = 1.2 \times 10^{-6}\ cm$

(m) $0.520\ km \times \frac{1000\ m}{km} \times \frac{100\ cm}{m} = 5.20 \times 10^{4}\ cm$

(n) $3.884\ Å \times \frac{10^{-10}\ m}{Å} \times \frac{1\ nm}{10^{-9}\ m} = 0.3884\ nm$

(o) $42.2\ in. \times \frac{2.54\ cm}{in.} = 107\ cm$

(p) $0.64\ mile \times \frac{5280\ ft}{mile} \times \frac{12\ in.}{ft} = 4.1 \times 10^{4}\ in.$

(q) $504\ miles \times \frac{1.61\ km}{mile} = 811\ km$

(r) $2.00 \text{ inches}^2 \times \left(\frac{2.54 \text{ cm}}{\text{inch}}\right)^2 = 12.9 \text{ cm}^2$

(s) $35.6 \text{ m} \times \frac{100 \text{ cm}}{\text{m}} \times \frac{1 \text{ in.}}{2.54 \text{ cm}} \times \frac{1 \text{ ft}}{12 \text{ in.}} = 117 \text{ ft.}$

(t) $16.5 \text{ km} \times \frac{1 \text{ mile}}{1.61 \text{ km}} = 10.2 \text{ miles}$

(u) $4.5 \text{ in.}^3 \times \left(\frac{2.54 \text{ cm}}{\text{in.}}\right)^3 \times \left(\frac{10 \text{ mm}}{\text{cm}}\right)^3 = 7.4 \times 10^4 \text{ mm}^3$

(v) $3.00 \text{ mile}^3 \times \left(\frac{1.61 \text{ km}}{\text{mile}}\right)^3 \times \left(\frac{1000 \text{ m}}{1 \text{ km}}\right)^3 \times \left(\frac{1000 \text{ mm}}{\text{m}}\right)^3$

$= 1.25 \times 10^{19} \text{ mm}^3$

9. Unit conversions.

(a) $10.68 \text{ g} \times \frac{1000 \text{ mg}}{\text{g}} = 1.068 \times 10^4 \text{ mg}$

(b) $6.8 \times 10^4 \text{ mg} \times \frac{1 \text{ g}}{1000 \text{ g}} \times \frac{1 \text{ kg}}{1000 \text{ g}} = 6.8 \times 10^{-2} \text{ kg}$

(c) $8.54 \text{ g} \times \frac{1 \text{ kg}}{1000 \text{ g}} = 8.54 \times 10^{-3} \text{ kg}$

(d) $42.8 \text{ kg} \times \frac{1000 \text{ g}}{\text{kg}} \times \frac{1 \text{ lb}}{454 \text{ g}} = 94.3 \text{ lbs}$

(e) $164 \text{ mg} \times \frac{1 \text{ g}}{1000 \text{ mg}} = 0.164 \text{ g}$

(f) $0.65 \text{ kg} \times \frac{1000 \text{ g}}{\text{kg}} \times \frac{1000 \text{ mg}}{\text{g}} = 6.5 \times 10^5 \text{ mg}$

(g) $5.5 \text{ kg} \times \frac{1000 \text{ g}}{\text{kg}} = 5.5 \times 10^3 \text{ g}$

(h) $95 \text{ lb} \times \frac{454 \text{ g}}{\text{lb}} = 4.3 \times 10^4 \text{ g}$

10. Unit conversions.

(a) $25.0 \text{ mL} \times \frac{1 \text{ L}}{1000 \text{ mL}} = 2.50 \times 10^{-2} \text{ L}$

(b) $22.4 \text{ L} \times \frac{1000 \text{ mL}}{\text{L}} = 2.24 \times 10^4 \text{ mL}$

(c) $3.5 \text{ qt} \times \frac{946 \text{ mL}}{\text{qt}} = 3.3 \times 10^3 \text{ mL}$

(d) $4.5 \times 10^4 \text{ ft}^3 \times \left(\frac{12 \text{ in.}}{\text{ft}}\right)^3 \times \left(\frac{2.54 \text{ cm}}{\text{in.}}\right)^3 \times \left(\frac{1 \text{ m}}{100 \text{ cm}}\right)^3$

$= 1.3 \times 10^3 \text{ m}^3$

(e) $0.468 \text{ L} \times \frac{1000 \text{ mL}}{\text{L}} = 468 \text{ mL}$

(f) $35.6\ \text{L} \times \frac{1\ \text{qt}}{0.946\ \text{L}} \times \frac{1\ \text{gal}}{4\ \text{qt}} = 9.41\ \text{gal}$

(g) $9.0\ \mu\text{L} \times \frac{1\ \text{L}}{10^6\ \mu\text{L}} \times \frac{1000\ \text{mL}}{\text{L}} = 9.0 \times 10^{-3}\ \text{mL}$

(h) $20.0\ \text{gal} \times \frac{4\ \text{qt}}{\text{gal}} \times \frac{0.946\ \text{L}}{\text{qt}} = 75.7\ \text{L}$

11. (a) $\frac{55\ \text{mi}}{\text{hr}} \times \frac{1.61\ \text{km}}{\text{mi}} = 89\ \text{km/hr}$

(b) $\frac{55\ \text{mi}}{\text{hr}} \times \frac{5280\ \text{ft}}{\text{mi}} \times \frac{1\ \text{hr}}{60\ \text{min}} \times \frac{1\ \text{min}}{60\ \text{sec}} = 81\ \text{ft/sec}$

12. (a) $\frac{100\ \text{m}}{9.82\ \text{s}} \times \frac{100\ \text{cm}}{\text{m}} \times \frac{1\ \text{in.}}{2.54\ \text{cm}} \times \frac{1\ \text{ft}}{12\ \text{in.}} = 33.4\ \text{ft/s}$

(b) $\frac{100\ \text{m}}{9.82\ \text{s}} \times \frac{1\ \text{km}}{1000\ \text{m}} \times \frac{1\ \text{mile}}{1.61\ \text{km}} \times \frac{60\ \text{s}}{\text{min.}} \times \frac{60\ \text{min.}}{\text{hr}} = 22.8\ \text{miles/hr}$

13. 1.00 lb = 454 g salt initially

$\frac{6.55\ \text{g salt}}{\text{student}} \times 24\ \text{students} = 157\ \text{g salt used}$

454 g - 157 g = 297 g salt remaining

14. (a) $\frac{13\ \text{mi}}{\text{s}} \times \frac{60\ \text{s}}{\text{min}} \times \frac{60\ \text{min}}{\text{hr}} = 4.7 \times 10^4\ \text{mi/hr}$

(b) $\frac{13\ \text{mi}}{\text{s}} \times \frac{60\ \text{s}}{\text{min}} \times \frac{60\ \text{min}}{\text{hr}} \times \frac{1.61\ \text{km}}{\text{mi}} = 7.5 \times 10^4\ \text{km/hr}$

15. $170\ \text{lb} \times \frac{454\ \text{g}}{\text{lb}} \times \frac{1\ \text{kg}}{1000\ \text{g}} = 77.2\ \text{kg}$

16. $5.0\ \text{grains} \times \frac{1\ \text{lb}}{7000\ \text{grains}} \times \frac{454\ \text{g}}{1\ \text{lb}} = 0.32\ \text{g}$

17. $93 \times 10^6\ \text{miles} \times \frac{1.61\ \text{km}}{\text{mile}} \times \frac{1000\ \text{m}}{\text{km}} \times \frac{100\ \text{cm}}{\text{m}} \times \frac{1\ \text{sec}}{3.00 \times 10^{10}\ \text{cm}}$

$= 5.0 \times 10^2\ \text{sec}$

18. $1\ \text{oz} \times \frac{1\ \text{lb}}{16\ \text{oz}} \times \frac{454\ \text{g}}{\text{lb}} \times \frac{1000\ \text{mg}}{\text{g}} = 3 \times 10^4\ \text{mg}$

19. $\frac{\$1.78}{10\ \text{lb}} \times \frac{1\ \text{lb}}{454\ \text{g}} \times \frac{1000\ \text{g}}{\text{kg}} = \$0.392/\text{kg}$

20. $227\ \text{g Au} \times \frac{1\ \text{lb}}{454\ \text{g}} \times \frac{14.58\ \text{oz}}{\text{lb}} \times \frac{\$345}{\text{oz}} = \$2.52 \times 10^3$

21. 1 condor weighs 21 lb; 1 hummingbird weighs 3.2 g

$$\frac{21\ \text{lb}}{3.2\ \text{g}} = \frac{21\ \text{lb} \times \frac{454\ \text{g}}{1\ \text{lb}}}{3.2\ \text{g}} = \frac{9534\ \text{g}}{3.2\ \text{g}} = 3.0 \times 10^3 \text{ times heavier}$$

22. $15.8\ \text{gal} \times \frac{4\ \text{qts}}{\text{gal}} \times \frac{0.946\ \text{L}}{\text{qt}} \times \frac{\$0.35}{\text{L}} = \$21$

23. $500\ \text{miles} \times \frac{1\ \text{gal}}{34\ \text{miles}} \times \frac{4\ \text{qt}}{\text{gal}} \times \frac{0.946\ \text{L}}{\text{qt}} = 56\ \text{L}$

24. $75\ \text{cm} \times 55\ \text{cm} \times 55\ \text{cm} \times \frac{1\ \text{L}}{1000\ \text{cm}^3} = 2.3 \times 10^2\ \text{L}$

25. $1.0\ \text{gal} \times \frac{4\ \text{qt}}{\text{gal}} \times \frac{946\ \text{mL}}{\text{qt}} \times \frac{20\ \text{drops}}{1.0\ \text{mL}} = 7.6 \times 10^4\ \text{drops}$

26. $42\ \text{gal} \times \frac{4\ \text{qt}}{\text{gal}} \times \frac{0.946\ \text{L}}{\text{qt}} = 1.6 \times 10^2\ \text{L}$

27. $5.0\ \mu\text{L} \times \frac{1\ \text{L}}{10^6\ \mu\text{L}} \times \frac{1000\ \text{mL}}{\text{L}} = 5.0 \times 10^{-3}\ \text{mL}$

28. $1.00\ \text{ft}^3 \times \left(\frac{12\ \text{in.}}{\text{ft}}\right)^3 \times \left(\frac{2.54\ \text{cm}}{\text{in.}}\right)^3 \times \frac{1\ \text{mL}}{\text{cm}^3} = 2.83 \times 10^4\ \text{mL}$

29. $\frac{200\ \text{cm}^3}{5\ \text{Å}} \times \frac{1\ \text{Å}}{10^{-8}\ \text{cm}} \times \left(\frac{1\ \text{m}}{100\ \text{cm}}\right)^2 = 4 \times 10^5\ \text{m}^2$

30. (a) $(162°\text{F} - 32°\text{F}) \times \frac{1°\text{C}}{1.8°\text{F}} = 72.2°\text{C}$

(b) $(0.0°\text{F} - 32°\text{F}) \times \frac{1°\text{C}}{1.8°\text{F}} = -18°\text{C}$

(c) $\text{K} = °\text{C} + 273 = -18 + 273 = 255\ \text{K}$

(d) $(-18°\text{C}) \times \frac{1.8°\text{F}}{1°\text{C}} + 32°\text{F} = -0.40°\text{F}$

(e) $(32°\text{C}) \times \frac{1.8°\text{F}}{1°\text{C}} + 32°\text{F} = 90°\text{F}$

(f) $(-8.6°\text{F} - 32°\text{F}) \times \frac{1°\text{C}}{1.8°\text{F}} = -23°\text{C}$

(g) $273°\text{C} + 273 = 546\ \text{K}$

(h) $212\ \text{K} - 273°\text{C} = -\ 61°\text{C}$

31. $37.0°\text{C} \times \frac{1.8°\text{F}}{1°\text{C}} + 32°\text{F} = 98.6°\text{F}$

32. $(-100°C) \times \frac{1.8°F}{1°C} + 32°F = -148°F$; $-100°C$ is $10°F$ colder than $-138°F$.

33. (a) At what temperature are Fahrenheit and Celsius scales equal?

$°F = °C$; $°F = (1.8 \times °F) + 32$; $°F - 1.8°F = 32$; $-0.8°F = 32$

$°F = \frac{32}{-0.8} = -40$; The two are identical at $-40°$.

(b) At what temperature are Fahrenheit and Celsius scales numerically equal but opposite in sign?

$°F = -°C$; $-°C = (1.8 \times °C) + 32$; $-2.8°C = 32$;

$°C = \frac{32}{-2.8} = -11.4°$

When the temperature is $-11.4°C$, the two scales are numerically equal but opposite in sign.

34. g × specific heat × Δt = calories; $\Delta t = 70°C - 20°C = 50°C$

$80\ g \times \frac{4.184\ J}{g\ °C} \times 50°C = 1.7 \times 10^4\ J$

35. $80\ g \times \frac{0.473\ J}{g\ °C} \times 50°C = 1.9 \times 10^3\ J$

$80\ g \times \frac{0.113\ cal}{g\ °C} \times 50°C = 4.5 \times 10^2\ cal$

36. Specific heat $= \frac{J}{g \times \Delta t}$; $\Delta t = 100°C - 22°C = 78°C$

Specific heat $= \frac{5.866\ kJ}{250\ g \times 78°C} \times \frac{1000\ J}{kJ} = 0.30\ J/g°C$

37. Heat lost by metal = heat gained by water

Change in temperature of 100 g of water = 4.0°C

Heat gained by water $= 100.0\ g \times \frac{4.184\ J}{g\ °C} \times 4.0°C = 1.7 \times 10^3\ J$

Therefore, heat gained by metal $= 1.7 \times 10^3\ J$

Specific heat of metal $= \frac{J}{g \times \Delta t} = \frac{1.7 \times 10^3\ J}{20.0\ g \times 174°C} = \frac{0.49\ J}{g\ °C}$

38. Heat lost by water = heat gained by water; final temperature = T;

Δt for hot water = $50°C - T$; Δt for cold water = $T - 10°C$;

$10\ g \times \frac{4.184\ J}{g\ °C} \times (50°C - T) = 50\ g \times \frac{4.184\ J}{g\ °C} \times (T - 10°C)$

$500 - 10T = 50T - 500$; $60T = 1000$; $T = 17°C$ (answer)

39. Heat lost by gold = heat gained by water; final temperature = T;

Δt for water = T - 22.0°C; Δt for gold = 427°C - T;

200 mL water = 200 g water

$$325 \text{ g} \times \frac{0.131 \text{ J}}{\text{g °C}} \times (427\text{°C} - T) = 200 \text{ g} \times \frac{4.184 \text{ J}}{\text{g °C}} \times (T - 22.0\text{°C})$$

$$1.82 \times 10^4 - 42.6T = 837T - 1.84 \times 10^4$$

$879.6T = 3.66 \times 10^4$; $T = 41.6$°C (answer)

40. Density = $\frac{\text{grams}}{\text{mL}} = \frac{78.26 \text{ g}}{50.00 \text{ mL}} = 1.565$ g/mL

41. Density = $\frac{39.9 \text{ g}}{12.8 \text{ mL}} = 3.12$ g/mL

42. 29.6 mL - 25.0 mL = 4.6 mL (volume of metal); $d = \frac{32.7 \text{ g}}{4.6 \text{ mL}} = 7.1$ g/mL

43. $500 \text{ mL} \times \frac{1.84 \text{ g}}{1 \text{ mL}} = 920$ g

44. 106.773 g - 42.817 g = 63.956 g (weight of liquid)

$\frac{63.956 \text{ g}}{50.0 \text{ mL}} = 1.28$ g/mL

45. $25.0 \text{ mL} \times \frac{13.6 \text{ g}}{\text{mL}} = 340$ g

46. $35.0 \text{ mL} \times \frac{0.789 \text{ g}}{\text{mL}} + 49.28 \text{ g} = 76.9$ g

47. Density = $\frac{\text{mass}}{\text{volume}}$

The cube with the largest volume will have the lowest density.

Cube A, lowest density, 1.74 g/mL - Magnesium

Cube B, middle density, 2.70 g/mL - Aluminum

Cube C, highest density, 10.5 g/mL - Silver

48. Al: $V = \frac{\text{mass}}{\text{density}} = 500 \text{ g} \times \frac{1 \text{ mL}}{2.70 \text{ g}} = 185$ mL

Au: mass = volume × density = $185 \text{ mL} \times \frac{19.3 \text{ g}}{\text{mL}} = 3.57 \times 10^3$ g

49. (a) d(water at 80°C) = $\frac{24.30 \text{ g}}{25.0 \text{ mL}} = 0.972$ g/mL

(b) $\frac{\text{volume}}{\text{gram}} = \frac{1.00 \text{ mL}}{0.972 \text{ g}} = 1.03$ mL/g

50. (a) Density = $\frac{g}{mL} = \frac{136\ g}{50.0\ mL} = 2.72\ g/mL$

(b) Specific gravity = $\frac{\text{density of solid}}{\text{density of water}} = \frac{2.72\ g/mL}{1.00\ g/mL}$
= 2.72 (unitless)

51. 150.50 g - 88.25 g = 62.25 g (grams of liquid)

$62.25\ g \times \frac{1\ mL}{1.25\ g} = 49.8\ mL$ (50 mL, volume of container)

52. $50\ g\ H_2O \times \frac{1.00\ mL}{1.00\ g} = 50\ mL$ water

50 g ethyl alcohol $\times \frac{1.00\ mL}{0.789\ g} = 63\ mL$ ethyl alcohol

Greater volume, ethyl alcohol

53. $2.00\ cm \times 15.0\ cm \times 6.00\ cm = 180\ cm^3$ (volume gold)

$\frac{3300\ g}{180\ cm^3} = 18.3\ g/cm^3$ (density of bar)

The bar is either not pure gold (density 19.3 g/cm^3) or is somewhat hollow inside.

54. $93.3\ kg \times \frac{1000\ g}{kg} \times \frac{1\ cm^3}{19.3\ g} = 4.83 \times 10^3\ cm^3$

55. Forgetful Freddie's dilemma.

Weight of cylinder: 89.450 g

Weight of slug: 15.434 g

Volume of slug: 30.7 mL - 25.0 mL = 5.7 mL

Volume of liquid: 25.0 mL

Weight of liquid: 125.934 g - 89.450 g -15.434 g = 21.050 g

Density of liquid = $\frac{21.050\ g}{25.0\ mL} = 0.842\ g/mL$

Density of slug = $\frac{15.434\ g}{5.7\ mL} = 2.7\ g/mL$

56. (a) 143 g acid + 500 g water = 643 g solution

$\frac{643\ g}{554\ mL} = 1.16\ g/mL$ = density. Hydrometers measure specific gravity (unitless); hence, the hydrometer will read 1.16.

(b) 143 g acid $\times \frac{1\ mL}{1.84\ g} = 77.7\ mL$ sulfuric acid

57. (a) Assume that the mass of 500 mL of water is 500 g. Then the total mass of the solution is 1000 g.

$$d(\text{solution}) = \frac{1000\ g}{717\ mL} = 1.39\ g/mL$$

Hydrometers measure specific gravity.

$$\text{Specific gravity} = \frac{d(\text{solution})}{d(\text{water})} = \frac{1.39\ g/mL}{1.00\ g/mL} = 1.39$$

Therefore the hydrometer will read 1.39.

(b) Volume of sulfuric acid added

$$d = \frac{g}{mL}; \quad mL = \frac{g}{d} = \frac{500\ g}{1.84\ g/mL} = 272\ mL\ \text{sulfuric acid}$$

58. $d(\text{gasoline}) = \dfrac{144\ g}{200\ mL} = 0.720\ g/mL$

$d(\text{water}) = 1.00\ g/mL$; sp gr (water) = 1.00

$$\text{sp gr (gasoline)} = \frac{d(\text{gasoline})}{d(\text{water})} = \frac{0.720\ g/mL}{1.00\ g/mL} = 0.720$$

Therefore, the specific gravity of gasoline is less than that of water.

B. 1. The water level is lower in the test tube than in the beaker. The trapped air occupying space in the tube prevents the water from rising higher.

2. Of the liquids listed in Table 3.1 alcohol, mercury and water are not mixtures.

3. Air is a mixture of gases and is the only gas in Table 3.1 that is not a pure substance.

4. Acetic acid is a solid at 10°C because its freezing point is 16.7°C.

5. Chlorine is a solid at 102 K because 102 K = −171°C and the freezing point of chlorine is −101.6°C.

6. The appearance of droplets of mercury in the upper part of the tube (Figure 3.7) is visible evidence of chemical change.

7. When the bulb is flashed, the magnesium metal reacts with oxygen gas to form the white solid, magnesium oxide (a new substance).

8. In the electrolysis of water the liquid is converted to hydrogen gas and oxygen gas.

C. 1. There are many examples of substances in the three physical states. Solid state: ice, aluminum, sugar, salt, etc. Liquid state: water, ethyl alcohol, mercury, benzene, etc. Gaseous state: oxygen, carbon dioxide, methane, nitrogen, etc.

2. (a) The attractive forces among the ultimate particles of a solid (atoms, ions, or molecules) are strong enough to hold these particles in fixed positions within the solid and thus maintain the solid in a definite shape. The attractive forces among the ultimate particles in a liquid (usually molecules) are sufficiently strong to hold them together (and prevent the liquid from rapidly becoming a gas) but are not strong enough to hold the particles in fixed positions (as in a solid).

(b) The ultimate particles in a liquid are quite closely packed (essentially in contact with each other) and thus the volume of the liquid is fixed. But the ultimate particles in a gas are relatively far apart and essentially independent of each other. Consequently the gas does not have a definite volume.

(c) In a gas the particles are relatively far apart and are easily compressed, but in a solid the particles are closely packed together and are thus virtually incompressible.

3. Three. At 16.7°C both solid and liquid acetic acid can be present and represents two phases. The presence of the odor of the acetic acid indicates gaseous molecules are present; hence, a third phase, gas, is also present.

4. Heterogeneous. Matter consisting, as in this example, of two or more physically distinct phases is heterogeneous.

5. No. A system is heterogeneous if two or more phases exist. Thus ice floating in water is a heterogeneous system even though only one substance (water) is present.

6. No. A system is homogeneous when only one phase is present. Thus a system containing sugar dissolved in water is homogeneous even though two substances are present.

7. Physical properties are the inherent characteristics of a substance that may be determined without altering the composition of that substance; they are associated with its physical existence. Chemical properties describe the ability of a substance to form new substances, either by reaction with other substances or by decomposition.

8. New substances are always formed in chemical changes but are never formed in purely physical changes.

9. Primarily physical: a, b, c, and e
Primarily chemical: d, and f

10. Primarily physical: b, c, and f
Primarily chemical: a, d, and e

11. Although the appearance of the platinum wire changed during the heating, the original appearance was restored when the wire cooled, and no change in the composition of the platinum could be detected.

12. The copper wire, like the platinum wire, changed to glowing red color when heated (physical change). upon cooling the original appearance of the copper was not restored, but a new substance, black copper(II) oxide, had appeared in its stead (chemical change).

13. When mercury(II) oxide is heated, the red powder changes color and produces liquid silvery globules of mercury and oxygen gas. Evidence of the oxygen is observed when the gas given off during the heating is observed to enhance the burning of a wood splint.

14. (a) Reactants: Copper and oxygen; Product: Copper(II) oxide
(b) Reactant: Mercury(II) oxide; Products: Mercury and oxygen.

15. In a chemical change, any change of mass that occurs is so small as to be undetectable in an experiment, though it can be calculated from the energy change. Since it is undetectable, we say that mass is neither lost nor gained.

16. Potential energy is energy of position. By the position of an object it has the potential of movement to a lower energy state. Kinetic energy is the energy matter possesses due to its motion.

17. The kinetic energy is converted to thermal energy (heat), chiefly in the brake system, and eventually dissipated in the atmosphere.

18. The transformation of kinetic energy to thermal energy (heat) is responsible for the fiery reentry of a space vehicle.

19. The energy equivalent of mass as given by Einstein's expression, energy = mass × speed of light squared, shows that a huge amount of energy is released when any measurable amount of matter is converted to energy. The mass equivalent of the energy released by a bulb flashing is too small to be detected by ordinary laboratory methods.

20. The following statements are correct: c,f,g,j,l,n,o,q, and t.

D. 1. (a) $118.0°C + 273 = 391\ K$ (b) $118.0°C \times \frac{1.8°F}{1°C} + 32°F = 244.4°F$

2. 15.0 g iron + 16.0 g sulfur + 18.5 g sand = 49.5 g total

$$\frac{15.0\ \text{g iron}}{49.5\ \text{g}} \times 100\% = 30.3\%\ \text{iron}$$

3. $$2.80\ \text{g copper} \times \frac{1.251\ \text{g copper(II) oxide}}{1.00\ \text{g copper}} = 3.50\ \text{g copper(II) oxide}$$

4. In section 3.6 we saw that 1.00 g copper combines with 0.251 g oxygen

$$\frac{1.00\ \text{g copper}}{0.251\ \text{g oxygen}} \times 5.50\ \text{g oxygen} = 21.9\ \text{g copper}$$

5. According to Section 3.7, 100 g mercury(II) oxide contains 92.6 g mercury.

$$85.0\ \text{g mercury(II) oxide} \times \frac{92.6\ \text{g mercury}}{100\ \text{g mercury(II) oxide}} = 78.7\ \text{g mercury}$$

6. $$\frac{6.40\ \text{g}}{60.0\ \text{g}} \times 100\% = 10.7\%\ \text{fat}$$

7. (a) 21.5 g magnesium oxide - 13.0 g magnesium = 8.5 g oxygen

(b) $$\frac{8.5\ \text{g oxygen}}{21.5\ \text{g magnesium oxide}} \times 100\% = 40\%\ \text{oxygen}$$

8. 1.00 g Al yields 4.94 g aluminum chloride ($AlCl_3$)

4.94 g $AlCl_3$ - 1.00 g Al = 3.94 g chlorine combined with 1.00 g Al

$$\frac{3.94\ \text{g chlorine}}{1.00\ \text{g aluminum}} \times 5.50\ \text{g aluminum} = 21.7\ \text{g chlorine}$$

21.7 g chlorine will combine with 5.50 g aluminum.

9. 1 ug mass = 9.2×10^{7} J = 1×10^{-6} g mass

(a) $5.5 \text{ g} \times \dfrac{9.2 \times 10^{7} \text{ J}}{1 \times 10^{-6} \text{ g}} = 5.1 \times 10^{14} \text{ J}$

$5.1 \times 10^{14} \text{ J} \times \dfrac{1. \text{ cal}}{4.184 \text{ J}} = 1.2 \times 10^{14} \text{ cal}$

(b) $5.1 \times 10^{14} \text{ J} \times \dfrac{1 \text{ gal.}}{1.27 \times 10^{6} \text{ J}} = 4.0 \times 10^{8} \text{ gal.}$

B. 1. Seven metals: Al, Fe, Ca, Na, K, Mg, Ti; Two nonmetals: O, H; One metalloid: Si

2. One metal: Ca; Five nonmetals: O, C, H, N, P; No metalloids

3. Au was derived from aurum, the Latin name for gold.

4. (a) The solid diatomic nonmetal is iodine.
(b) The liquid diatomic nonmetal is bromine.

5. When in binary compounds with metals, the name of iodine is modified to iodide and that of bromine, to bromide.

6. Silicon, 25.67% by weight; hydrogen, 0.87% by weight.

$$\frac{\text{weight of all silicon atoms}}{\text{weight of all hydrogen atoms}} = \frac{25.67}{0.87} = 30$$

Since each silicon atom weighs 28 times the weight of each hydrogen atom and the total weight of silicon is 30 times that of hydrogen, there are more atoms of silicon.

7. There are 83 metals, 18 nonmetals, and 7 metalloids, on the basis of 108 elements.

C. Review questions.

1. The symbol stands for the element itself, for one atom of the element, and may also represent a particular quantity of the element, the mole.

2. See Table 4.3.

3. (a) Si represents a silicon atom while SI would represent one sulfur atom (S) combined with one iodine atom (I).

(b) Pb represents a lead atom while PB represents one atom of phosphorus (P) combined with one atom of boron (B).

(c) 4 P represents four individual phosphorus atoms, whereas P_4 represents a molecule of phosphorus which consists of four phosphorus atoms.

4. Sodium, Na: Potassium, K; Iron, Fe; Silver, Ag; Tin, Sn; Antimony, Sb; Gold, Au; Mercury, Hg; Lead, Pb; etc.

5. Hydrogen, H; Boron, B; Carbon, C; Nitrogen, N; Oxygen, O; Fluorine, F; Phosphorus, P; Sulfur, S; Potassium, K; Vanadium, V; Yttrium, Y; Iodine, I; Tungsten, W; Uranium, U.

6. The elements are the basic building blocks of all substances, and cannot be resolved into simpler substances by chemical means. Compounds are pure substances composed of two or more elements. Compounds have a definite composition by weight.

7. In a mixture, the components are not chemically combined in a definite ratio, so they can exist in any ratio. Thus mixtures do not have a definite composition.

8. The two general types of compounds are molecular and ionic compounds. Molecular compounds consist of molecules which are uncharged units of a compound formed by the union of two or more atoms. An ion is a positively or negatively charged atom or group of atoms. Ionic compounds consist of aggregates of positive and negative ions which are held together by the attractive forces of their opposite charges.

9. A compound is identified and distinguished from other compounds by its characteristic physical and chemical properties.

10. Elements present in each compound:

(a)	KI	potassium, iodine
(b)	Na_2CO_3	sodium, carbon, oxygen
(c)	Al_2O_3	aluminum, oxygen
(d)	$CaBr_2$	calcium, bromine
(e)	CCl_4	carbon, chlorine
(f)	$MgBr_2$	magnesium, bromine
(g)	HNO_3	hydrogen, nitrogen, oxygen
(h)	$BaSO_4$	barium, sulfur, oxygen
(i)	$AlPO_4$	aluminum, phosphorus, oxygen
(j)	$HC_2H_3O_2$	hydrogen, carbon, oxygen

11. Formulas from data of number of atoms of each element:

(a)	Zinc oxide	ZnO	(b)	Potassium chlorate	$KClO_3$
(c)	Sodium hydroxide	NaOH	(d)	Aluminum bromide	$AlBr_3$
(e)	Calcium fluoride	CaF_2	(f)	Lead(II) chromate	$PbCrO_4$
(g)	Ethyl alcohol	C_2H_6O	(h)	Benzene	C_6H_6

12. (a) 2 atoms hydrogen; 1 atom oxygen
(b) 1 atom aluminum; 3 atoms bromine
(c) 2 atoms sodium; 1 atom sulfur; 4 atoms oxygen
(d) 1 atom nickel; 2 atoms nitrogen; 6 atoms oxygen
(e) 12 atoms carbon; 22 atoms hydrogen; 11 atoms oxygen

13. Number of atoms:

(a) KF, 2 atoms
(b) $CaCO_3$, 5 atoms
(c) N_2, 2 atoms
(d) $Ba(ClO_3)_2$, 9 atoms
(e) $K_2Cr_2O_7$, 11 atoms
(f) $NaC_2H_3O_2$, 8 atoms
(g) CCl_2F_2, 5 atoms
(h) $Al_2(SO_4)_3$, 17 atoms
(i) $(NH_4)_2C_2O_4$, 16 atoms

14. (a) H_2, 2 atoms (b) H_2O, 3 atoms (c) H_2SO_4, 7 atoms

15. A cation is a positively charged ion. An anion is a negatively charged ion.

16. The elements which exist as diatomic molecules are: hydrogen, H_2; nitrogen, N_2; oxygen, O_2; fluorine, F_2; chlorine, Cl_2; bromine, Br_2; and iodine, I_2.

17. Number of oxygen atoms in each expression:

(a) 4 H_2O; 4
(b) 3 $CuSO_4$; 12
(c) H_2O_2; 2
(d) 3 $Fe(OH)_3$; 9
(e) $Al(ClO_3)_3$; 9

18. Number of hydrogen atoms in each expression:

(a) 5 H_2; 10
(b) 2 $Ba(C_2H_3O_2)_2$; 12
(c) 2 $C_6H_{12}O_6$; 24
(d) 2 $HC_2H_3O_2$; 8

19. In a homogeneous mixture there is only one phase between the components of the mixture. Two or more phases exist in a heterogeneous mixture.

20. Classification as E, element; C, compound; or M, mixture: (a) air, M; (b) oxygen, E; (c) sodium chloride, C; (d) platinum, E; (e) wine, M; (f) iodine, E; (g) sulfuric acid (solution of H_2SO_4 in H_2O), M; or sulfuric acid (100%, no water), C; (h) crude oil, M.

21. Classification as E, element; C, compound; or M, mixture: (a) paint, M; (b) salt, C; (c) copper, E; (d) beer, M; (e) sulfuric acid, C; (f) silver, E; (g) milk, M; (h) sodium hydroxide, C.

22. The original white solid must have been a compound, because it decomposed to two different compounds. The change to a colorless gas could have been just a physical change of an element (vaporization) except for the appearance of the yellow solid.

23. Properties of metals: solid at room temperature (except mercury); high luster; good conductors of heat and electricity; malleable and ductile; do not tend to react with other metals to form compounds, but do react with many nonmetals to form compounds. Nonmetals exist as solids, liquids, and gases. They are nonlustrous, generally poor conductors of heat and electricity, and react with one another to form molecular compounds.

24. Diatomic molecules: (a) H_2, (b) HCl, (c) NO

25. Naming compounds:

(a)	AgBr	silver bromide	(b)	HI	hydrogen iodide
(c)	$MgBr_2$	magnesium bromide	(d)	CaS	calcium sulfide
(e)	NaF	sodium fluoride	(f)	K_2O	potassium oxide
(g)	LiCl	lithium chloride	(h)	BN	boron nitride
(i)	$BaCl_2$	barium chloride	(j)	Al_2S_3	aluminum sulfide

All of these names end in <u>ide</u>, the characteristic name-ending of binary compounds.

26. (a) Five silver atoms, 5 Ag; (b) four hydrogen molecules, 4 H_2; (c) three water molecules, 3 H_2O

27. Balancing equations:

(a) $H_2 + Cl_2 \rightarrow 2\ HCl$

(b) $Zn + CuSo_4 \rightarrow Cu + ZnSO_4$

(c) $HCl + NaOH \rightarrow NaCl + H_2O$

(d) $2\ Ca + O_2 \rightarrow 2\ CaO$

(e) $Fe + 2\ HCl \rightarrow FeCl_2 + H_2$

(f) $2\ P + 3\ I_2 \rightarrow 2\ PI_3$

(g) $MgO + 2\ HCl \rightarrow MgCl_2 + H_2O$

(h) $2\ HNO_3 + Ba(OH)_2 \rightarrow Ba(NO_3)_2 + 2\ H_2O$

(i) $2\ BiCl_3 + 3\ H_2S \rightarrow 6\ HCl + Bi_2S_3$

(j) $Mg_3N_2 + 6\ H_2O \rightarrow 3\ Mg(OH)_2 + 2\ NH_3$

28. Balancing equations:

(a) $2\ H_2O_2 \rightarrow 2\ H_2O + O_2$

(b) $2\ KClO_3 \rightarrow 2\ KCl + 3\ O_2$

(c) $2\ KNO_3 \rightarrow 2\ KNO_2 + O_2$

(d) $2\ Na_2O_2 + 2\ H_2O \rightarrow 4\ NaOH + O_2$

(e) $2\ H_2O \xrightarrow[H_2SO_4]{\text{electrical energy}} 2\ H_2 + O_2$

29. Balancing equations:

(a) $Zn + 2\ HCl \rightarrow ZnCl_2 + H_2$

(b) $2\ Al + 3\ H_2SO_4 \rightarrow Al_2(SO_4)_3 + 3\ H_2$

(c) $2\ Na + 2\ H_2O \rightarrow 2\ NaOH + H_2$

(d) $C + H_2O(\text{steam}) \rightarrow CO + H_2$

(e) $3\ Fe + 4\ H_2O \rightarrow Fe_3O_4 + 4\ H_2$

30. The following statements are correct: c,e,g,j,k,m,n,p,r,u,v,w,y.

D. Review Problems.

1. 39.3% Na means that there are 39.3 g Na in 100 g NaCl(salt).

$$25.0 \text{ g salt} \times \frac{39.3 \text{ g Na}}{100.0 \text{ g salt}} = 9.83 \text{ g Na}$$

2. 100 g of red brass contains 10 g of Zn and 90 g Cu.

Total mass of the mixture is 140 g after the 40 g of Zn is added.

$$\frac{40 \text{ g Zn} + 10 \text{ g Zn}}{140 \text{ g brass}} \times 100\% = 36\% \text{ Zn}$$

$$\frac{90 \text{ g Cu}}{140 \text{ g brass}} \times 100\% = 64\% \text{ Cu}$$

3. $$\frac{12.0 \text{ g Ca}}{\text{g CaO}} \times 100\% = 71.5\% \text{ Ca}$$

$$\text{g CaO} = \frac{12.0 \text{ g Ca} \times 100\%}{71.5\% \text{ Ca}} = 16.8 \text{ g CaO}$$

Alternate solution: 71.5% Ca in CaO means that 100 g CaO contains 71.5 g Ca. Then converting 12.0 g Ca to grams CaO we have

$$12.0 \text{ g Ca} \times \frac{100 \text{ g CaO}}{71.5 \text{ g Ca}} = 16.8 \text{ g CaO}$$

4. First calculate the mass of CCl_4 and CBr_4 from their volumes and densities.

$2.50 \text{ mL } CCl_4 \times 1.595 \text{ g/mL} = 3.99 \text{ g } CCl_4$

$3.50 \text{ mL } CBr_4 \times 3.420 \text{ g/mL} = 12.0 \text{ g } CBr_4$

Now calculate the density of the mixture from the total mass and total volume.

Total volume = 2.50 mL + 3.50 mL = 6.00 mL

$$\text{Density} = \frac{g}{mL} = \frac{(3.99 + 12.0)\ g}{6.00\ mL} = 2.66\ g/mL$$

5. $12.5\ mL\ Hg \times \frac{13.6\ g}{mL} = 170\ g\ Hg$ (which also equals the mass of Br_2)

$170\ g\ Br_2 \times \frac{1\ mL}{3.12\ g} = 54.5\ mL\ Br_2$

54.5 mL Br_2 has the same mass as 12.5 mL Hg

6. (a) 14.5 g S - 2.90 g S = 11.6 g S reacted

11.6 g S reacts with 14.5 g Ca to give 26.1 g CaS

$\frac{11.6\ g\ S}{14.5\ g\ CaS} \times 100\% = 44.4\%\ S$

(b) A Ca atom has a greater mass than a S atom because, in the compound CaS, there are equal numbers of Ca and S atoms but the mass of the Ca (14.5 g) is greater than the mass of the S (11.65 g).

(c) $30.0\ g\ Ca \times \frac{11.6\ g\ S}{14.5\ g\ Ca} = 24.0\ g\ S$

7. Total mass = 60 g Au + 8 g Ag + 12 g Cu = 80 g total

$\frac{60\ g\ Au}{80\ g\ total} \times 24\ carat = 18\ carat$

8. Consider that each C atom weighs 12 units; then each H atom weighs 1 unit. Total weight of one formula unit CH_4 = 12 + 4 = 16

$\%\ C = \frac{12\ units}{16\ units} \times 100\% = 75\%\ C$

9. $9420\ g\ white\ gold \times \frac{90\ g\ Au}{100\ g\ white\ Au} = 8.5 \times 10^3\ g\ Au$

10. The ratio of Cu to Ni is 75 kg Cu/25 kg Ni.

$450\ kg\ Ni \times \frac{75\ kg\ Cu}{25\ kg\ Ni} = 1350\ kg\ Cu$ required to combine with 450 kg Ni. There is not sufficient Cu to combine with 450 kg Ni, so we calculate how much Ni is required to combine with 1180 kg of Cu.

$1180\ kg\ Cu \times \frac{25\ kg\ Ni}{75\ kg\ Cu} = 393$ (or 3.9×10^2) kg Ni required to combine with 1180 kg Cu. 1180 kg Cu + 393 kg Ni = 1573 (or 1.6×10^3) kg alloy can be produced. There will be 57 kg of Ni left over (450 kg - 393 kg).

B. 1. Atomic numbers: (a) copper = 29; (b) nitrogen = 7; (c) phosphorus = 15; (d) radium = 88; (e) zinc = 30.

2. The neutron is about 1840 times heavier than an electron.

3. Gold nuclei carry a large positive charge, are massive, but very small (relative to the rest of the atom). Like charges repel each other. Consequently, when the stream of positively charged alpha particles was directed at the foil, some particles were deflected at large angles by passing very close to a gold nucleus. Other alpha particles, traveling directly toward a gold nucleus, were literally bounced back.

4. Thirty two. $(2n^2 = 2 \times 4^2 = 32)$

5. 9 orbitals: one s, three p, and five d orbitals

6. See Figure 5.5 in Section 5.8.

7.

	Atom	Nucleus	Electron shells n = 1	2	3	4	5
(a)	$^{14}_{7}N$	7p 7n	$2e^-$	$5e^-$			
(b)	$^{35}_{17}Cl$	17p 18n	$2e^-$	$8e^-$	$7e^-$		
(c)	$^{65}_{30}Zn$	30p 35n	$2e^-$	$8e^-$	$18e^-$	$2e^-$	
(d)	$^{91}_{40}Zr$	40p 51n	$2e^-$	$8e^-$	$18e^-$	$10e^-$	$2e^-$
(e)	$^{127}_{53}I$	53p 74n	$2e^-$	$8e^-$	$18e^-$	$18e^-$	$7e^-$

8. :Ċ· Mg: Ȧl: :C̤̈l· K·

9. The significance of the designation $3d^7$ is that the 3 indicates the principal energy level or shell, the d indicates the kind of orbital, and the 7 indicates the number of electrons in that orbital.

10.

	1s	2s	$2p_x$	$2p_y$	$2p_z$	3s	$3p_x$	$3p_y$	$3p_z$	4s	3d	3d	3d
Si	↑↓	↑↓	↑↓	↑↓	↑↓	↑↓	↑	↑					
S	↑↓	↑↓	↑↓	↑↓	↑↓	↑↓	↑↓	↑	↑				
Ar	↑↓	↑↓	↑↓	↑↓	↑↓	↑↓	↑↓	↑↓	↑↓				
V	↑↓	↑↓	↑↓	↑↓	↑↓	↑↓	↑↓	↑↓	↑↓	↑↓	↑	↑	↑

11. All noble gases except helium have eight electrons in their outer energy level (ns^2 np^6).

C. Review questions.

1. All three particles differ in charge. In terms of unit charge, the proton is +1, the electron -1, and the neutron is 0. The masses of the proton and neutron are nearly equal, but the electron weighs only 1/1837 as much as the proton.

2. In the atom, the protons and neutrons are tightly packed in the nucleus. The electrons move in the space outside the nucleus. The area in which an electron moves is dictated by its energy level. The electrons are located in orbitals outside the nucleus.

3. The nucleus contains practically all the mass of the atom.

4. From his experiments with the scattering of alpha particles by a gold foil, Rutherford concluded:

   (a) the nucleus of the atom must contain most of the mass, because it would require collision with a rather dense object to bounce the alpha particles back toward their source.

   (b) the nucleus of the atom is positively charged because the nature of the deflections of alpha particles indicates a large positive charge concentrated in a small space; and

   (c) the atom consists mostly of empty space since most of the alpha particles passed through the gold foil with little or no deflection.

5. Contributions to atomic theory:

   (a) Dalton: contributed the idea that each element is composed of tiny particles called atoms which are unique for each element, and that compounds are formed from simple combinations of these atoms.

   (b) Thomson: discovered the electron, and found that the mass of the proton was 1837 times that of the electron.

   (c) Rutherford: discovered that the positive charge and almost all the mass of the atom were concentrated in a small nucleus, with the bulk of the atom being empty space.

   (d) Chadwick: discovered the neutron.

   (e) Bohr: devised a theory in which electrons could possess only certain specified energies, which led to the concept of electron orbits.

6. In a Bohr orbit, electrons were thought to revolve around the nucleus of an atom like the planets revolve around the sun. An orbital is a region in space outside the nucleus where we can expect to find an electron.

7. The spectral lines represent quanta of energy, visible as light at characteristic wave lengths, that are emitted as electrons fall from higher to lower energy levels within the hydrogen atoms.

8. The following are correct statements: a,b,c,f,g, and i.

9. An atomic orbital is a region in space about the nucleus of an atom in which we may expect to find an electron.

10. A second electron can enter an orbital already containing one electron only if all other orbitals of the same sublevel contain one electron.

11. The ground state of an atom is the stable state in which it normally exists. This means that all electrons are in the lowest available energy levels. The excited state is when one or more of the electrons are excited to a higher energy level (which would be vacant in the ground state). The absorption of light or energy from an electrical discharge are ways in which electrons are excited above the ground state.

12. Both 1s and 2s orbitals are spherically symmetrical around the nucleus. The 2s orbital is a larger sphere than the 1s orbital, and thus the 2s electrons are located farther from the nucleus than the 1s electrons.

13. The letters s p d f are used to designate the energy sublevels.

14. Subenergy levels in order of increasing energy: 1s, 2s, 2p, 3s, 3p, 4s, 3d, 4p.

15. In any electron shell there can be two s electrons. Any shell, except the first, can hold six p electrons and any shell, except the first and second, can hold ten d electrons.

16. Numbers of protons in nuclei: H, 1; B, 5; F, 9; Sc, 21; Ag, 47; U, 92; Br, 35; Sb, 51; Pb, 82.

17. Electron structures:

B $1s^22s^22p^1$

Ti $1s^22s^22p^63s^23p^64s^23d^2$

Zn $1s^22s^22p^63s^23p^64s^23d^{10}$

Br $1s^22s^22p^63s^23p^64s^23d^{10}4p^5$

Sr $1s^22s^22p^63s^23p^64s^23d^{10}4p^65s^2$

18. The eleventh electron of sodium is located in the third energy level because the first and second energy levels are completely filled. The eleventh electron must occupy the available orbital of lowest

energy, and that is the 3s orbital.

19. The last electron of potassium is located in the 4s sublevel of the fourth energy level. The previous electrons filled the available orbitals through 3s and 3p. The 4s level is available, and is of lower energy than the 3d level, which is also available.

20. (a) $1s^2 2s^2 2p^6 3s^2$; Mg

(b) $1s^2 2s^2 2p^5$; F

(c) $1s^2 2s^2 2p^6 3s^2 3p^6 4s^2 3d^8$; Ni

(d) $1s^2 2s^2 2p^6 3s^2 3p^6 4s^2 3d^5$; Mn

(e) $1s^2 2s^2 2p^6 3s^2 3p^6 4s^2 3d^{10} 4p^6 5s^1 4d^5$; Mo

21.

	Number of electrons	Structure
(a)	8	$1s^2 2s^2 2p^4$
(b)	11	$1s^2 2s^2 2p^6 3s^1$
(c)	17	$1s^2 2s^2 2p^6 3s^2 3p^5$
(d)	23	$1s^2 2s^2 2p^6 3s^2 3p^6 4s^2 3d^3$ or $[Ar] 4s^2 3d^3$
(e)	28	$[Ar] 4s^2 3d^8$
(f)	34	$[Ar] 4s^2 3d^{10} 4p^4$

22.

	Number of electrons	Structure
(a)	9	$1s^2 2s^2 2p^5$
(b)	26	$1s^2 2s^2 2p^6 3s^2 3p^6 4s^2 3d^6$ or $[Ar] 4s^2 3d^6$
(c)	31	$[Ar] 4s^2 3d^{10} 4s^1$
(d)	39	$[Kr] 5s^2 4d^1$
(e)	52	$[Kr] 5s^2 4d^{10} 5p^4$

23. (a) $[Ar] 4s^2 3d^1$; Sc (b) $[Ar] 4s^2 3d^{10} 4p^6$; Kr

(c) $[Kr] 5s^2 4d^{10} 5p^2$; Sn (d) $[Xe] 6s^1$; Cs

24. (a) Sulfur (b) Nickel

25.

		Nucleus			
(a)	$^{27}_{13}Al$	13p 14n	$2e^-$	$8e^-$	$3e^-$

(b) $^{51}_{23}V$ 23p 28n $2e^-$ $8e^-$ $11e^-$ $2e^-$

26. Through chemical changes many of the elements attain an electron structure of eight electrons in their outer shell. The noble gases have a very stable electron structure which is eight electrons (except He) in their outer shell.

27. Electron-dot structures:

He: Ḃ: ·Ö: Na·   ·Ṡi: :Är: Ġa: Ca: ·Br: :Kr:

28. The following statements are correct: a,c,d,e,f,h,i,j,k,l,n,p,q,r.

29. Isotopes are alike in having the same number of protons and electrons. They differ in the number of neutrons, and therefore in their atomic masses.

30. Names of the hydrogen isotopes: $^{1}_{1}H$, protium; $^{2}_{1}H$, deuterium; $^{3}_{1}H$, tritium

31. All three isotopes of hydrogen, $^{1}_{1}H$, $^{2}_{1}H$, and $^{3}_{1}H$, have the same number of protons (1) and electrons (1). They differ in number of neutrons, having 0, 1, and 2, respectively.

32. Cr, chromium

33. (a) Nuclear charge = 201 - 121 = 80 (number of protons)

(b) Hg, mercury

34. The composition of the six naturally occurring isotopes of calcium, having mass numbers of 40, 42, 43, 44, 46, and 48: all six isotopes have 20 protons and 20 electrons. The numbers of neutrons are, respectively, 20, 22, 23, 24, 26, and 28. The protons and neutrons are in the nucleus; the electrons are not. The electron structure, $1s^22s^22p^63s^23p^64s^2$, is the same for each isotope.

35. (a) $^{79}_{35}Br$, 35p, 44n, $35e^-$ (b) $^{131}_{56}Ba$, 56p, 75n, $56e^-$

(c) $^{238}_{92}U$, 92p, 146n, $92e^-$ (d) $^{56}_{26}Fe$, 26p, 30n, $26e^-$

36. Z = atomic number; A = mass number

37. (a) $^{55}_{26}Fe$ (b) $^{26}_{12}Mg$ (c) $^{6}_{3}Li$ (d) $^{29}_{14}Si$ (e) $^{188}_{79}Au$

38. (a) $^{59}_{27}Co$ (b) $^{184}_{74}W$

39. The following statements are correct: b,c,f,g,h,k.

40. Atomic weights of elements are not whole numbers for two reasons: (1) the neutron and proton are not exactly the same weight, and neither is exactly 1.00 amu; and (2) most elements have more than one naturally occurring isotope with different atomic weights and the atomic weight of the element is the average of all the atoms of that element.

41. The isotopic mass of $^{12}_{6}C$ is an exact number by definition, and will always be unless redefined. The masses of other isotopes, such as copper-63, will not be exact whole numbers, for reasons given in the answer to Question 40.

42. Question 34 listed naturally occurring isotopes of calcium with mass numbers of 40, 42, 43, 44, 46, and 48. The most abundant is clearly the first one, with mass number 40, because the atomic weight of calcium is 40.08, showing that there must be only small amounts of the other isotopes.

43. The following statements are correct: b,d.

D. Review problems.

1. Powers of 10.

(a) $510{,}000 = 5.1 \times 10^5$ (b) $0.000274 = 2.74 \times 10^{-4}$

(c) $(0.0001)^2 = 0.00000001 = 1 \times 10^{-8}$

(d) $(8.0)^3 = 5.1 \times 10^2$

2. Number of electrons $= 2n^2$

n	Number of electrons	n	Number of electrons
1	$2(1)^2 = 2$	4	$2(4)^2 = 32$
2	$2(2)^2 = 8$	5	$2(5)^2 = 50$
3	$2(3)^2 = 18$	6	$2(6)^2 = 72$

3.

	Atomic Number	Mass Number	Symbol of Element	Number of Protons	Number of Neutrons
(a)	8	16	O	8	8
(b)	28	58	Ni	28	30
(c)	80	199	Hg	80	119

4. Calculate by comparing the mass of the unknown isotope to that of the mass of one atom of C-12, 1.9927×10^{-23} amu.

$$\frac{2.18 \times 10^{-22}\ \text{amu}}{1.9927 \times 10^{-23}\ \text{amu (C)}} \times 12.0\ \text{amu C-12} = 131\ \text{amu (atomic mass of the unknown)}$$

5. ^{107}Ag: 106.9041 amu × 0.5182 = 55.40 amu

^{109}Ag: 108.9047 amu × 0.4818 = 52.47 amu

Atomic weight = 107.87 amu

6. ^{24}Mg: 23.985 amu × 0.7899 = 18.95 amu

^{25}Mg: 24.986 amu × 0.1000 = 2.50 amu

^{26}Mg: 25.983 amu × 0.1101 = 2.86 amu

Atomic weight = 24.31 amu

B. 1. Elements in the s block all have one or two s electrons in their outermost energy levels.

2. 8 = O; 16 = S; 34 = Se; 52 = Te; 84 = Po. Each of these elements is in Group VIA and each has six electrons (2-s and 4-p) in its outermost energy level.

3. (a) Germanium (Ge). (b) Mendeleev's predicted density for eka-silicon was 5.5 g/mL; the density of germanium is 5.35 g/mL.

4. The size of atoms decreases from left to right in the third period.

5. The symbols of the family of elements having seven electrons in their outer energy level are: F, Cl, Br, I, and At.

6. The symbols of the alkali metals in order of increasing atomic size are: Li, Na, K, Rb, Cs, and Fr.

7. The greatest number of elements in a single period is 32, and only the sixth period has this number.

8. The last electron is placed in an outer energy level for Group A elements and in an inner level for Group B elements.

C. Review Questions.

1. Dobereiner's triads were elements in groups of three. The three elements in each triad were similar in properties with the center element reflecting an average of the other two in some respects. These observations indicated a recurrence of properties among some elements and pointed toward periodicity among the elements.

2. Newlands' Law of Octaves is based on the fact that up to element 21, where d electrons first appear, properties do repeat with each eighth element. This is because an s orbital contains 2 electrons and three p orbitals contain 6 electrons; so for every 8 electrons, the element will have the same number of valence electrons, and that is what causes similar properties. Newlands' law repeated every seventh element, like the musical scales, which fit because the noble gases were not known at the time.

3. Newlands' Law of Octaves does not extend as far as the element bromine. The appearance of d electrons, beginning with element 21, Sc, disrupts the sequence of eight.

4. Periodicity refers to the fact that as the elements are arranged consecutively in order of increasing atomic number a periodic recurrence of similarities in the properties of the elements is observed.

5. Vacancies in Mendeleev's periodic table indicated the existence of elements not yet discovered at that time. From the position of the vacancy in the periodic table, expected properties of the missing

element could be proposed.

6. Three major differences have been made to the periodic table since Mendeleev's time: (1) a new family of elements, the noble gases, was discovered and added, (2) the elements are arranged by atomic number, not atomic weight, and (3) elements having atomic numbers greater than 92 have been discovered and fitted into the table.

7. Moseley's work indicated the periodic arrangement of the elements should be based on the atomic numbers of the elements rather than on the atomic weights of the elements.

8. The current statement of the periodic law is: The properties of the chemical elements recur periodically when the elements are arranged in increasing order of their atomic numbers.

9. Pairs of elements which are out of sequence with respect to atomic weights are: Ar, and K, Co and Ni, Te and I, Th and Pa, U and Np, Es and Fm, and, Md and No.

10. Going from left to right in any period of elements the atomic number increases by one from one element to the next, and the atomic radius generally decreases. Each period (except period 1) starts with an alkali metal and ends with a noble gas. There is a trend in properties of the elements changing from metallic to nonmetallic from the beginning to the end of the period.

11. Elements in a group all have the same number of outershell (valence) electrons.

12. The alkali metals all have only one electron in their outer energy level.

13. Zinc, cadmium, and mercury each has two electrons in its outer energy level. Also each of the three has eighteen electrons in the next to outermost energy level.

14. Electron-dot formulas: $Cs\cdot$ $Ba:$ $\dot{T}l:$ $\cdot\dot{P}b:$ $\cdot\underset{\cdot}{\ddot{P}}o:$ $\cdot\underset{\cdot\cdot}{\ddot{A}}t:$ $:\underset{\cdot\cdot}{\ddot{R}}n:$

    Each of these is a representative element and has the same number of electrons in its outer shell as its periodic group number.

15. Elements in the same family of elements: (a) and (g); each has 1 electron in its outer energy level and belongs to Group IA. (b) and (d); each has 6 electrons in its outer energy level and belongs to Group VIA.

16. Elements in the same family of elements: (a) and (f); each has 8 electrons in its outer energy level and belongs to the noble gas family. (e) and (h) belong to the IB family.

17. Elements 12, Mg, and 38, Sr, are in the same family as Ca, element 20. Consequently, the physical and chemical properties of Mg and Sr will resemble those of Ca.

18. Oxygen and sulfur are grouped in the same family because each has six electrons in its outer energy level.

19. Classification of elements:

    Metals: potassium, plutonium, molybdenum, tungsten.

    Nonmetals: sulfur, iodine.

    Metalloids: germanium, antimony.

20. A d electron first appears in period 4 and in group IIIB.

21. Group IIIA elements contain 3 electrons in their outer shell. Group IIIB elements are transition elements. They contain two electrons in their outer shell; the third electron is in an inner d shell rather than in the valence shell.

22. Transition elements are located in Groups IB, IIIB, IVB, VB, VIB, VIIB, and VIII.

23. Transition elements differ from the representative elements in having a partially filled set of d or f orbitals.

24. (a) K is larger than Na; (b) Na is larger than Mg; (c) O is larger than F; (d) I is larger than Br; (e) Zr is larger than Ti.

25. The smallest atomic radius in each of these groups is: IA, H; IIA, Be; IIIA, B; IVA, C; VA, N; VIA, O; VIIA, F.

26. In going down any group of the periodic table, each successive element has one additional shell of electrons which is located farther from the nucleus causing the atom to be larger.

27.

Group	IA	IIA	IIIA	IVA	VA	VIA	VIIA
	E·	E:	Ė:	·Ė:	·Ė: (dot below)	·Ë̤:	·Ë: (two dots below)

28. NaH, Na_2O; CaH_2, CaO; AlH_3, Al_2O_3; SnH_4, SnO_2; SbH_3, Sb_2O_5; H_2Se, SeO_3; HCl, Cl_2O_7.

29. Yes, since Group IA formed compounds such as NaCl, KCl, and since the number of outer shell electrons is the greatest single factor in the way elements combine.

30. Predicted formulas: tin(II) bromide, $SnBr_2$; Germanium(II) bromide, $GeBr_2$.

31. Lithium sulfate, Li_2SO_4; potassium sulfate, K_2SO_4; rubidium sulfate, Rb_2SO_4; cesium sulfate, Cs_2SO_4; francium sulfate, Fr_2SO_4.

32. Magnesium bromide, $MgBr_2$; strontium bromide, $SrBr_2$; barium bromide, $BaBr_2$.

33. Discovery of isotopes was an indication that atomic weight was not as fundamental a characteristic of an element as atomic number. The atomic weight of an element is the average mass of all its isotopes and mainly reflects the mass of the most abundant isotope. Thus the value of the atomic weight of an element could be less than that of another element of lower atomic number.

34. The atomic numbers of the elements with six electrons in their outer electron shell are 8, 16, 34, 52, and 84.

35. The outer energy level of element 87 consists of a 7s orbital which is half filled (contains only one electron). The element has seven energy levels.

36. If element 36 is a noble gas, element 35 would be in Group VIIA (the halogens) and element 37 would be in Group IA (alkali metals).

37. Each students' answer should be an individual composition.

38. The radii of the five atoms, from smallest to largest, are Cl, Mg, Na, K, Rb.

39. The following statements are correct: a,d,e,g,h,i,k,n,o,p,t,v.

B. 1. More energy is required for neon because neon has a very stable outer electron structure consisting of an octet of electrons in filled orbitals (noble gas electron structure). Sodium, an alkali metal, has a relatively unstable outer electron structure with a single electron in an unfilled orbital. The sodium electron is also farther away from the nucleus and is shielded by more inner electron shells than are neon outer shell electrons.

2. When a third electron is removed from beryllium, it must come from a very stable electron shell structure corresponding to that of the noble gas, helium. In addition, the third electron must be removed from a +2 beryllium ion, which increases the difficulty of removing it.

3. The first ionization energy decreases from top to bottom because in the successive alkali metals the outermost electron is farther away from the nucleus and is shielded from the positive nucleus by an additional electron shell.

4. The first ionization energy decreases from top to bottom because the outermost electrons in the successive noble gases are farther away from the nucleus and are shielded by an additional inner electron shell.

5. Elements of lowest ionization energies and lowest electronegativities are located in the lower left of the periodic table. Elements of highest ionization energies and highest electronegativities are located in the upper right of the periodic table. (The noble gases have not been assigned electronegativities, but they do have ionization energies.)

6. A magnesium atom is larger because the ion has one less electron shell and because the 12 positive nuclear charges are also able to attract 10 electrons more closely than they do 12 electrons.

7. The bromine atom is smaller because the ion has one more electron in the outer shell and because the 35 positive nuclear charges are not able to attract 36 electrons as closely as they do 35 electrons.

8. (a) H_2O; H+, O- (b) NaF; Na+, F- (c) NH_3; H+, N-
(d) PbS; Pb+, S- (e) NO; N+, O- (f) CH_4; H+, C-
(g) HCl; H+, Cl- (h) LiH; Li+, H- (i) CCl_4; C+, Cl-
(j) IBr; I+, Br- (k) MgH_2; Mg+, H- (l) OF_2; O+, F-

9. (a) Sodium and chlorine; principally ionic
(b) Carbon and hydrogen; principally covalent
(c) Chlorine and carbon; principally covalent
(d) Calcium and oxygen; principally ionic
(e) Hydrogen and sulfur; principally covalent
(f) Barium and oxygen; principally ionic
(g) Fluorine and fluorine; principally covalent
(h) Potassium and fluorine; principally ionic

C. Review questions.

1. While the number of electrons in the outermost energy level is the same for all elements in a family, the energy level in which they are located is not the same. As one goes down (top to bottom) a family of elements the valence electrons are successively located in energy levels farther removed from the nucleus. The farther the energy level is from the nucleus the greater will be the energy of the electrons in that energy level and the less will be the additional energy required to remove the electron from the atom.

2. Barium and beryllium are in the same family. The electron to be removed from barium is, however, located in an energy level farther removed form the nucleus than is the energy level holding the electron in beryllium. Hence, it requires less energy to remove the electron from barium than to remove the electron from beryllium and barium has a lower ionization energy than beryllium. (see answer to question C.1)

3. The first electron removed from a sodium atom is the one valence electron, which is shielded from most of the nuclear charge by the electrons of lower levels. To remove a second electron from a sodium atom requires breaking into the noble gas structure and this requires much more energy than that required to remove the first electron. The fact that the second electron would be removed from a positively charged ion also contributes to the energy required.

4. Electron affinity is the amount of energy released or absorbed when an atom gains an electron to form a negative ion. Ionization energy is the amount of energy necessary to remove one electron from an atom or an ion.

5. Magnesium has the electron structure $1s^2 2s^2 2p^6 3s^2$ while the structure for chlorine is $1s^2 2s^2 2p^6 3s^2 3p^5$. When these two elements react each magnesium atom loses its $3s^2$ electrons, one to each of two chlorine atoms. The electron structures that result for both the magnesium and the chlorine atoms are noble gas structures.

6. (a) $F + 1e^- \rightarrow F^-$ (b) $Ca \rightarrow Ca^{2+} + 2e^-$

7. 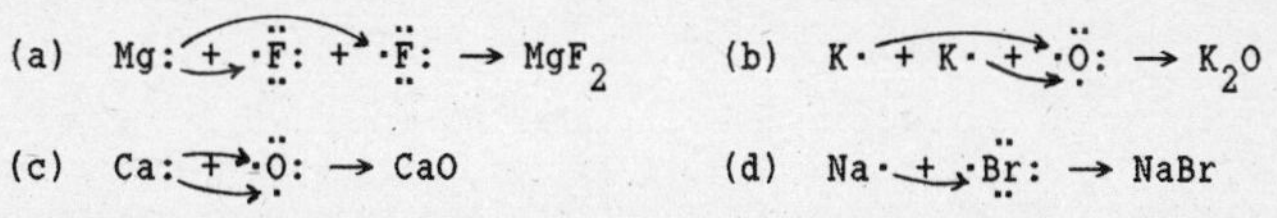

8. Valence electrons are the electrons in the outermost energy level of an atom.

9. Valence electrons: H, 1; K, 1; Mg, 2; He, 2; Al, 3; Si, 4; N, 5; P, 5; O, 6; Cl, 7.

10. Noble gas structures: (a) Calcium atom, lose 2 electrons; (b) Sulfur atom, gain 2 electrons; (c) Helium atom, none (already

noble gas structure); (d) Chloride ion, none (already noble gas structure); (e) Nitrogen atom, gain 3 electrons or lose 5 electrons; (f) Potassium atom, lose 1 electron.

11. By losing one electron a potassium atom acquires a noble gas structure and becomes a K^+ ion. To become a K^{2+} ion requires loss of a second electron and breaking into the noble gas structure of the K^+ ion. This requires too much energy to generally occur.

12. An aluminum ion has a +3 charge because it has lost 3 electrons in acquiring a noble gas structure.

13. (a) A potassium atom is larger than a potassium ion because the atom has one more electron containing energy level than the ion.

    (b) A bromide ion is larger than a bromine atom because it has, within the same principal energy level, one more electron than a bromine atom.

    (c) A magnesium ion, Mg^{2+}, is larger than an aluminum ion, Al^{3+}. Both ions have the same number of electrons (are isoelectronic), with identical distribution of these electrons in their energy levels. The increased nuclear charge of the Al^{3+} ion (over the Mg^{2+} ion) pulls the electrons in the Al^{3+} ion closer to its nucleus, making it, the Al^{3+} ion, smaller than the Mg^{2+} ion.

    (d) The Fe^{2+} ion is larger than the Fe^{3+} ion because the Fe^{2+} ion has one more electron than the Fe^{3+} ion. Both ions have the same nuclear charge.

14. Lewis dot structures: (a) $Na\cdot$ (b) $:\underset{\cdot\cdot}{\overset{\cdot\cdot}{Br}}:^{-}$ (c) $:\underset{\cdot\cdot}{\overset{\cdot\cdot}{O}}:^{2-}$

    (d) $\dot{Ga}:$ (e) Ga^{3+}

15. Sodium chloride exists as aggregates of sodium and chloride ions. An electron is transferred from a sodium atom to a chlorine atom forming these ions. Substances with ionic bonds do not exist as molecules. The ions are held together by ionic bonds.

16. A covalent bond results from the sharing of a pair of electrons between two bonding atoms; whereas, an ionic bond results from the transfer of an electron (or electrons) from one atom to another.

17. In a coordinate covalent bond the electrons being shared are donated by one of the atoms, whereas, in an ordinary covalent bond electrons shared come from both atoms.

18. Compounds with ionic bonding: (b) NaCl, (c) MgO, and (f) $BaCl_2$.

    Compounds with covalent bonding: (a) H_2O, (d) Br_2, (e) HCl,

(g) NH_3, and (h) SO_2.

19. Predictions are based on difference in electronegativities. (a) Na and N, ionic; (b) N and S, covalent; (c) Br and I, covalent; (d) H and Si, covalent; (e) O and F, covalent.

20. (a) H:H (b) :N:::N: (c) :Cl:Cl: (d) :O::O: (e) :Br:Br:

21.
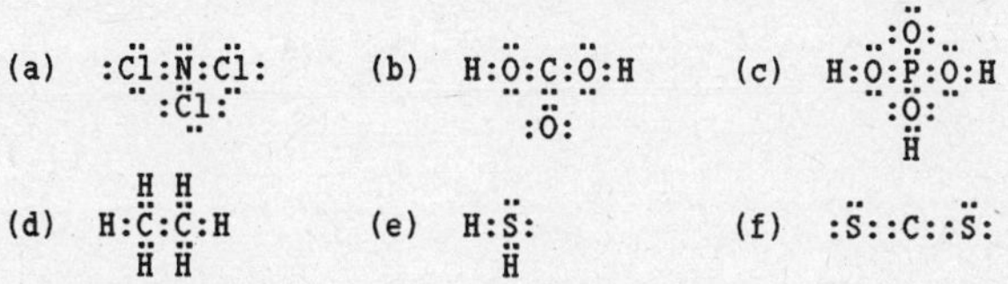

22. The structure written as Na:O:Na indicates covalent bonds between Na and O. However, the difference in electronegativity between Na and O, 2.6, clearly indicates that the bonding is ionic.

23. The four most electronegative elements are: F, O, N, and Cl.

24. Lewis-dot structures

(a) Ba^{2+} (b) Al^{3+} (c) $:I:^-$ (d) $:S:^{2-}$

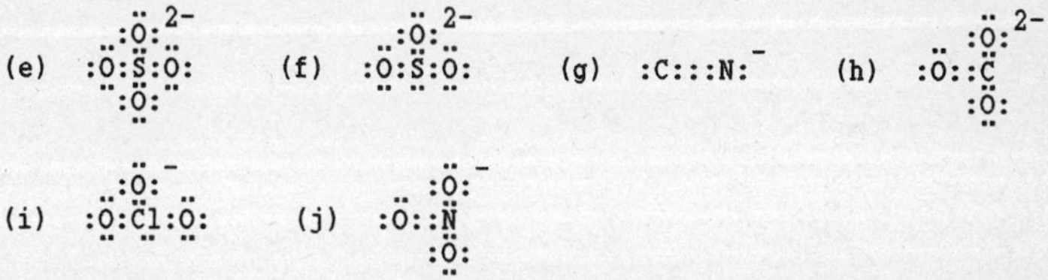

25. Chloric acid

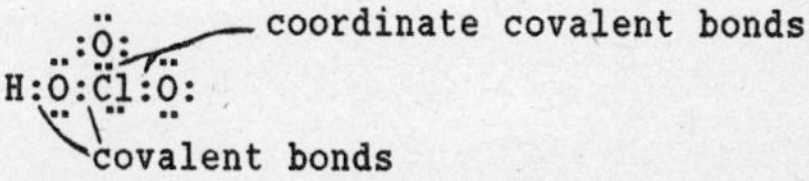

26. 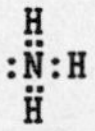 All three nitrogen-hydrogen bonds are covalent bonds. It is possible for NH_3 to form one coordinate covalent bond.

27. Electronegativities, highest to lowest: F, O, S, H, Mg, Cs.

28. Polar molecules are: (a) H_2O, (b) HBr, and (f) NH_3
Nonpolar molecules are: (c) CF_4, (d) F_2, and (e) CO_2.

29. A symmetrical molecule will be nonpolar even though it may contain polar bonds. Polar bonds in a symmetrical molecule will be so oriented that the center of gravity of the positive charge and the negative charge will coincide.

30. Both the CO_2 and the CO molecules contain polar carbon-oxygen bonds; however, the CO_2 molecule is a symmetrical molecule and the two carbon-oxygen dipoles, being equal but in exactly opposite directions, cancel. The single dipole in the CO molecule makes the CO molecule polar.

31. The following statements are correct: a,b,e,h,i,j,l,o,q,r,s,t,u,v, w,x,z,bb,cc,gg,hh, and jj.

1. Formulas:
   (a) Hydrogen compounds: LiH, CH_4, HI, NH_3, SrH_2
   (b) Oxygen compounds: CaO, SiO_2, Br_2O, Al_2O_3, P_2O_5
   (c) Iodine compounds: LiI, AlI_3, CI_4, SrI_2, SI_2

2. Fromulas:
   (a) Sodium, Na^+, and chlorate, ClO_3^-: $NaClO_3$
   (b) Hydrogen, H^+, and sulfate, SO_4^{2-}: H_2SO_4
   (c) Tin(II), Sn^{2+}, and acetate, $C_2H_3O_2^-$: $Sn(C_2H_3O_2)_2$
   (d) Copper(I), Cu^+, and oxide, O^{2-}: Cu_2O
   (e) Zinc, Zn^{2+}, and bicarbonate, HCO_3^-: $Zn(HCO_3)_2$

3. Oxidation numbers of underlined elements.
   (a) $\underline{Mn}CO_3$ +2   (b) $\underline{Sn}F_4$ +4   (c) $K\underline{N}O_3$ +5
   (d) $K\underline{Mn}O_4$ +7   (e) $Ba\underline{C}O_3$ +4   (f) $\underline{P}Cl_3$ +3
   (g) $\underline{W}Cl_5$ +5   (h) $K_2\underline{Cr}_2O_7$ +6

4. Oxidation numbers of underlined elements.
   (a) $\underline{In}I_3$ +3   (b) $K\underline{Cl}O_3$ +5   (c) $Na_2\underline{S}O_4$ +6
   (d) $\underline{C}_2H_5OH$ -2   (e) $Mg(\underline{N}O_3)_2$ +5   (f) $\underline{Sn}O_2$ +4
   (g) $\underline{Fe}_2(CO_3)_3$ +3   (h) $Na\underline{Cl}O_4$ +7

5. Formulas of compounds.
   (a) Na and I: NaI (b) Ba and F: BaF_2 (c) Al and O: Al_2O_3
   (d) K and S: K_2S (e) Cs and Cl: $CsCl$ (f) Sr and Br: $SrBr_2$

6. Formulas of compounds.
   (a) Ba and O: BaO (b) H and S: H_2S (c) Ga and Cl: $GaCl_3$
   (d) Be and Br: $BeBr_2$ (e) Li and Si: Li_4Si (f) Mg and P: Mg_3P_2

7. No. It means their oxidation numbers, while of opposite sign, are of the same numerical magnitude.

8. Compounds of X and Z. (a) X_2Z (b) XZ_4 (c) XZ

9. Cation formulas: Sodium, Na^+; magnesium, Mg^{2+}; aluminum, Al^{3+}; copper(II), Cu^{2+}; iron(II), Fe^{2+}; ferric, Fe^{3+}; lead(II), Pb^{2+}; silver, Ag^+; cobalt(II), Co^{2+}; barium, Ba^{2+}; hydrogen, H^+;

mercury(II), Hg^{2+}; tin(II), Sn^{2+}; chromium(III), Cr^{3+}; stannic, Sn^{4+}; manganese(II), Mn^{2+}; bismuth(III), Bi^{3+}.

10. Anion formulas: chloride, Cl^-; bromide, Br^-; fluoride, F^-; iodide, I^-; cyanide, CN^-; oxide, O^{2-}; hydroxide, OH^-; sulfide, S^{2-}; sulfate, SO_4^{2-}; bisulfate, HSO_4^-; bisulfite, HSO_3^-; chromate, CrO_4^{2-}; carbonate, CO_3^{2-}; acetate, $C_2H_3O_2^-$; chlorate, ClO_3^-; permanganante, MnO_4^-; oxalate, $C_2O_4^{2-}$.

11. Complete the table:

	Br^-	O^{2-}	NO_3^-	PO_4^{3-}	CO_3^{2-}
K^+	KBr	K_2O	KNO_3	K_3PO_4	K_2CO_3
Mg^{2+}	$MgBr_2$	MgO	$Mg(NO_3)_2$	$Mg_3(PO_4)_2$	$MgCO_3$
Al^{3+}	$AlBr_3$	Al_2O_3	$Al(NO_3)_3$	$AlPO_4$	$Al_2(CO_3)_3$
Zn^{2+}	$ZnBr_2$	ZnO	$Zn(NO_3)_2$	$Zn_3(PO_4)_2$	$ZnCO_3$
H^+	HBr	H_2O	HNO_3	H_3PO_4	H_2CO_3

12. Complete the table

	SO_4^{2-}	Cl^-	AsO_4^{3-}	$C_2H_3O_2^-$	CrO_4^{2-}
NH_4^+	$(NH_4)_2SO_4$	NH_4Cl	$(NH_4)_3AsO_4$	$NH_4C_2H_3O_2$	$(NH_4)_2CrO_4$
Ca^{2+}	$CaSO_4$	$CaCl_2$	$Ca_3(AsO_4)_2$	$Ca(C_2H_3O_2)_2$	$CaCrO_4$
Fe^{3+}	$Fe_2(SO_4)_3$	$FeCl_3$	$FeAsO_4$	$Fe(C_2H_3O_2)_3$	$Fe_2(CrO_4)_3$
Ag^+	Ag_2SO_4	$AgCl$	Ag_3AsO_4	$AgC_2H_3O_2$	Ag_2CrO_4
Cu^{2+}	$CuSO_4$	$CuCl_2$	$Cu_3(AsO_4)_2$	$Cu(C_2H_3O_2)_2$	$CuCrO_4$

13. Usage of prefixes and suffixes in naming inorganic compounds:

ide: always indicates a binary compound (only two elements) except for hydroxides, cyanides, and ammonium compounds

ous: used in an acid to designate the lower oxidation state of the element other than hydrogen and oxygen; also used for the lower oxidation state of a multivalent metal.

hypo: used as a prefix in acids or salts when the oxidation state is lower than that of the -ous acid or the -ite salt.

per: used as a prefix in acids or salts when the oxidation state is greater than that of the -ic acid or the -ate salt.

ite: the suffix of a salt derived from an -ous acid.

ate: the suffix of a salt derived from an -ic acid.

Roman numerals: In the Stock System roman numerals are used in naming compounds that contain metals that may exist in more than one oxidation state. The oxidation number of the metal is indicated by a roman numeral written in parentheses immediately after the name of the metal.

14. Nonmetal binary compound formulas:

(a) carbon monoxide, CO
(b) sulfur trioxide, SO_3
(c) carbon tetrabromide, CBr_4
(d) phosphorus trichloride, PCl_3
(e) nitrogen dioxide, NO_2
(f) dinitrogen pentoxide, N_2O_5

15. Naming binary nonmetal compounds:

(a) CO_2, carbon dioxide
(b) N_2O, dinitrogen oxide
(c) PCl_5, phosphorus pentachloride
(d) CCl_4, carbon tetrachloride
(e) SO_2, sulfur dioxide
(f) N_2O_4, dinitrogen tetroxide
(g) P_2O_5, diphosphorus pentoxide
(h) OF_2, oxygen difluoride
(i) NF_3, nitrogen trifluoride
(j) CS_2, carbon disulfide

16. (a) K_2O, potassium oxide
(b) NH_4Br, ammonium bromide
(c) CaI_2, calcium iodide
(d) $BaCO_3$, barium carbonate
(e) Na_3PO_4, sodium phosphate
(f) Al_2O_3, aluminum oxide
(g) $Zn(NO_3)_2$, zinc nitrate
(h) Ag_2SO_4, silver sulfate

17. (a) sodium nitrate, $NaNO_3$
(b) magnesium fluoride, MgF_2
(c) barium hydroxide, $Ba(OH)_2$
(d) ammonium sulfate, $(NH_4)_2SO_4$
(e) silver carbonate, Ag_2CO_3
(f) calcium phosphate, $Ca_3(PO_4)_2$

18. (a) $CuCl_2$, copper(II) chloride; cupric chloride
(b) $CuBr$, copper(I) bromide; cuprous bromide
(c) $Fe(NO_3)_2$, iron(II) nitrate; ferrous nitrate
(d) $FeCl_3$, iron(III) chloride; ferric chloride
(e) SnF_2, tin(II) fluoride; stannous fluoride
(f) $HgCO_3$, mercury(II) carbonate; mercuric carbonate
(g) $As(C_2H_3O_2)_3$, arsenic(III) acetate; arsenous acetate
(h) TiI_3, titanium(III) iodide; titanous iodide

19. Formulas:

(a) tin(IV) bromide, $SnBr_4$
(b) copper(I) sulfate, Cu_2SO_4
(c) ferric carbonate, $Fe_2(CO_3)_3$
(d) mercuric nitrite, $Hg(NO_2)_2$

(e) titanic sulfide, TiS_2 (f) iron(II) acetate, $Fe(C_2H_3O_2)_2$

20. Acid formulas:
(a) hydrochloric acid, HCl (b) chloric acid, $HClO_3$
(c) nitric acid, HNO_3 (d) carbonic acid, H_2CO_3
(e) sulfurous acid, H_2SO_3 (f) phosphoric acid, H_3PO_4

21. Naming acids:
(a) HNO_2, nitrous acid (b) H_2SO_4, sulfuric acid
(c) $H_2C_2O_4$, oxalic acid (d) HBr, hydrobromic acid
(e) H_3PO_3, phosphorous acid (f) $HC_2H_3O_2$, acetic acid
(g) HF, hydrofluoric acid (h) $HBrO_3$, bromic acid

22. Formulas of acids:
(a) acetic acid, $HC_2H_3O_2$ (b) hydrofluoric acid, HF
(c) hypochlorous acid, $HClO$ (d) boric acid, H_3BO_3
(e) nitrous acid, HNO_2 (f) hydrosulfuric acid, H_2S

23. Naming acids:
(a) H_3PO_4, phosphoric acid (b) H_2CO_3, carbonic acid
(c) HIO_3, iodic acid (d) HCl, hydrochloric acid
(e) $HClO$, hypochlorous acid (f) HNO_3, nitric acid
(g) HI, hydriodic acid (h) $HClO_4$, perchloric acid

24. Naming compounds:
(a) $Ba(NO_3)_2$, barium nitrate (b) $NaC_2H_3O_2$, sodium acetate
(c) PbI_2, lead(II) iodide (d) $MgSO_4$, magnesium sulfate
(e) $CdCrO_4$, cadmium chromate (f) $BiCl_3$, bismuth(III) chloride
(g) NiS, nickel(II) sulfide (h) $Sn(NO_3)_4$, tin(IV) nitrate
(i) $Ca(OH)_2$, calcium hydroxide

25. Write formulas for:
(a) silver sulfite, Ag_2SO_3 (b) cobalt(II) bromide, $CoBr_2$
(c) tin(II) hydroxide, $Sn(OH)_2$ (d) aluminum sulfate, $Al_2(SO_4)_3$
(e) manganese(II) fluoride, MnF_2 (f) ammonium carbonate, $(NH_4)_2CO_3$
(g) chromium(III) oxide, Cr_2O_3 (h) cupric chloride, $CuCl_2$
(i) potassium permanganate, $KMnO_4$ (j) barium nitrite, $Ba(NO_2)_2$
(k) sodium peroxide, Na_2O_2 (l) ferrous sulfate, $FeSO_4$
(m) potassium dichromate, $K_2Cr_2O_7$
(n) bismuth(III) chromate, $Bi_2(CrO_4)_3$

26. Write formulas for:

(a) sodium chromate, Na_2CrO_4

(b) magnesium hydride, MgH_2

(c) nickel(II) acetate, $Ni(C_2H_3O_2)_2$

(d) calcium chlorate, $Ca(ClO_3)_2$

(e) lead(II) nitrate, $Pb(NO_3)_2$

(f) potassium dihydrogen phosphate, KH_2PO_4

(g) manganese(II) hydroxide, $Mn(OH)_2$

(h) cobalt(II) bicarbonate, $Co(HCO_3)_2$

(i) sodium hypochlorite, NaClO

(j) arsenic(V) carbonate, $As_2(CO_3)_5$

(k) chromium(III) sulfite, $Cr_2(SO_3)_3$

(l) antimony(III) sulfate, $Sb_2(SO_4)_3$

(m) sodium oxalate, $Na_2C_2O_4$

(n) potassium thiocyanate, KSCN

27. Write the name of each salt and its formula and the name of the acid from which the salt may be derived.

	Salt Formula	Salt Name	Acid Formula	Acid Name
(a)	$ZnSO_4$	zinc sulfate	H_2SO_4	sulfuric acid
(b)	$HgCl_2$	mercury(II) chloride	HCl	hydrochloric acid
(c)	$CuCO_3$	copper(II) carbonate	H_2CO_3	carbonic acid
(d)	$Cd(NO_3)_2$	cadmium nitrate	HNO_3	nitric acid
(e)	$Al(C_2H_3O_2)_3$	aluminum acetate	$HC_2H_3O_2$	acetic acid
(f)	CoF_2	cobalt(II) fluoride	HF	hydrofluoric acid
(g)	$Cr(ClO_3)_3$	chromium(III) chlorate	$HClO_3$	chloric acid
(h)	Ag_3PO_4	silver phosphate	H_3PO_4	phosphoric acid
(i)	NiS	nickel(II) sulfide	H_2S	hydrosulfuric acid
(j)	$BaCrO_4$	barium chromate	H_2CrO_4	chromic acid
(k)	$Ca(HSO_4)_2$	calcium hydrogen sulfate	H_2SO_4	sulfuric acid
(l)	$As_2(SO_3)_3$	arsenic(III) sulfite	H_2SO_3	sulfurous acid
(m)	$Sn(NO_2)_2$	tin(II) nitrite	HNO_2	nitrous acid
(n)	$FeBr_3$	iron(III) bromide	HBr	hydrobromic acid
(o)	$KHCO_3$	potassium hydrogen carbonate	H_2CO_3	carbonic acid
(p)	$BiAsO_4$	bismuth(III) arsenate	H_3AsO_4	arsenic acid

(q)	$Fe(BrO_3)_2$	iron(II) bromate	$HBrO_3$	bromic acid
(r)	$(NH_4)_2HPO_4$	ammonium monohydrogen phosphate	H_3PO_4	phosphoric acid
(s)	NaClO	sodium hypochlorite	HClO	hypochlorous acid
(t)	$KMnO_4$	potassium permanganate	$HMnO_4$	permanganic acid

28. Write formulas for:
    (a) Baking soda, $NaHCO_3$
    (b) Lime, CaO
    (c) Epsom salts, $MgSO_4 \cdot 7H_2O$
    (d) Muriatic acid, HCl
    (e) Vinegar, $HC_2H_3O_2$
    (f) Potash, K_2CO_3
    (g) Lye, NaOH
    (h) Quicksilver, Hg
    (i) Fool's gold, FeS_2
    (j) Saltpeter, $NaNo_3$
    (k) Limestone, $CaCO_3$
    (l) Cane sugar, $C_{12}H_{22}O_{11}$
    (m) Milk of magnesia, $Mg(OH)_2$
    (n) Washing soda, $Na_2CO_3 \cdot 10H_2O$
    (o) Grain alcohol, C_2H_5OH

29. Calcium cyanide, $Ca(CN)_2$; Ammonium bromide, NH_4Br; Lithium cyanide, LiCN.

30. The following statements are correct: a,d,f,g,i,j,l,m,o,p,r,u,w,x,y, and aa.

B. Review questions.

1. A mole is defined as the amount of a substance that contains the same number of units as the number of atoms in exactly 12 grams of $^{12}_{6}C$. The number of atoms in 12 grams of $^{12}_{6}C$, and hence, the number of units in a mole, is 6.022×10^{23}.

2. A mole of gold atoms contains the same number of atoms as a mole of potassium atoms. In as much as the mass of a gold atom is 197.0 compared to 39.1 for a potassium atom a mole of gold atoms will weigh more than a mole of potassium atoms.

3. The number of atoms in one mole of potassium atoms is the same as the number of atoms in one mole of gold atoms.

4. 1 mole gold = 6.022×10^{23} atoms of gold.

   1 mole potassium = 6.022×10^{23} atoms of potassium

   One gold atom contains 79 electrons and one potassium atom contains 19 electrons. Therefore, a mole of gold atoms contains more electrons than a mole of potassium atoms.

5. No. Avogadro's number is defined as the number of carbon-12 atoms in 12 grams of carbon-12. Defining an atom of carbon-12 as weighing 50 amu simply amounts to changing the size of the atomic mass unit (amu).

6. Avogadro's number is 6.022×10^{23}.

7. A mole of anything (atoms, molecules, ants) contains Avogadro's number of these units.

8. (a) A mole of oxygen atoms (O) contains 6.022×10^{23} atoms.

   (b) A mole of oxygen molecules (O_2) contains 6.022×10^{23} molecules.

   (c) A mole of oxygen molecules (O_2) contains 1.204×10^{24} atoms.

   (d) A mole of oxygen atoms (O) weighs 16.0 grams.

   (e) A mole of oxygen molecules (O_2) weighs 32.0 grams.

9. The following statements are correct: a,b,c,d, and h.

10. Formula weight and molecular weight are determined in the same way. If the substance is molecular, its formula weight is its molecular weight. But for those substances not molecular, then the formula weight merely represents the total mass of all the atoms in the formula.

11. One molecular weight of sulfuric acid, H_2SO_4, contains 6.022×10^{23} molecules.

$$\text{Atoms present} = \frac{6.022 \times 10^{23} \text{ molecules}}{\text{mol } H_2SO_4} \times \frac{7 \text{ atoms}}{\text{molecule}} =$$

$$\frac{4.215 \times 10^{24} \text{ atoms}}{\text{mol } H_2SO_4}$$

12. A mole is the amount of a compound which represents 6.022×10^{23} molecules or formula units and weighs (a) the molecular weight in grams, or (b) the formula weight in grams.

13. One mole of NaCl is a correct term, meaning the amount of NaCl which weighs 58.5 g, the formula weight in grams. But the term molecular weight implies the weight of Avogadro's number of molecules, which is incorrect for an ionic substance like sodium chloride. There are no molecules of sodium chloride.

14. In calculating the empirical formula of a compound from its percentage composition, we choose to start with 100 grams of the compound because the grams of an element in 100 g of the compound is conveniently the same number as the percentage of that element in the compound. For instance, if you have 100 g of a substance which is 25.3% carbon, then that sample contains 25.3 g of carbon.

15. The following statements are correct: a,b,e,g,h,k,l, and n.

C. Problems.

1. Formula weight; molecular weight

(a)	CsBr	1 Cs	132.9	(b)	K_2SO_4	2 K	78.2
		1 Br	79.9			1 S	32.1
			212.8			4 O	64.0
							174.3
(c)	$Ba(NO_3)_2$	1 Ba	137.3	(d)	C_3H_7OH	3 C	36.0
		2 N	28.0			8 H	8.0
		6 O	96.0			1 O	16.0
			261.3				60.0
(e)	$H_2C_2O_4$	2 H	2.0	(f)	Fe_2O_3	2 Fe	111.6
		2 C	24.0			3 O	48.0
		4 O	64.0				159.6
			90.0				
(g)	$C_{18}H_{36}O_{18}$	18 C	216.0	(h)	$Al_2(CrO_4)_3$	2 Al	54.0
		36 H	36.0			3 Cr	156.0
		18 O	288.0			12 O	144.0
			540.0				354.0
(i)	$(NH_4)_2CO_3$	2 N	28.0	(j)	$CaCl_2 \cdot 2H_2O$	1 Ca	40.1
		8 H	8.0			2 Cl	71.0
		1 C	12.0			4 H	4.0
		3 O	48.0			2 O	32.0
			96.0				147.1

2. Formula weights.

(a) KOH

1 K	39.1
1 O	16.0
1 H	1.0
	56.1

(b) Ag_2SO_4

2 Ag	215.8
1 S	32.1
4 O	64.0
	311.9

(c) $CrCl_3$

1 Cr	52.0
3 Cl	106.5
	158.5

(d) $Cu(NO_3)_2$

1 Cu	63.5
2 N	28.0
6 O	96.0
	187.5

(e) $Sr(HCO_3)_2$

1 Sr	87.6
2 H	2.0
2 C	24.0
6 O	96.0
	209.6

(f) $C_6H_5NO_2$

6 C	72.0
5 H	5.0
1 N	14.0
2 O	32.0
	123.0

(g) $C_{18}H_{36}O_{18}$

18 C	216.0
36 H	36.0
18 O	288.0
	540.0

(h) $K_3Fe(CN)_6$

3 K	117.3
1 Fe	55.8
6 C	72.0
6 N	84.0
	329.1

(i) $Mg_3(AsO_4)_2$

3 Mg	72.9
2 As	149.8
8 O	128.0
	350.7

(j) C_4H_9I

4 C	48.0
9 H	9.0
1 I	126.9
	183.9

3. Moles of atoms.

(a) $22.5 \text{ g Zn} \times \frac{1 \text{ mol}}{65.4 \text{ g}} = 0.344 \text{ mol Zn}$

(b) $0.688 \text{ g Mg} \times \frac{1 \text{ mol}}{24.3 \text{ g}} = 2.83 \times 10^{-2} \text{ mol Mg}$

(c) $4.5 \times 10^{22} \text{ atoms Cu} \times \frac{1 \text{ mol Cu}}{6.022 \times 10^{23} \text{ atoms Cu}} = 7.5 \times 10^{-2} \text{ mol Cu}$

(d) $382 \text{ g Co} \times \frac{1 \text{ mol}}{58.9 \text{ g}} = 6.49 \text{ mol Co}$

(e) $0.055 \text{ g Sn} \times \frac{1 \text{ mol}}{118.7 \text{ g}} = 4.6 \times 10^{-4} \text{ mol Sn}$

(f) $8.4 \times 10^{24} \text{ molecules } N_2 \times \frac{2 \text{ atoms N}}{1 \text{ molecule } N_2} \times \frac{1 \text{ mol N atoms}}{6.022 \times 10^{23} \text{ atoms N}}$

$= 28 \text{ mol N atoms}$

4. Number of moles

(a) $25.0 \text{ g NaOH} \times \frac{1 \text{ mol NaOH}}{40.0 \text{ g NaOH}} = 0.625 \text{ mol NaOH}$

(b) $44.0 \text{ g } Br_2 \times \frac{1 \text{ mol } Br_2}{159.8 \text{ g } Br_2} = 0.275 \text{ mol } Br_2$

(c) $0.684 \text{ g } MgCl_2 \times \frac{1 \text{ mol } MgCl_2}{95.3 \text{ g } MgCl_2} = 7.18 \times 10^{-3} \text{ mol } MgCl_2$

(d) $14.8 \text{ g } CH_3OH \times \frac{1 \text{ mol } CH_3OH}{32.0 \text{ g } CH_3OH} = 0.463 \text{ mol } CH_3OH$

(e) $2.88 \text{ g } Na_2SO_4 \times \frac{1 \text{ mol } Na_2SO_4}{142.1 \text{ g } Na_2SO_4} = 2.03 \times 10^{-2} \text{ mol } Na_2SO_4$

(f) $4.20 \text{ lb } ZnI_2 \times \frac{454 \text{ g}}{1 \text{ lb}} \times \frac{1 \text{ mol } ZnI_2}{319.2 \text{ g } ZnI_2} = 5.97 \text{ mol } ZnI_2$

5. Number of grams.

(a) $0.550 \text{ mol Au} \times \frac{197.0 \text{ g Au}}{1 \text{ mol Au}} = 108 \text{ g Au}$

(b) $15.8 \text{ mol } H_2O \times \frac{18.0 \text{ g } H_2O}{\text{mol } H_2O} = 284 \text{ g } H_2O$

(c) $12.5 \text{ mol } Cl_2 \times \frac{71.0 \text{ g } Cl_2}{\text{mol } Cl_2} = 888 \text{ g } Cl_2$

(d) $3.15 \text{ mol } NH_4NO_3 \times \frac{80.0 \text{ g } NH_4NO_3}{\text{mol } NH_4NO_3} = 252 \text{ g } NH_4NO_3$

(e) $4.25 \times 10^{-4} \text{ mol } H_2SO_4 \times \frac{98.1 \text{ g } H_2SO_4}{\text{mol } H_2SO_4} = 0.0417 \text{ g } H_2SO_4$

(f) $4.5 \times 10^{22} \text{ molecules } CCl_4 \times \frac{1 \text{ mol } CCl_4}{6.022 \times 10^{23} \text{ molecules}} \times \frac{154 \text{ g } CCl_4}{\text{mol } CCl_4}$

$= 11.5 \text{ g } CCl_4$

(g) $0.00255 \text{ mol Ti} \times \frac{47.9 \text{ g Ti}}{\text{mol Ti}} = 0.122 \text{ g Ti}$

(h) $1.5 \times 10^{16} \text{ atoms S} \times \frac{32.1 \text{ g S}}{6.022 \times 10^{23} \text{ atoms}} = 8.0 \times 10^{-7} \text{ g S}$

6. Number of molecules.

(a) $1.26 \text{ mol } O_2 \times \frac{6.022 \times 10^{23} \text{ molecules } O_2}{\text{mol } O_2}$

$= 7.59 \times 10^{23} \text{ molecules } O_2$

(b) $0.56 \text{ mol } C_6H_6 \times \frac{6.022 \times 10^{23} \text{ molecules } C_6H_6}{\text{mol } C_6H_6}$

$= 3.4 \times 10^{23}$ molecules C_6H_6

(c) $16.0 \text{ g } CH_4 \times \dfrac{6.022 \times 10^{23} \text{ molecules } CH_4}{16.0 \text{ g } CH_4}$

$= 6.02 \times 10^{23}$ molecules CH_4

(d) $1000 \text{ g HCl} \times \dfrac{6.022 \times 10^{23} \text{ molecules HCl}}{36.5 \text{ g HCl}}$

$= 1.65 \times 10^{25}$ molecules HCl

7. Number of grams.

(a) $1 \text{ atom Pb} \times \dfrac{207.2 \text{ g Pb}}{6.022 \times 10^{23} \text{ atoms Pb}} = 3.441 \times 10^{-22} \text{ g Pb}$

(b) $1 \text{ atom Ag} \times \dfrac{107.9 \text{ g Ag}}{6.022 \times 10^{23} \text{ atoms Ag}} = 1.791 \times 10^{-22} \text{ g Ag}$

(c) $1 \text{ molecule } H_2O \times \dfrac{18.0 \text{ g } H_2O}{6.022 \times 10^{23} \text{ molecules } H_2O}$

$= 2.99 \times 10^{-23}$ g H_2O

(d) $1 \text{ molecule } C_3H_5(NO_3)_3 \times \dfrac{227 \text{ g } C_3H_5(NO_3)_3}{6.022 \times 10^{23} \text{ molecules } C_3H_5(NO_3)_3}$

$= 3.77 \times 10^{-22}$ g $C_3H_5(NO_3)_3$

8. (a) $8.66 \text{ mol Cu} \times \dfrac{63.5 \text{ g Cu}}{1 \text{ mol Cu}} = 550 \text{ g Cu}$

(b) $125 \text{ mol Au} \times \dfrac{197.0 \text{ g Au}}{1 \text{ mol Au}} \times \dfrac{1 \text{ kg}}{1000 \text{ g}} = 24.6 \text{ kg Au}$

(c) $10 \text{ atoms C} \times \dfrac{1 \text{ mol C}}{6.022 \times 10^{23} \text{ atoms C}} = 1.7 \times 10^{-23} \text{ mol C}$

(d) $5000 \text{ molecules } CO_2 \times \dfrac{1 \text{ mol } CO_2}{6.022 \times 10^{23} \text{ molecules } CO_2}$

$= 8.3 \times 10^{-21}$ mol CO_2

(e) $28.4 \text{ g S} \times \dfrac{1 \text{ mol S}}{32.1 \text{ g S}} = 0.885 \text{ mol S}$

(f) $2.50 \text{ kg NaCl} \times \dfrac{1000 \text{ g}}{\text{kg}} \times \dfrac{1 \text{ mol NaCl}}{58.5 \text{ g NaCl}} = 42.7 \text{ mol NaCl}$

(g) $42.4 \text{ g Mg} \times \frac{6.022 \times 10^{23} \text{ atoms Mg}}{24.3 \text{ g Mg}} = 1.05 \times 10^{24} \text{ atoms Mg}$

(h) $485 \text{ mL } Br_2 \times \frac{3.12 \text{ g}}{\text{mL}} \times \frac{1 \text{ mol } Br_2}{159.8 \text{ g } Br_2} = 9.47 \text{ mol } Br_2$

9. One mole of carbon disulfide (CS_2) contains:

(a) 6.022×10^{23} molecules CS_2

(b) $6.022 \times 10^{23} \text{ molecules } CS_2 \times \frac{1 \text{ C atom}}{1 \text{ molecule } CS_2}$

$= 6.022 \times 10^{23}$ C atoms

(c) $6.022 \times 10^{23} \text{ molecules } CS_2 \times \frac{2 \text{ S atoms}}{1 \text{ molecule } CS_2}$

$= 1.204 \times 10^{24}$ S atoms

(d) $6.022 \times 10^{23} \text{ atoms} + 1.204 \times 10^{24} \text{ atoms} = 1.806 \times 10^{24} \text{ atoms}$

10. $0.350 \text{ mol } P_4 \times \frac{6.022 \times 10^{23} \text{ molecules } P_4}{\text{mol } P_4} \times \frac{4 \text{ atoms P}}{\text{molecule } P_4}$

$= 8.43 \times 10^{23}$ atoms P

11. Conversion: g K $\rightarrow$ mol K $\rightarrow$ mol Na $\rightarrow$ g Na

$10.0 \text{ g K} \times \frac{1 \text{ mol K}}{39.1 \text{ g K}} \times \frac{1 \text{ mol Na}}{1 \text{ mol K}} \times \frac{23.0 \text{ g Na}}{\text{mol Na}} = 5.88 \text{ g Na}$

12. $\frac{1.79 \times 10^{-23} \text{ g}}{\text{atom}} \times \frac{6.022 \times 10^{23} \text{ atoms}}{\text{atomic weight}} = 10.8 \text{ g/atomic weight}$

13. $6.022 \times 10^{23} \text{ sheets} \times \frac{4.60 \text{ cm}}{500 \text{ sheets}} \times \frac{1 \text{ m}}{100 \text{ cm}} = 5.54 \times 10^{19} \text{ m}$

14. $\frac{6.022 \times 10^{23} \text{ dollars}}{5.0 \times 10^{9} \text{ people}} = 1.2 \times 10^{14} \text{ dollars/person}$

15. (a) $1 \text{ mi}^3 \times \left(\frac{5280 \text{ ft}}{\text{mile}}\right)^3 \times \left(\frac{12.0 \text{ in.}}{\text{ft}}\right)^3 \times \left(\frac{2.54 \text{ cm}}{\text{in.}}\right)^3 \times \frac{20 \text{ drops}}{1.0 \text{ cm}^3}$

$= 8.3 \times 10^{16} \text{ drops/mi}^3$

(b) $6.022 \times 10^{23} \text{ drops} \times \frac{1 \text{ mi}^3}{8.3 \times 10^{16} \text{ drops}} = 7.3 \times 10^{6} \text{ mi}^3$ (cubic miles)

16. 1 mol Ag = 107.9 g

(a) $107.9 \text{ g Ag} \times \frac{1 \text{ cm}^3 \text{ Ag}}{10.5 \text{ g Ag}} = 10.3 \text{ cm}^3 \text{ Ag}$

(b) $10.3 \text{ cm}^3 = \text{volume of cube} = (\text{side})^3$

$\text{side} = \sqrt[3]{10.3 \text{ cm}^3} = 2.18 \text{ cm}$

17. How many atoms of oxygen in:

(a) $16.0 \text{ g } O_2 \times \frac{1 \text{ mol O}}{16.0 \text{ g}} \times \frac{6.022 \times 10^{23} \text{ atoms}}{\text{mol O}} = 6.02 \times 10^{23} \text{ atoms O}$

(b) $0.622 \text{ mol MgO} \times \frac{1 \text{ mol O}}{\text{mol MgO}} \times \frac{6.022 \times 10^{23} \text{ atoms}}{\text{mol O}}$

$= 3.75 \times 10^{23} \text{ atoms O}$

(c) $6.00 \times 10^{22} \text{ molecules } C_6H_{12}O_6 \times \frac{6 \text{ atoms O}}{\text{molecule } C_6H_{12}O_6}$

$= 3.60 \times 10^{23} \text{ atoms O}$

(d) $5.0 \text{ mol } MnO_2 \times \frac{2 \text{ mol O}}{\text{mol } MnO_2} \times \frac{6.022 \times 10^{23} \text{ atoms O}}{\text{mol O}}$

$= 6.0 \times 10^{24} \text{ atoms O}$

(e) $250 \text{ g } MgCO_3 \times \frac{1 \text{ mol } MgCO_3}{84.3 \text{ g } MgCO_3} \times \frac{3 \text{ mol O}}{\text{mol } MgCO_3} \times \frac{6.022 \times 10^{23} \text{ atoms O}}{\text{mol O}}$

$= 5.36 \times 10^{24} \text{ atoms O}$

(f) $5.0 \times 10^{18} \text{ molecules } H_2O \times \frac{1 \text{ atom O}}{\text{molecule } H_2O} = 5.0 \times 10^{18} \text{ atoms O}$

18. Calculate the number of grams of:

(a) silver in 25.0 g AgBr

$25.0 \text{ g AgBr} \times \frac{107.9 \text{ g Ag}}{187.8 \text{ g AgBr}} = 14.4 \text{ g Ag}$

(b) chlorine in 5.00 g $PbCl_2$

$5.00 \text{ g } PbCl_2 \times \frac{71.0 \text{ g Cl}}{278.2 \text{ g } PbCl_2} = 1.28 \text{ g Cl}$

(c) nitrogen in 6.34 mol $(NH_4)_3PO_4$

$6.34 \text{ mol } (NH_4)_3PO_4 \times \frac{42.0 \text{ g N}}{\text{mol } (NH_4)_3PO_4} = 266 \text{ g N}$

(d) oxygen in 8.45×10^{22} molecules SO_3

$8.45 \times 10^{22} \text{ molecules } SO_3 \times \frac{1 \text{ mol } SO_3}{6.022 \times 10^{23} \text{ molecules}} \times \frac{48.0 \text{ g O}}{\text{mol } SO_3}$

$= 6.74 \text{ g O}$

(e) hydrogen in 45.0 g C_3H_8O

$$45.0 \text{ g } C_3H_8O \times \frac{8.0 \text{ g H}}{60.0 \text{ g } C_3H_8O} = 6.00 \text{ g H}$$

19. $1.00 \text{ L} \times \frac{1000 \text{ mL}}{\text{L}} \times \frac{1.55 \text{ g sol'n}}{1.00 \text{ mL sol'n}} \times \frac{0.650 \text{ g } H_2SO_4}{1.00 \text{ g sol'n}} \times \frac{1 \text{ mol } H_2SO_4}{98.1 \text{ g } H_2SO_4}$

$= 10.3 \text{ mol } H_2SO_4$

20. $100 \text{ mL sol'n} \times \frac{1.42 \text{ g}}{\text{mL sol'n}} \times \frac{0.720 \text{ g } HNO_3}{1.00 \text{ g sol'n}} \times \frac{1 \text{ mol } HNO_3}{63.0 \text{ g } HNO_3} = 1.62 \text{ mol } HNO_3$

21. (a) First determine the formula weight of each compound. CO_2, 44.0 g; O_2, 32.0 g; H_2O, 18.0 g; CH_3OH, 32.0 g. The 1.00 g sample with the lowest formula weight will contain the most molecules. The H_2O will contain the most molecules.

(b) By inspection the H_2O or CH_3OH will contain the most atoms.

$$1.00 \text{ g } H_2O \times \frac{1 \text{ mol}}{18.0 \text{ g}} \times \frac{3 \times 6.022 \times 10^{23} \text{ atoms}}{\text{mol}}$$

$= 1.00 \times 10^{23}$ atoms

$$1.00 \text{ g } CH_3OH \times \frac{1 \text{ mol}}{32.0 \text{ g}} \times \frac{6 \times 6.022 \times 10^{23} \text{ atoms}}{\text{mol}}$$

$= 1.13 \times 10^{23}$ atoms

The 1.00 g sample of CH_3OH contains the most atoms.

22. 1 mol Fe_2S_3 = 207.9 g Fe_2S_3 = 6.022×10^{23} formula units.

$$\text{g } Fe_2S_3 = 6.022 \times 10^{23} \text{ atoms} \times \frac{207.9 \text{ g } Fe_2S_3}{6.022 \times 10^{23} \text{ formula units}}$$

$$\times \frac{1 \text{ formula unit}}{5 \text{ atoms}} = 41.58 \text{ g } Fe_2S_3$$

23. Percentage composition

(a) NaBr

Na	23.0	$\frac{23.0}{102.9} \times 100\% = 22.4\%$ Na
Br	79.9	$\frac{79.9}{102.9} \times 100\% = 77.6\%$ Br
	102.9	

(b) $KHCO_3$

K	39.1	$\frac{39.1}{100.1} \times 100\% = 39.1\%$ K
H	1.0	$\frac{1.0}{100.1} \times 100\% = 1.0\%$ H
C	12.0	$\frac{12.0}{100.1} \times 100\% = 12.0\%$ C
3 O	48.0	$\frac{48.0}{100.1} \times 100\% = 48.0\%$ O
	100.1	

(c) $FeCl_3$

Fe	55.8	$\frac{55.8}{162.3} \times 100\% = 34.4\%$ Fe
3 Cl	106.5	$\frac{106.5}{162.3} \times 100\% = 65.6\%$ Cl
	162.3	

(d) $SiCl_4$

Si	28.1	$\frac{28.1}{170.1} \times 100\% = 16.5\%$ Si
4 Cl	142.0	$\frac{142.0}{170.1} \times 100\% = 83.5\%$ Cl
	170.1	

(e) $Al_2(SO_4)_3$

2 Al	54.0	$\frac{54.0}{342.3} \times 100\% = 15.8\%$ Al
3 S	96.3	$\frac{96.3}{342.3} \times 100\% = 28.1\%$ S
12 O	192.0	$\frac{192.0}{342.3} \times 100\% = 56.1\%$ O
	342.3	

(f) $AgNO_3$

Ag	107.9	$\frac{107.9}{169.9} \times 100\% = 63.5\%$ Ag
N	14.0	$\frac{14.0}{169.9} \times 100\% = 8.2\%$ N
3 O	48.0	$\frac{48.0}{169.9} \times 100\% = 28.3\%$ O
	169.9	

24. Percentage composition

(a) $ZnCl_2$

Zn	65.4	$\frac{65.4}{136.4} \times 100\% = 47.9\%$ Zn
2 Cl	71.0	$\frac{71.0}{136.4} \times 100\% = 52.1\%$ Cl
	136.4	

(b) $NH_4C_2H_3O_2$

N	14.0	$\frac{14.0}{77.0} \times 100\% = 18.2\%$ N
7 H	7.0	$\frac{7.0}{77.0} \times 100\% = 9.1\%$ H
2 C	24.0	$\frac{24.0}{77.0} \times 100\% = 31.2\%$ C
2 O	32.0	$\frac{32.0}{77.0} \times 100\% = 41.6\%$ O
	77.0	

(c) MgP_2O_7

Mg	24.3	$\frac{24.3}{198.3} \times 100\% = 12.3\%$ Mg
2 P	62.0	$\frac{62.0}{198.3} \times 100\% = 31.3\%$ P
7 O	112.0	$\frac{112.0}{198.3} \times 100\% = 56.5\%$ O
	198.3	

(d) $(NH_4)_2SO_4$

2 N	28.0	$\frac{28.0}{132.1} \times 100\% = 21.2\%$ N
8 H	8.0	$\frac{8.0}{132.1} \times 100\% = 6.1\%$ H
S	32.1	$\frac{32.1}{132.1} \times 100\% = 24.3\%$ S
4 O	64.0	$\frac{64.0}{132.1} \times 100\% = 48.4\%$ O
	132.1	

(e) $Fe(NO_3)_3$

Fe	55.8	$\frac{55.8}{241.8} \times 100\% = 23.1\%$ Fe
3 N	42.0	$\frac{42.0}{241.8} \times 100\% = 17.4\%$ N
9 O	144.0	$\frac{144.0}{241.8} \times 100\% = 59.6\%$ O
	241.8	

(f) ICl_3

I	126.9	$\frac{126.9}{233.4} \times 100\% = 54.4\%$ I
3 Cl	106.5	$\frac{106.5}{233.4} \times 100\% = 45.6\%$ Cl
	233.4	

25. Calculate the percentage of iron in the following compounds:

(a) FeO

Fe	55.8	$\frac{55.8}{71.8} \times 100\% = 77.7\%$ Fe
O	16.0	
	71.8	

(b) Fe_2O_3

2 Fe	111.6	$\frac{111.6}{159.6} \times 100\% = 69.9\%$ Fe
3 O	48.0	
	159.6	

(c) Fe_3O_4

3 Fe	167.4	$\frac{167.4}{231.4} \times 100\% = 72.3\%$ Fe
4 O	64.0	
	231.4	

(d) $K_4Fe(CN)_6$

4 K	156.4	$\frac{55.8}{368.2} \times 100\% = 15.2\%$ Fe
1 Fe	55.8	
6 C	72.0	
6 N	84.0	
	368.2	

26. Percentage chlorine.

(a) KCl

1 K	39.1	$\frac{35.5}{74.6} \times 100\% = 47.6\%$ Cl
1 Cl	35.5	
	74.6	

(b) $BaCl_2$

1 Ba	137.3	$\frac{71.0}{208.3} \times 100\% = 34.1\%$ Cl
2 Cl	71.0	
	208.3	

(c) $SiCl_4$

1 Si	28.1	$\frac{142.0}{170.1} \times 100\% = 83.5\%$ Cl
4 Cl	142.0	
	170.1	

(d) LiCl

1 Li	6.9	$\frac{35.5}{42.4} \times 100\% = 83.7\%$ Cl	
1 Cl	35.5		
	42.4		

Highest % Cl is LiCl; lowest % Cl is $BaCl_2$

27. 14.2 g oxide
 6.20 g P
 8.00 g oxygen

$$\frac{6.20 \text{ g P}}{14.20 \text{ g}} \times 100\% = 43.7\% \text{ P}$$

$$\frac{8.00 \text{ g O}}{14.20 \text{ g}} \times 100\% = 56.3\% \text{ O}$$

28. Percent composition of ethylene chloride.

 6.00 g C
 1.00 g H
 17.75 g Cl
 24.75 g total

$$\frac{6.00 \text{ g C}}{24.75 \text{ g}} \times 100\% = 24.2\% \text{ C}$$

$$\frac{1.00 \text{ g H}}{24.75 \text{ g}} \times 100\% = 4.0\% \text{ H}$$

$$\frac{17.75 \text{ g Cl}}{24.75 \text{ g}} \times 100\% = 71.72\% \text{ Cl}$$

29. From the formula, 2 Li (13.8 g) combine with 1 S (32.1 g).

$$\frac{13.8 \text{ g Li}}{32.1 \text{ g S}} \times 20.0 \text{ g S} = 8.60 \text{ g Li will combine with } 20.0 \text{ g S}$$

30. Calculate the percentage of

(a) Hg in $HgCO_3$

1 Hg	200.6	$\frac{200.6}{260.6} \times 100\% = 76.98\%$ Hg
1 C	12.0	
3 O	48.0	
	260.6	

(b) O in $Ca(ClO_3)_2$

1 Ca	40.1	$\frac{96.0}{207.1} \times 100\% = 46.4\%$ O
2 Cl	71.0	
6 O	96.0	
	207.1	

(c) N in $C_{10}H_{14}N_2$

2 N	28.0	$\frac{28.0}{162.0} \times 100\% = 17.3\%$ N
10 C	120.0	
14 H	14.0	
	162.0	

(d) Mg in $C_{55}H_{72}MgN_4O_5$

Mg	24.3	$\frac{24.3}{892.3} \times 100\% = 2.72\%$ Mg
55 C	660.0	
72 H	72.0	
4 N	56.0	
5 O	80.0	
	892.3	

31. (a) H_2O (b) N_2O_3 (c) equal (d) $KClO_3$ (e) $KHSO_4$ (f) Na_2CrO_4

32. Calculation of empirical formulas from percentage compositions.

(a) Step 1. Express each element as grams/100 g material

63.6% N = 63.6 g N/100 g material

36.4% O = 36.4 g O/100 g material

Step 2. Calculate relative moles of each element

$$63.6 \text{ g N} \times \frac{1 \text{ mol N}}{14.0 \text{ g N}} = 4.54 \text{ mol N}$$

$$36.4 \text{ g O} \times \frac{1 \text{ mol O}}{16.0 \text{ g O}} = 2.28 \text{ mol O}$$

Step 3. Change these moles to whole numbers by dividing each by the smaller number.

$$\frac{4.54 \text{ mol N}}{2.28} = 1.99 \text{ mol N}; \quad \frac{2.28 \text{ mol O}}{2.28} = 1.00 \text{ mol O}$$

The simplest ratio of N:O is 2:1

The empirical formula is, therefore, N_2O

(b) 46.7% N, 53.3% O; 46.7 g N, 53.3 g O

$$46.7 \text{ g N} \times \frac{1 \text{ mol N}}{14.0 \text{ g N}} = 3.34 \text{ mol N}$$

$$53.3 \text{ g O} \times \frac{1 \text{ mol O}}{16.0 \text{ g O}} = 3.33 \text{ mol O}$$

$$\frac{3.34 \text{ mol N}}{3.33} = 1.00 \text{ mol N}; \quad \frac{3.33 \text{ mol O}}{3.33} = 1.00 \text{ mol O}$$

The empirical formula is NO

(c) 25.9% N, 74.1% O; 25.9 g N, 74.1 g O

$$25.9 \text{ g N} \times \frac{1 \text{ mol N}}{14.0 \text{ g N}} = 1.85 \text{ mol N}$$

$$74.1 \text{ g O} \times \frac{1 \text{ mol O}}{16.0 \text{ g O}} = 4.63 \text{ mol O}$$

$$\frac{1.85 \text{ mol N}}{1.85} = 1.00 \text{ mol N}; \quad \frac{4.63 \text{ mol O}}{1.85} = 2.5 \text{ mol O}$$

Since these values are not whole numbers, multiply each by 2 to change them to whole numbers.

1.00 mol N × 2 = 2 mol N; 2.5 mol O × 2 = 5 mol O

The empirical formula is N_2O_5

(d) 43.4% Na, 11.3% C, 45.3% O; 43.4 g Na, 11.3 g C, 45.3 g O

$$43.4 \text{ g Na} \times \frac{1 \text{ mol Na}}{23.0 \text{ g}} = 1.89 \text{ mol Na}$$

$$11.3 \text{ g C} \times \frac{1 \text{ mol C}}{12.0 \text{ g}} = 0.942 \text{ mol C}$$

$$45.3 \text{ g O} \times \frac{1 \text{ mol O}}{16.0 \text{ g}} = 2.83 \text{ mol O}$$

$$\frac{1.89}{0.942} = 2.01 \text{ mol Na}; \quad \frac{0.942}{0.942} = 1.00 \text{ mol C}; \quad \frac{2.83}{0.942} = 3.00 \text{ mol O}$$

The empirical formula is Na_2CO_3

(e) 18.8% Na, 29.0% Cl, 52.3% O; 18.8 g Na, 29.0 g Cl, 52.3 g O

$18.8 \text{ g Na} \times \frac{1 \text{ mol Na}}{23.0 \text{ g}} = 0.817 \text{ mol Na}$; $29.0 \text{ g Cl} \times \frac{1 \text{ mol Cl}}{35.5 \text{ g}}$

$= 0.817 \text{ mol Cl}$; $52.3 \text{ g O} \times \frac{1 \text{ mol O}}{16.0 \text{ g}} = 3.27 \text{ mol O}$

$\frac{0.817}{0.817} = 1.00$ mol Na and Cl; $\frac{3.27}{0.817} = 4.00 \text{ mol O}$

The empirical formula is $NaClO_4$

(f) 72.02% Mn, 27.98% O; 72.02 g Mn, 27.98 g O

$72.02 \text{ g Mn} \times \frac{1 \text{ mol Mn}}{54.9 \text{ g}} = 1.31 \text{ mol Mn}$

$27.98 \text{ g O} \times \frac{1 \text{ mol O}}{16.0 \text{ g}} = 1.75 \text{ mol O}$

$\frac{1.31 \text{ mol Mn}}{1.31} = 1.00 \text{ mol Mn}$; $\frac{1.75 \text{ mol O}}{1.31} = 1.34 \text{ mol O}$

Multiply both values by 3 to give whole numbers.

$1.00 \times 3 = 3.00$ mol Mn; $1.34 \times 3 = 4.02$ mol O

The empirical formula is Mn_3O_4

33. Calculation of empirical formulas from percentage compositions.

(a) 64.1% Cu, 35.9% Cl; 64.1 g Cu, 35.9 g Cl

$64.1 \text{ g Cu} \times \frac{1 \text{ mol Cu}}{63.5 \text{ g}} = 1.01 \text{ mol Cu}$

$35.9 \text{ g Cl} \times \frac{1 \text{ mol Cl}}{35.5 \text{ g}} = 1.01 \text{ mol Cl}$

The empirical formula is CuCl

(b) 47.2% Cu, 52.8% Cl; 47.2 g Cu, 52.8 g Cl

$47.2 \text{ g Cu} \times \frac{1 \text{ mol Cu}}{63.5 \text{ g}} = 0.743 \text{ mol Cu}$

$52.8 \text{ g Cl} \times \frac{1 \text{ mol Cl}}{35.5 \text{ g}} = 1.49 \text{ mol Cl}$

$Cu_{\frac{0.743}{0.743}}Cl_{\frac{1.49}{0.743}} = Cu_{1.00}Cl_{2.00}$

The empirical formula is $CuCl_2$

(c) 51.9% Cr, 48.1% S; 51.9 g Cr, 48.1 g S

$51.9 \text{ g Cr} \times \frac{1 \text{ mol Cr}}{52.0 \text{ g}} = 0.998 \text{ mol Cr}$

$48.1 \text{ g S} \times \frac{1 \text{ mol S}}{32.1 \text{ g}} = 1.50 \text{ mol S}$

$$Cr_{\frac{0.998}{0.998}}S_{\frac{1.50}{0.998}} = Cr_{1.00}S_{1.50} = Cr_{2.00}S_{3.00}$$

The empirical formula is Cr_2S_3

(d) 55.3% K, 14.6% P, 30.1% O; 55.3 g K, 14.6 g P, 30.1 g O

$$55.3\ g\ K \times \frac{1\ mol\ K}{39.1\ g} = 1.41\ mol\ K$$

$$14.6\ g\ P \times \frac{1\ mol\ P}{31.0\ g} = 0.471\ mol\ P$$

$$30.1\ g\ O \times \frac{1\ mol\ O}{16.0\ g} = 1.88\ mol\ O$$

$$K_{\frac{1.41}{0.471}}P_{\frac{0.471}{0.471}}O_{\frac{1.88}{0.471}} = K_{2.99}P_{1.00}O_{3.99}$$

The empirical formula is K_3PO_4

(e) 38.9% Ba, 29.4% Cr, 31.7% O; 38.9 g Ba, 29.4 g Cr, 31.7 g O

$$38.9\ g\ Ba \times \frac{1\ mol\ Ba}{137.3\ g} = 0.283\ mol\ Ba$$

$$29.4\ g\ Cr \times \frac{1\ mol\ Cr}{52.0\ g} = 0.565\ mol\ Cr$$

$$31.7\ g\ O \times \frac{1\ mol\ O}{16.0\ g} = 1.98\ mol\ O$$

$$Ba_{\frac{0.283}{0.283}}Cr_{\frac{0.565}{0.283}}O_{\frac{1.98}{0.283}} = Ba_{1.00}Cr_{1.99}O_{7.00}$$

The empirical formula is $BaCr_2O_7$

(f) 3.99% P, 82.3% Br, 13.7% Cl; 3.99 g P, 82.3 g Br, 13.7 g Cl

$$3.99\ g\ P \times \frac{1\ mol\ P}{31.0\ g} = 0.129\ mol\ P$$

$$82.3\ g\ Br \times \frac{1\ mol\ Br}{79.9\ g} = 1.03\ mol\ Br$$

$$13.7\ g\ Cl \times \frac{1\ mol\ Cl}{35.5\ g} = 0.386\ mol\ Cl$$

$$P_{\frac{0.129}{0.129}}Br_{\frac{1.03}{0.129}}Cl_{\frac{0.386}{0.129}} = P_{1.00}Br_{7.98}Cl_{2.99}$$

The empirical formula is PBr_8Cl_3

34. Empirical formula

$$3.996\ g\ Sn \times \frac{1\ mol\ Sn}{118.7\ g} = 0.0337\ mol\ Sn$$

$$1.077\ g\ O \times \frac{1\ mol\ O}{16.0\ g} = 0.0673\ mol\ O$$

$Sn_{\frac{0.0337}{0.0337}}O_{\frac{0.0673}{0.0337}} = Sn_{1.00}O_{2.00}$; the empirical formula is SnO_2

35. Empirical formula

5.454 g product - 3.054 g V = 2.400 g O

$3.054 \text{ g V} \times \frac{1 \text{ mol V}}{50.9 \text{ g}} = 0.0600 \text{ mol V}$; $2.400 \text{ g O} \times \frac{1 \text{ mol O}}{16.0 \text{ g}}$

= 0.1500 mol O

$V_{\frac{0.0600}{0.0600}}O_{\frac{0.150}{0.0600}} = V_{1.00}O_{2.50} = V_{2.00}O_{5.00}$; Empirical formula is V_2O_5

36. According to the formula, 1 mol (65.4 g) Zn combines with 1 mol (32.1 g) S.

$19.5 \text{ g Zn} \times \frac{32.1 \text{ g S}}{65.4 \text{ g Zn}} = 9.57 \text{ g S}$

19.5 g Zn require 9.57 g S for complete reaction. Therefore, there is not sufficient S present (9.40 g) to react with the Zn.

37. Molecular weight, 110; 65.45% C, 5.45% H, 29.09% O

$65.45 \text{ g C} \times \frac{1 \text{ mol C}}{12.0 \text{ g}} = 5.45 \text{ mol C}$

$5.45 \text{ g H} \times \frac{1 \text{ mol H}}{1.0 \text{ g}} = 5.45 \text{ mol H}$

$29.09 \text{ g O} \times \frac{1 \text{ mol O}}{16.0 \text{ g}} = 1.82 \text{ mol O}$

$C_{\frac{5.45}{1.82}}H_{\frac{5.45}{1.82}}O_{\frac{1.82}{1.82}} = C_{2.99}H_{2.99}O_{1.00} = C_3H_3O$ (empirical formula)

Empirical formula weight = 55.0 (36.0 + 3.0 + 16.0)

$\frac{\text{molecular weight}}{\text{empirical formula weight}} = \frac{110}{55} = 2$

The molecular formula is twice that of the empirical formula.

Molecular formula = $(C_3H_3O)_2 = C_6H_6O_2$

38. Molecular weight, 180; 40.4% C, 6.7% H, 53.3% O

$40.0 \text{ g C} \times \frac{1 \text{ mol C}}{12.0 \text{ g}} = 3.33 \text{ mol C}$

$6.7 \text{ g H} \times \frac{1 \text{ mol H}}{1.0 \text{ g}} = 6.7 \text{ mol H}$; $53.3 \text{ g O} \times \frac{1 \text{ mol O}}{16.0 \text{ g}} = 3.33 \text{ mol O}$

$C_{\frac{3.33}{3.33}}H_{\frac{6.7}{3.33}}O_{\frac{3.33}{3.33}} = C_{1.00}H_{2.00}O_{1.00}$

Empirical formula is CH_2O; empirical formula weight = 30.0

$$\frac{\text{Molecular weight}}{\text{Empirical formula weight}} = \frac{180}{30.0} = 6$$

Molecular formula = $(CH_2O)_6 = C_6H_{12}O_6$

39. Aspirin; molecular weight, 180; 60.0% C, 4.48% H, 35.5% O

$60.0\ g\ C \times \frac{1\ mol\ C}{12.0\ g} = 5.00\ mol\ C$; $4.48\ g\ H \times \frac{1\ mol\ H}{1.0\ g} = 4.48\ mol\ H$

$35.5\ g\ O \times \frac{1\ mol\ O}{16.0\ g} = 2.22\ mol\ O$; $C_{\frac{5.00}{2.22}}H_{\frac{4.48}{2.22}}O_{\frac{2.22}{2.22}} = C_{2.25}H_{2.02}O_{1.00}$

Multiply mole subscripts by 4 to get: $C_{9.00}H_{8.08}O_{4.00}$. Empirical formula, $C_9H_8O_4$; empirical formula weight, 180.

Formula weight = molecular weight, therefore, molecular formula is the same as the empirical formula, $C_9H_8O_4$.

40. Grams of oxygen in 8.50 g $Al_2(SO_4)_3$. First find the percent of oxygen in $Al_2(SO_4)_3$.

2 Al	54.0	$\frac{192.0}{342.3} \times 100\% = 56.1\%\ O$
3 S	96.3	
12 O	192.0	
	342.3	Now take 56.1% of 8.5 g.

$8.50\ g \times 0.561 = 4.77\ g\ O$ in $8.50\ g\ Al_2(SO_4)_3$

41. Empirical formula of gallium arsenide

48.2% Ga, 51.8% As; 48.2 g Ga, 51.8 g As

$48.2\ g\ Ga \times \frac{1\ mol\ Ga}{69.7\ g} = 0.692\ mol\ Ga$

$51.8\ g\ As \times \frac{1\ mol\ As}{74.9\ g} = 0.692\ mol\ As$

The empirical formula is GaAs

42. (a) 7.79% C, 92.21% Cl; 7.79 g C, 92.21 g Cl

$7.79\ g\ C \times \frac{1\ mol\ C}{12.0\ g} = 0.649\ mol\ C$

$92.21\ g\ Cl \times \frac{1\ mol\ Cl}{35.5\ g} = 2.60\ mol\ Cl$

$C_{\frac{0.649}{0.649}}Cl_{\frac{2.60}{0.649}} = C_{1.00}Cl_{4.01}$

The empirical formula is CCl_4

Empirical formula weight = 154 = molecular weight; therefore the molecular formula is the same as the empirical formula, CCl_4.

(b) 10.13% C, 89.87% Cl; 10.13 g C, 89.87 g Cl

$$10.13 \text{ g C} \times \frac{1 \text{ mol C}}{12.0 \text{ g}} = 0.844 \text{ mol C}$$

$$89.87 \text{ g Cl} \times \frac{1 \text{ mol Cl}}{35.5 \text{ g}} = 2.53 \text{ mol Cl}$$

$C_{\frac{0.844}{0.844}}Cl_{\frac{2.53}{0.844}} = C_{1.00}Cl_{3.00}$; Empirical formula is CCl_3

Empirical formula weight = 118.5; molecular weight = 237

$\frac{237}{118.5} = 2$; Molecular formula = $(CCl_3)_2 = C_2Cl_6$

(c) 25.26% C, 74.74% Cl; 25.26 g C, 74.74 g Cl

$$25.26 \text{ g C} \times \frac{1 \text{ mol C}}{12.0 \text{ g}} = 2.11 \text{ mol C}$$

$$74.74 \text{ g Cl} \times \frac{1 \text{ mol Cl}}{35.5 \text{ g}} = 2.11 \text{ mol Cl}$$

Empirical formula is CCl

Empirical formula weight = 47.5; molecular weight = 285

$\frac{285}{47.5} = 6$; Molecular formula = $(CCl)_6 = C_6Cl_6$

(d) 11.25% C, 88.75% Cl; 11.25 g C, 88.75 g Cl

$$11.25 \text{ g C} \times \frac{1 \text{ mol C}}{12.0 \text{ g}} = 0.938 \text{ mol C}$$

$$88.75 \text{ g Cl} \times \frac{1 \text{ mol Cl}}{35.5 \text{ g}} = 2.50 \text{ mol Cl}$$

$C_{\frac{0.938}{0.938}}Cl_{\frac{2.50}{0.938}} = C_{1.00}Cl_{2.67}$

Multiply subscripts by 3 to give empirical formula, C_3Cl_8.

Empirical formula weight = 320; molecular weight = 320; therefore, the molecular and empirical formulas are the same, C_3Cl_8.

C. Review Questions.

1. Balanced equations.

(a) $2\ H_2 + O_2 \rightarrow 2\ H_2O$

(b) $H_2 + Br_2 \rightarrow 2\ HBr$

(c) $2\ C + Fe_2O_3 \rightarrow 2\ Fe + 3\ CO$

(d) $2\ H_2O_2 \rightarrow 2\ H_2O + O_2$

(e) $Ba(ClO_3)_2 \xrightarrow{\Delta} BaCl_2 + 3\ O_2$

(f) $H_2SO_4 + 2\ NaOH \rightarrow 2\ H_2O + Na_2SO_4$

(g) $2\ NH_4I + Cl_2 \rightarrow 2\ NH_4Cl + I_2$

(h) $CrCl_3 + 3\ AgNO_3 \rightarrow Cr(NO_3)_3 + 3\ AgCl$

(i) $Al_2(CO_3)_3 \xrightarrow{\Delta} Al_2O_3 + 3\ CO_2$

(j) $4\ Al + 3\ C \rightarrow Al_4C_3$

2. Classification of reactions in Question C.1.

(a) Combination (b) Combination (c) Single displacement
(d) Decomposition (e) Decomposition (f) Double displacement
(g) Single displacement (h) Double displacement
(i) Decomposition (j) Combination

3. Balanced equations:

(a) $2\ SO_2 + O_2 \rightarrow 2\ SO_3$

(b) $4\ Al + 3\ MnO_2 \rightarrow 3\ Mn + 2\ Al_2O_3$

(c) $2\ Na + 2\ H_2O \rightarrow 2\ NaOH + H_2$

(d) $2\ AgNO_3 + Ni \rightarrow Ni(NO_3)_2 + 2\ Ag$

(e) $Bi_2S_3 + 6\ HCl \rightarrow 2\ BiCl_3 + 3\ H_2S$

(f) $2\ PbO_2 \xrightarrow{\Delta} 2\ PbO + O_2$

(g) $2\ LiAlH_4 \rightarrow 2\ LiH + 2\ Al + 3\ H_2$

(h) $2\ KI + Br_2 \rightarrow 2\ KBr + I_2$

(i) $2\ K_3PO_4 + 3\ BaCl_2 \rightarrow 6\ KCl + Ba_3(PO_4)_2$

4. Balanced equations:

(a) $2\ MnO_2 + CO \rightarrow Mn_2O_3 + CO_2$

(b) $2\ Mg_3N_2 + 12\ H_2O \rightarrow 6\ Mg(OH)_2 + 4\ NH_3$

(c) $4\ C_3H_5(NO_3)_3 \xrightarrow{\Delta} 12\ CO_2 + 10\ H_2O + 6\ N_2 + O_2$

(d) $4\ FeS_2 + 11\ O_2 \rightarrow 2\ Fe_2O_3 + 8\ SO_2$

(e) $2\ Cu(NO_3)_2 \xrightarrow{\Delta} 2\ CuO + 4\ NO_2 + O_2$

(f) $3\ NO_2 + H_2O \rightarrow 2\ HNO_3 + NO$

(g) $2\ Al + 3\ H_2SO_4 \rightarrow Al_2(SO_4)_3 + 3\ H_2$

(h) $4\ HCN + 5\ O_2 \rightarrow 2\ N_2 + 4\ CO_2 + 2\ H_2O$

(i) $2\ B_5H_9 + 12\ O_2 \rightarrow 5\ B_2O_3 + 9\ H_2O$

(j) $4\ NH_3 + 5\ O_2 \rightarrow 4\ NO + 6\ H_2O$

5. Change the following word equations into formula equations and balance them:

(a) Copper + Sulfur $\xrightarrow{\Delta}$ Copper(I) sulfide

$2\ Cu + S \xrightarrow{\Delta} Cu_2S$

(b) Phosphoric acid + Calcium hydroxide → Calcium phosphate + water

$2\ H_3PO_4 + 3\ Ca(OH)_2 \rightarrow Ca_3(PO_4)_2 + 6\ H_2O$

(c) Silver oxide $\xrightarrow{\Delta}$ Silver + Oxygen

$2\ Ag_2O \xrightarrow{\Delta} 4\ Ag + O_2$

(d) Iron(III) chloride + Sodium hydroxide →

$FeCl_3 + 3\ NaOH \rightarrow$

Iron(III) hydroxide + Sodium Chloride

$Fe(OH)_3 + 3\ NaCl$

(e) Nickel(II) phosphate + Sulfuric acid →

$Ni_3(PO_4)_2 + 3\ H_2SO_4 \rightarrow$

Nickel(II) sulfate + Phosphoric acid

$3\ NiSO_4 + 2\ H_3PO_4$

(f) Zinc carbonate + Hydrochloric acid →

$ZnCO_3 + 2\ HCl \rightarrow$

Zinc chloride + Water + Carbon dioxide

$ZnCl_2 + H_2O + CO_2$

(g) Silver nitrate + Aluminum chloride → Silver chloride + Aluminum nitrate

$3\ AgNO_3 + AlCl_3 \rightarrow 3\ AgCl + Al(NO_3)_3$

6. Change the following word equations into formula equations and balance them:

(a) Water → Hydrogen + Oxygen

$2\ H_2O \rightarrow 2\ H_2 + O_2$

(b) Acetic acid + Potassium hydroxide → Potassium acetate + Water

$HC_2H_3O_2 + KOH \rightarrow KC_2H_3O_2 + H_2O$

(c) Phosphorus + Iodine → Phosphorus triiodide

$2\ P + 3\ I_2 \rightarrow 2\ PI_3$

(d) Aluminum + Copper(II) sulfate → Copper + Aluminum sulfate

$2\ Al + 3\ CuSO_4 \rightarrow 3\ Cu + Al_2(SO_4)_3$

(e) Ammonium sulfate + Barium chloride → Ammonium chloride + Barium sulfate

$(NH_4)_2SO_4 + BaCl_2 \rightarrow 2\ NH_4Cl + BaSO_4$

(f) Sulfur tetrafluoride + Water → Sulfur dioxide + Hydrogen fluoride

$SF_4 + 2\ H_2O \rightarrow SO_2 + 4\ HF$

(g) Chromium(III) carbonate → Chromium(III) oxide + Carbon dioxide

$Cr_2(CO_3)_3 \rightarrow Cr_2O_3 + 3\ CO_2$

7. Complete and balance (combination reactions).

(a) $4\ K + O_2 \xrightarrow{\Delta} 2\ K_2O$

(b) $2\ Al + 3\ Cl_2 \rightarrow 2\ AlCl_3$

(c) $SO_3 + H_2O \rightarrow H_2SO_4$

(d) $CaO + H_2O \rightarrow Ca(OH)_2$

8. Complete and balance (decomposition reactions).

(a) $2\ HgO \xrightarrow{\Delta} 2\ Hg + O_2$

(b) $2\ NaClO_3 \xrightarrow{\Delta} 2\ NaCl + 3\ O_2$

(c) $MgCO_3 \xrightarrow{\Delta} MgO + CO_2$

(d) $2\ PbO_2 \rightarrow 2\ PbO + O_2$

9. Complete and balance (single displacement reactions).

(a) $Zn + H_2SO_4 \rightarrow ZnSO_4 + H_2(g)$

(b) $AlI_3 + 3\ Cl_2 \rightarrow 2\ AlCl_3 + 3\ I_2$

(c) $Mg + 2\ AgNO_3 \rightarrow Mg(NO_3)_2 + 2\ Ag$

(d) $2\ Al + 3\ CoSO_4 \rightarrow Al_2(SO_4)_3 + 3\ Co$

10. Predicting single displacement reactions.

(a) $Ag + H_2SO_4(aq) \rightarrow$ no reaction

(b) $Cl_2 + 2\ NaBr(aq) \rightarrow Br_2 + 2\ NaCl(aq)$

(c) $Mg + ZnCl_2(aq) \rightarrow Zn + MgCl_2(aq)$

(d) $Pb + 2\ AgNO_3(aq) \rightarrow 2\ Ag + Pb(NO_3)_2$

(e) $Cu + FeCl_3(aq) \rightarrow$ no reaction

(f) $H_2 + Al_2O_3(s) \xrightarrow{\Delta}$ no reaction

(g) $2\ Al + 6\ HBr(aq) \rightarrow 3\ H_2 + 2\ AlBr_3$

(h) $I_2 + HCl(aq) \rightarrow$ no reaction

11. Complete and balance (double displacement reactions).

(a) $ZnCl_2 + 2\ KOH \rightarrow Zn(OH)_2 + 2\ KCl$

(b) $CuSO_4 + H_2S \rightarrow CuS + H_2SO_4$

(c) $3\ Ca(OH)_2 + 2\ H_3PO_4 \rightarrow Ca_3(PO_4)_2 + 6\ H_2O$

(d) $2\ (NH_4)_3PO_4 + 3\ Ni(NO_3)_2 \rightarrow Ni_3(PO_4)_2 + 6\ NH_4NO_3$

(e) $Ba(OH)_2 + 2\ HNO_3 \rightarrow Ba(NO_3)_2 + 2\ H_2O$

(f) $(NH_4)_2S + 2\ HCl \rightarrow 2\ NH_4Cl + H_2S$

12. Complete and balance.

(a) $H_2 + I_2 \rightarrow 2\ HI$

(b) $CaCO_3 \rightarrow CaO + CO_2$

(c) $Mg + H_2SO_4 \rightarrow MgSO_4 + H_2$

(d) $FeCl_2 + 2\ NaOH \rightarrow Fe(OH)_2 + 2\ NaCl$

(e) $SO_2 + H_2O \rightarrow H_2SO_3$

(f) $SO_3 + H_2O \rightarrow H_2SO_4$

(g) $Ca + 2\ H_2O \rightarrow Ca(OH)_2 + H_2$

(h) $2\ Bi(NO_3)_3 + 3\ H_2S \rightarrow Bi_2S_3 + 6\ HNO_3$

13. Complete and balance.

(a) $2\ Ba + O_2 \rightarrow 2\ BaO$

(b) $2\ NaHCO_3 \xrightarrow{\Delta} Na_2CO_3 + H_2O + CO_2$

(c) $Ni + CuSO_4 \rightarrow NiSO_4 + Cu$

(d) $MgO + 2\ HCl \rightarrow MgCl_2 + H_2O$

(e) $H_3PO_4 + 3\ KOH \rightarrow K_3PO_4 + 3\ H_2O$

(f) $C + O_2 \rightarrow CO_2$

(g) $2\ Al(ClO_3)_3 \xrightarrow{\Delta} 2\ AlCl_3 + 9\ O_2$

(h) $CuBr_2 + Cl_2 \rightarrow CuCl_2 + Br_2$

(i) $2\ SbCl_3 + 3\ (NH_4)_2S \rightarrow Sb_2S_3 + 6\ NH_4Cl$

(j) $2\ NaNO_3 \rightarrow 2\ NaNO_2 + O_2$

14. The major purpose of balancing equations is to determine the ratios of materials consumed and produced by the reaction in conformance with the law of conservation of mass.

15. The numbers in front of the formulas in a balanced equation represent the relative number of moles or molecules (or formula units) of the various substances consumed or produced in the reaction.

16. Chemical reactions.

(a) $MgBr_2 + 2\ AgNO_3 \rightarrow Mg(NO_3)_2 + 2\ AgBr$

One mole of $MgBr_2$ reacts with two moles of $AgNO_3$ to yield one mole of $Mg(NO_3)_2$ and two moles of AgBr.

(b) $N_2 + 3\ H_2 \rightarrow 2\ NH_3$

One mole of N_2 reacts with 3 moles of H_2 to yield 2 moles of NH_3.

(c) $C_3H_7OH + 9\ O_2 \rightarrow 6\ CO_2 + 8\ H_2O$

Two moles of C_3H_7OH react with nine moles of O_2 to yield six moles of CO_2 and eight moles of H_2O.

17. Interpret the meaning of the following equations, in terms of moles:

(a) $2\ Na + Cl_2 \rightarrow 2\ NaCl + 822\ kJ$

Two moles of Na react with one mole of Cl_2 to form 2 moles of NaCl and 822 kilojoules of heat are liberated. The reaction is exothermic.

(b) $PCl_5 + 92.9\ kJ \rightarrow PCl_3 + Cl_2$

One mole of PCl_5 gas absorbs 92.9 kJ of heat to form one mole of PCl_3 gas and one mole of Cl_2. The reaction is endothermic.

18. Balance equations, including heat term.

(a) $CaO + H_2O \rightarrow Ca(OH)_2 + 65.3\ kJ$

(b) $2\ Al_2O_3 + 3260\ kJ \rightarrow 4\ Al + 3\ O_2$

19. Combustion of hydrocarbons

(a) $2\ C_2H_6 + 7\ O_2 \rightarrow 4\ CO_2 + 6\ H_2O$

(b) $2\ C_6H_6 + 15\ O_2 \rightarrow 12\ CO_2 + 6\ H_2O$

(c) $C_7H_{16} + 11\ O_2 \rightarrow 7\ CO_2 + 8\ H_2O$

20. The following statements are correct: a,d,e,f,h,i,j,l,n,o,p, and q.

B. Review problems

1. Calculate the number of moles in each of the following quantities:

(a) $25.0 \text{ g } KNO_3 \times \frac{1 \text{ mol}}{101.1 \text{ g}} = 0.247 \text{ mol } KNO_3$

(b) $10.8 \text{ g } Ca(NO_3)_2 \times \frac{1 \text{ mol}}{164.1 \text{ g}} = 0.0658 \text{ mol } Ca(NO_3)_2$

(c) $5.4 \times 10^2 \text{ g } (NH_4)_2C_2O_4 \times \frac{1 \text{ mol}}{124.0 \text{ g}} = 4.4 \text{ mol } (NH_4)_2C_2O_4$

(d) $2.10 \text{ kg } NaHCO_3 \times \frac{1000 \text{ g}}{\text{kg}} \times \frac{1 \text{ mol}}{84.0 \text{ g}} = 25.0 \text{ mol } NaHCO_3$

(e) $525 \text{ mg } ZnCl_2 \times \frac{1 \text{ g}}{1000 \text{ mg}} \times \frac{1 \text{ mol}}{136.4 \text{ g}} = 3.85 \times 10^{-3} \text{ mol } ZnCl_2$

(f) $56 \text{ mmol NaOH} \times \frac{1 \text{ mol}}{1000 \text{ mmol}} = 0.056 \text{ mol NaOH}$

(g) $9.8 \times 10^{24} \text{ molecules } CO_2 \times \frac{1 \text{ mol}}{6.022 \times 10^{23} \text{ molecules}} = 16 \text{ mol } CO_2$

(h) $250 \text{ mL } C_2H_5OH \times \frac{0.789 \text{ g}}{\text{mL}} \times \frac{1 \text{ mol}}{46.0 \text{ g}} = 4.29 \text{ mol } C_2H_5OH$

(i) $16.8 \text{ mL } H_2SO_4 \text{ sol'n} \times \frac{1.727 \text{ g sol'n}}{1 \text{ mL sol'n}} \times \frac{0.800 \text{ g } H_2SO_4}{1 \text{ g sol'n}}$

$\times \frac{1 \text{ mol } H_2SO_4}{98.1 \text{ g } H_2SO_4} = 0.237 \text{ mol } H_2SO_4$

2. Calculate the number of grams in each of the following:

(a) $2.55 \text{ mol } Fe(OH)_3 \times \frac{106.8 \text{ g}}{\text{mol}} = 272 \text{ g } Fe(OH)_3$

(b) $0.00844 \text{ mol } NiSO_4 \times \frac{154.8 \text{ g}}{\text{mol}} = 1.31 \text{ g } NiSO_4$

(c) $125 \text{ kg } CaCO_3 \times \frac{1000 \text{ g}}{\text{kg}} = 1.25 \times 10^5 \text{ g } CaCO_3$

(d) $0.0600 \text{ mol } HC_2H_3O_2 \times \frac{60.0 \text{ g}}{\text{mol}} = 3.60 \text{ g } HC_2H_3O_2$

(e) $10.5 \text{ mol } NH_3 \times \frac{17.0 \text{ g}}{\text{mol}} = 179 \text{ g } NH_3$

(f) $0.725 \text{ mol } Bi_2S_3 \times \frac{514.3 \text{ g}}{\text{mol}} = 373 \text{ g } Bi_2S_3$

(g) $72 \text{ mmol HCl} \times \frac{1 \text{ mol}}{1000 \text{ mmol}} \times \frac{36.5 \text{ g}}{\text{mol}} = 2.6 \text{ g HCl}$

(h) $4.50 \times 10^{21} \text{ molecules } C_6H_{12}O_6 \times \frac{1 \text{ mol}}{6.022 \times 10^{23} \text{ molecules}}$

$\times \frac{180.0\ g}{mol} = 1.35\ g\ C_6H_{12}O_6$

(i) $500\ mL\ Br_2 \times \frac{3.119\ g}{mL} = 1.56 \times 10^3\ g\ Br_2$

(j) $75\ mL\ K_2CrO_4\ sol'n \times \frac{1.175\ g\ sol'n}{mL\ sol'n} \times \frac{0.200\ g\ K_2CrO_4}{1.0\ g\ sol'n}$

$= 18\ g\ K_2CrO_4$

3. Which contains the larger number of molecules?

(a) 10.0 g H_2O or 10.0 g H_2O_2

Because H_2O has a lower molecular weight than H_2O_2, 10.0 g of H_2O contains more moles, and, therefore, more molecules than 10.0 g of H_2O_2.

(b) 25.0 g HCl or 85.0 g $C_6H_{12}O_6$

$25.0\ g\ HCl \times \frac{1\ mol}{36.5\ g} \times \frac{6.022 \times 10^{23}\ molecules}{mol}$

$= 4.12 \times 10^{23}$ molecules HCl

$85.0\ g\ C_6H_{12}O_6 \times \frac{1\ mol}{180.0\ g} \times \frac{6.022 \times 10^{23}\ molecules}{mol}$

$= 2.84 \times 10^{23}$ molecules $C_6H_{12}O_6$

More molecules of HCl

4. Given the equation $2\ C_3H_7OH + 9\ O_2 \rightarrow 6\ CO_2 + 8\ H_2O$ set up the correct mole ratio for:

(a) CO_2 to C_3H_7OH $\qquad \frac{6\ mol\ CO_2}{2\ mol\ C_3H_7OH}$

(b) C_3H_7OH to O_2 $\qquad \frac{2\ mol\ C_3H_7OH}{9\ mol\ O_2}$

(c) O_2 to CO_2 $\qquad \frac{9\ mol\ O_2}{6\ mol\ CO_2}$

(d) H_2O to C_3H_7OH $\qquad \frac{8\ mol\ H_2O}{2\ mol\ C_3H_7OH}$

(e) CO_2 to H_2O $\qquad \frac{6\ mol\ CO_2}{8\ mol\ H_2O}$

(f) H_2O to O_2 $\qquad \dfrac{8 \text{ mol } H_2O}{9 \text{ mol } O_2}$

5. $3\ CaCl_2 + 2\ H_3PO_4 \rightarrow Ca_3(PO_4)_2 + 6\ HCl$ Write mole ratios for

(a) $CaCl_2$ to $Ca_3(PO_4)_2$ $\qquad \dfrac{3 \text{ mol } CaCl_2}{1 \text{ mol } Ca_3(PO_4)_2}$

(b) HCl to H_3PO_4 $\qquad \dfrac{6 \text{ mol HCl}}{2 \text{ mol } H_3PO_4}$

(c) $CaCl_2$ to H_3PO_4 $\qquad \dfrac{3 \text{ mol } CaCl_2}{2 \text{ mol } H_3PO_4}$

(d) $Ca_3(PO_4)_2$ to H_3PO_4 $\qquad \dfrac{1 \text{ mol } Ca_3(PO_4)_2}{2 \text{ mol } H_3PO_4}$

(e) HCl to $Ca_3(PO_4)_2$ $\qquad \dfrac{6 \text{ mol HCl}}{1 \text{ mol } Ca_3(PO_4)_2}$

(f) H_3PO_4 to HCl $\qquad \dfrac{2 \text{ mol } H_3PO_4}{6 \text{ mol HCl}}$

6. Moles of Cl_2:

$4\ HCl + O_2 \rightarrow 2\ Cl_2 + 2\ H_2O$

mol HCl $\rightarrow$ mol Cl_2

$$5.60 \text{ mol HCl} \times \frac{2 \text{ mol } Cl_2}{4 \text{ mol HCl}} = 2.80 \text{ mol } Cl_2$$

7. Grams of NaOH:

$Ca(OH)_2 + Na_2CO_3 \rightarrow 2\ NaOH + CaCO_3$

g $Ca(OH)_2$ $\rightarrow$ mol $Ca(OH)_2$ $\rightarrow$ mol NaOH $\rightarrow$ g NaOH

$$500 \text{ g } Ca(OH)_2 \times \frac{1 \text{ mol } Ca(OH)_2}{74.1 \text{ g}} \times \frac{2 \text{ mol NaOH}}{1 \text{ mol } Ca(OH)_2} \times \frac{40.0 \text{ g NaOH}}{1 \text{ mol NaOH}}$$

= 540 g NaOH

8. $Al_4C_3 + 12\ H_2O \rightarrow 4\ Al(OH)_3 + 3\ CH_4$

(a) g Al_4C_3 $\rightarrow$ mol Al_4C_3 $\rightarrow$ mol H_2O

$$100 \text{ g } Al_4C_3 \times \frac{1 \text{ mol } Al_4C_3}{144.0 \text{ g}} \times \frac{12 \text{ mol } H_2O}{1 \text{ mol } Al_4C_3} = 8.33 \text{ mol } H_2O$$

(b) mol CH_4 $\rightarrow$ mol $Al(OH)_3$

$$0.600 \text{ mol } CH_4 \times \frac{4 \text{ mol } Al(OH)_3}{3 \text{ mol } CH_4} = 0.800 \text{ mol } Al(OH)_3$$

9. First we need to write the balanced equation for the reaction.

$3\ Zn + 2\ H_3PO_4 \rightarrow Zn_3(PO_4)_2 + 3\ H_2$

Grams of $Zn_3(PO_4)_2$:

g Zn $\rightarrow$ mol Zn $\rightarrow$ mol $Zn_3(PO_4)_2$ $\rightarrow$ g $Zn_3(PO_4)_2$

$$10.0 \text{ g Zn} \times \frac{1 \text{ mol Zn}}{65.4 \text{ g}} \times \frac{1 \text{ mol } Zn_3(PO_4)_2}{3 \text{ mol Zn}} \times \frac{386.2 \text{ g } Zn_3(PO_4)_2}{1 \text{ mol}}$$

$= 19.7$ g $Zn_3(PO_4)_2$

10. $4\ FeS_2 + 11\ O_2 \rightarrow 2\ Fe_2O_3 + 8\ SO_2$

(a) $1.00 \text{ mol } FeS_2 \times \frac{2 \text{ mol } Fe_2O_3}{4 \text{ mol } FeS_2} = 0.500 \text{ mol } Fe_2O_3$

(b) $4.50 \text{ mol } FeS_2 \times \frac{11 \text{ mol } O_2}{4 \text{ mol } FeS_2} = 12.4 \text{ mol } O_2$

(c) $1.55 \text{ mol } Fe_2O_3 \times \frac{8 \text{ mol } SO_2}{2 \text{ mol } Fe_2O_3} = 6.20 \text{ mol } SO_2$

(d) $0.512 \text{ mol } FeS_2 \times \frac{8 \text{ mol } SO_2}{4 \text{ mol } FeS_2} \times \frac{64.1 \text{ g } SO_2}{\text{mol } SO_2} = 65.6 \text{ g } SO_2$

(e) $40.6 \text{ g } SO_2 \times \frac{1 \text{ mol } SO_2}{64.1 \text{ g}} \times \frac{11 \text{ mol } O_2}{8 \text{ mol } SO_2} = 0.871 \text{ mol } O_2$

(f) $221 \text{ g } Fe_2O_3 \times \frac{1 \text{ mol } Fe_2O_3}{159.6 \text{ g } Fe_2O_3} \times \frac{4 \text{ mol } FeS_2}{2 \text{ mol } Fe_2O_3} \times \frac{120 \text{ g } FeS_2}{\text{mol } FeS_2}$

$= 332$ g FeS_2

11. The balanced equation is

$MnO_2(s) + 4\ HCl(aq) \rightarrow Cl_2(g) + MnCl_2(aq) + 2\ H_2O$

mol MnO_2 $\rightarrow$ mol HCl

$$1.05 \text{ mol } MnO_2 \times \frac{4 \text{ mol HCl}}{1 \text{ mol } MnO_2} = 4.20 \text{ mol HCl}$$

12. The reaction is $Zn + 2\ HCl \rightarrow ZnCl_2 + H_2$

Did you balance the equation?

180 g Zn - 35 g Zn = 145 g Zn reacted with the HCl

(a) g Zn → mol Zn → mol H_2

$$145 \text{ g Zn} \times \frac{1 \text{ mol Zn}}{65.4 \text{ g Zn}} \times \frac{1 \text{ mol } H_2}{\text{mol Zn}} = 2.22 \text{ mol } H_2$$

(b) g Zn → mol Zn → mol HCl → g HCl

$$145 \text{ g Zn} \times \frac{1 \text{ mol Zn}}{65.4 \text{ g Zn}} \times \frac{2 \text{ mol HCl}}{1 \text{ mol Zn}} \times \frac{36.5 \text{ g HCl}}{\text{mol HCl}} = 162 \text{ g HCl}$$

13. Balanced equation: $Fe_2O_3 + 3\ C \rightarrow 2\ Fe + 3\ CO$

kg Fe_2O_3 → kmol Fe_2O_3 → kmol Fe → kg Fe

$$125 \text{ kg } Fe_2O_3 \times \frac{1 \text{ kmol } Fe_2O_3}{159.6 \text{ kg}} \times \frac{2 \text{ kmol Fe}}{1 \text{ kmol } Fe_2O_3} \times \frac{55.8 \text{ kg Fe}}{\text{kmol Fe}}$$

= 87.4 kg Fe

14. Balanced equation: $3\ Fe + 4\ H_2O \rightarrow Fe_3O_4 + 4\ H_2$

g Fe_3O_4 → mol Fe_3O_4 → mol H_2O → g H_2O

$$375 \text{ g } Fe_3O_4 \times \frac{1 \text{ mol } Fe_3O_4}{231.4 \text{ g } Fe_3O_4} \times \frac{4 \text{ mol } H_2O}{1 \text{ mol } Fe_3O_4} \times \frac{18.0 \text{ g } H_2O}{\text{mol } H_2O}$$

= 117 g H_2O

g Fe_3O_4 → mol Fe_3O_4 → mol Fe → g Fe

$$375 \text{ g } Fe_3O_4 \times \frac{1 \text{ mol } Fe_3O_4}{231.4 \text{ g } Fe_3O_4} \times \frac{3 \text{ mol Fe}}{1 \text{ mol } Fe_3O_4} \times \frac{55.8 \text{ g Fe}}{\text{mol Fe}} = 271 \text{ g Fe}$$

15. Reaction: $2\ C_2H_6 + 7\ O_2 \rightarrow 4\ CO_2 + 6\ H_2O$

(a) mol C_2H_6 → mol O_2

$$15.0 \text{ mol } C_2H_6 \times \frac{7 \text{ mol } O_2}{2 \text{ mol } C_2H_6} = 52.5 \text{ mol } O_2$$

(b) g H_2O → mol H_2O → mol CO_2 → g CO_2

$$8.00 \text{ g } H_2O \times \frac{1 \text{ mol } H_2O}{18.0 \text{ g } H_2O} \times \frac{4 \text{ mol } CO_2}{6 \text{ mol } H_2O} \times \frac{44.0 \text{ g } CO_2}{\text{mol } CO_2} = 13.0 \text{ g } CO_2$$

(c) g C_2H_6 → mol C_2H_6 → mol CO_2 → g CO_2

$$75.0 \text{ g } C_2H_6 \times \frac{1 \text{ mol } C_2H_6}{30.0 \text{ g } C_2H_6} \times \frac{4 \text{ mol } CO_2}{2 \text{ mol } C_2H_6} \times \frac{44.0 \text{ g } CO_2}{\text{mol } CO_2} = 220 \text{ g } CO_2$$

16. Limiting reactant calculations. In each reaction the amount of the second reagent needed for reaction with the given amount of the first reagent will be calculated.

(a) $KOH + HNO_3 \rightarrow KNO_3 + H_2O$

16.0 g   12.0 g

$$16.0\ g\ KOH \times \frac{1\ mol\ KOH}{56.1\ g\ KOH} \times \frac{1\ mol\ HNO_3}{1\ mol\ KOH} \times \frac{63.0\ g\ HNO_3}{mol\ HNO_3}$$

= 18.0 g HNO_3; 16.0 g of KOH will react with 18.0 g of HNO_3. The amount of HNO_3 supplied (12.0 g) is less than the amount needed, so HNO_3 is the limiting reactant and the KOH is in excess.

(b) $2\ NaOH + H_2SO_4 \rightarrow Na_2SO_4 + 2\ H_2O$

10.0 g   10.0 g

$$10.0\ g\ NaOH \times \frac{1\ mol\ NaOH}{40.0\ g\ NaOH} \times \frac{1\ mol\ H_2SO_4}{2\ mol\ NaOH} \times \frac{98.1\ g\ H_2SO_4}{mol\ H_2SO_4}$$

= 12.3 g H_2SO_4

10.0 g of NaOH will react with 12.3 g of H_2SO_4.

The amount of H_2SO_4 supplied (10.0 g) is less than the amount needed, so the H_2SO_4 is the limiting reactant and the NaOH is in excess.

(c) $2\ Bi(NO_3)_3 + 3\ H_2S \rightarrow Bi_2S_3 + 6\ HNO_3$

50.0 g   6.0 g

$$50.0\ g\ Bi(NO_3)_3 \times \frac{1\ mol\ Bi(NO_3)_3}{395.0\ g\ Bi(NO_3)_3} \times \frac{3\ mol\ H_2S}{2\ mol\ Bi(NO_3)_3}$$

$$\times \frac{34.1\ g\ H_2S}{mol\ H_2S} = 6.48\ g\ H_2S$$

50.0 g of $Bi(NO_3)_3$ will react with 6.48 g of H_2S. The amount of H_2S supplied (6.0 g) is less than needed, so the H_2S is the limiting reactant and the $Bi(NO_3)_3$ is in excess.

(d) $3\ Fe + 4\ H_2O \rightarrow Fe_3O_4 + 4\ H_2$

40.0 g   16.0 g

$$40.0\ g\ Fe \times \frac{1\ mol\ Fe}{55.8\ g\ Fe} \times \frac{4\ mol\ H_2O}{3\ mol\ Fe} \times \frac{18.0\ g\ H_2O}{mol\ H_2O} = 17.2\ g\ H_2O$$

40.0 g of Fe will react with 17.2 g of H_2O. The amount of H_2O supplied (16.0 g) is less than needed, so the H_2O is the limiting reactant and the Fe is in excess.

17. $Fe(s) + CuSO_4(aq) \rightarrow Cu(s) + FeSO_4$

2.0 mol 3.0 mol

(a) 2.0 mol of Fe will react with 2.0 mol of $CuSO_4$ to yield 2.0 mol of Cu and 2.0 mol of $FeSO_4$. 1.0 mol of $CuSO_4$ remains unreacted. At the end of the reaction there will be 2.0 mol Cu, 2.0 mol $FeSO_4$, and 1.0 mol $CuSO_4$.

(b) Fe and $CuSO_4$ react in equal molar amounts.

$$20.0 \text{ g Fe} \times \frac{1 \text{ mol Fe}}{55.8 \text{ g}} = 0.358 \text{ mol Fe}$$

$$40.0 \text{ g } CuSO_4 \times \frac{1 \text{ mol } CuSO_4}{159.6 \text{ g } CuSO_4} = 0.251 \text{ mol } CuSO_4$$

$CuSO_4$ is limiting, Fe is in excess.

0.358 mol Fe - 0.251 mol $CuSO_4$ = 0.107 mol Fe unreacted.

Relocate items as shown

$$\underset{0.251 \text{ mol}}{Fe} + \underset{0.251 \text{ mol}}{CuSO_4} \rightarrow \underset{0.251 \text{ mol}}{Cu} + \underset{0.251 \text{ mol}}{FeSO_4}$$

At the end of the reaction, 0.251 mol Cu, 0.251 mol $FeSO_4$, and 0.107 mol Fe will be present. Convert these values to grams.

$$0.251 \text{ mol Cu} \times \frac{63.5 \text{ g Cu}}{\text{mol}} = 15.9 \text{ g Cu produced}$$

$$0.251 \text{ mol } FeSO_4 \times \frac{151.9 \text{ g } FeSO_4}{\text{mol}} = 38.1 \text{ g } FeSO_4 \text{ produced}$$

$$0.107 \text{ mol Fe} \times \frac{55.8 \text{ g Fe}}{\text{mol}} = 5.97 \text{ g Fe unreacted}$$

18. Limiting reactant calculations.

$C_3H_8 + 5\ O_2 \rightarrow 3\ CO_2 + 4\ H_2O$

(a) Reaction between 6.0 moles C_3H_8 and 6.0 moles O_2

$$6.0 \text{ mol } C_3H_8 \times \frac{3 \text{ mol } CO_2}{1 \text{ mol } C_3H_8} = 18 \text{ mol } CO_2$$

$$6.0 \text{ mol } O_2 \times \frac{3 \text{ mol } CO_2}{5 \text{ mol } O_2} = 3.6 \text{ mol } CO_2$$

The O_2 is the limiting reactant; 3.6 mol CO_2 produced.

(b) Reaction between 4.0 moles C_3H_8 and 30.0 moles O_2

$$4.0 \text{ mol } C_3H_8 \times \frac{3 \text{ mol } CO_2}{1 \text{ mol } C_3H_8} = 12 \text{ mol } CO_2$$

$$30.0 \text{ mol } O_2 \times \frac{3 \text{ mol } CO_2}{5 \text{ mol } O_2} = 18.0 \text{ mol } CO_2$$

C_3H_8 is the limiting reactant; 12 mol CO_2 produced.

(c) Reaction between 20.0 moles C_3H_8 and 4.0 moles O_2

According to the equation, 1 mol C_3H_8 reacts with 5 mol O_2.

O_2 is clearly the limiting reactant.

$$4.0 \text{ mol } O_2 \times \frac{3 \text{ mol } CO_2}{5 \text{ mol } O_2} = 2.4 \text{ mol } CO_2 \text{ produced}$$

(d) Reaction between 2.0 moles C_3H_8 and 20.0 moles O_2

According to the equation, 2 mol C_3H_8 will react with 10 mol O_2.

Therefore, C_3H_8 is the limiting reactant and 10.0 mol O_2 will remain unreacted.

$$2.0 \text{ mol } C_3H_8 \times \frac{3 \text{ mol } CO_2}{1 \text{ mol } C_3H_8} = 6.0 \text{ mol } CO_2$$

$$2.0 \text{ mol } C_3H_8 \times \frac{4 \text{ mol } H_2O}{1 \text{ mol } C_3H_8} = 8.0 \text{ mol } H_2O$$

When the reaction is completed, 6.0 mol CO_2, 8.0 mol H_2O, and 10.0 mol O_2 will be in the container.

(e) Reaction between 30.0 g C_3H_8 and 30.0 g O_2

$$30.0 \text{ g } C_3H_8 \times \frac{1 \text{ mol } C_3H_8}{44.0 \text{ g } C_3H_8} \times \frac{3 \text{ mol } CO_2}{1 \text{ mol } C_3H_8} \times \frac{44.0 \text{ g } CO_2}{1 \text{ mol } CO_2} = 90.0 \text{ g } CO_2$$

$$30.0 \text{ g } O_2 \times \frac{1 \text{ mol } O_2}{32.0 \text{ g } O_2} \times \frac{3 \text{ mol } CO_2}{5 \text{ mol } O_2} \times \frac{44.0 \text{ g } CO_2}{1 \text{ mol } CO_2} = 24.8 \text{ g } CO_2$$

O_2 is the limiting reactant; 24.8 g CO_2 produced

(f) Reaction between 25.0 g C_3H_8 and 100 g O_2

$$25.0 \text{ g } C_3H_8 \times \frac{1 \text{ mol } C_3H_8}{44.0 \text{ g } C_3H_8} \times \frac{3 \text{ mol } CO_2}{1 \text{ mol } C_3H_8} \times \frac{44.0 \text{ g } CO_2}{1 \text{ mol } CO_2} = 75.0 \text{ g } CO_2$$

$$100 \text{ g } O_2 \times \frac{1 \text{ mol } O_2}{32.0 \text{ g } O_2} \times \frac{3 \text{ mol } CO_2}{5 \text{ mol } O_2} \times \frac{44.0 \text{ g } CO_2}{1 \text{ mol } CO_2} = 82.5 \text{ g } CO_2$$

C_3H_8 is the limiting reactant; 75.0 g CO_2 produced

(g) Reaction between 20.0 g C_3H_8 and 300 g O_2.

$$20.0 \text{ g } C_3H_8 \times \frac{1 \text{ mol } C_3H_8}{44.0 \text{ g } C_3H_8} \times \frac{3 \text{ mol } CO_2}{1 \text{ mol } C_3H_8} \times \frac{44.0 \text{ g } CO_2}{1 \text{ mol } CO_2} = 60.0 \text{ g } CO_2$$

$$300 \text{ g } O_2 \times \frac{1 \text{ mol } O_2}{32.0 \text{ g } O_2} \times \frac{3 \text{ mol } CO_2}{5 \text{ mol } O_2} \times \frac{44.0 \text{ g } CO_2}{1 \text{ mol } CO_2} = 248 \text{ g } CO_2$$

C_3H_8 is the limiting reactant; 60.0 g CO_2 produced

19. Limiting reactant calculations

$2 \text{ Al} + 3 \text{ Br}_2 \rightarrow 2 \text{ AlBr}_3$

Reaction between 25.0 g Al and 100 g Br_2: Determine which reactant will produce the least amount of $AlBr_3$.

$$25.0 \text{ g Al} \times \frac{1 \text{ mol Al}}{27.0 \text{ g}} \times \frac{2 \text{ mol AlBr}_3}{2 \text{ mol Al}} \times \frac{266.7 \text{ g AlBr}_3}{\text{mol AlBr}_3} = 247 \text{ g AlBr}_3$$

$$100 \text{ g Br}_2 \times \frac{1 \text{ mol Br}_2}{159.8 \text{ g}} \times \frac{2 \text{ mol AlBr}_3}{3 \text{ mol Br}_2} \times \frac{266.7 \text{ g AlBr}_3}{\text{mol AlBr}_3} = 111 \text{ g AlBr}_3$$

Br_2 is limiting; 111 g $AlBr_3$ is the theoretical yield of product.

$$\text{Percentage yield} = \frac{64.2 \text{ g AlBr}_3}{111 \text{ g AlBr}_3} \times 100\% = 57.8\%$$

20. Limiting reactant calculation

$CO(g) + 2 \text{ H}_2(g) \rightarrow CH_3OH(l)$

Reaction between 30.0 g CO and 10.0 g H_2: Determine which reactant will produce the least amount of CH_3OH.

$$30.0 \text{ g CO} \times \frac{1 \text{ mol CO}}{28.0 \text{ g}} \times \frac{1 \text{ mol CH}_3\text{OH}}{1 \text{ mol CO}} \times \frac{32.0 \text{ g CH}_3\text{OH}}{\text{mol}} = 34.3 \text{ g CH}_3\text{OH}$$

$$10.0 \text{ g H}_2 \times \frac{1 \text{ mol H}_2}{2.0 \text{ g}} \times \frac{1 \text{ mol CH}_3\text{OH}}{2 \text{ mol H}_2} \times \frac{32.0 \text{ g CH}_3\text{OH}}{\text{mol}} = 80.0 \text{ g CH}_3\text{OH}$$

30.0 g CO is limiting; 10.0 g H_2 is in excess; 34.3 g CH_3OH will be produced.

Now calculate the grams of H_2 that react with 30.0 g CO and subtract this amount from 10.0 g H_2 to obtain the grams of H_2 unreacted.

$$30.0 \text{ g CO} \times \frac{1 \text{ mol CO}}{28.0 \text{ g}} \times \frac{2 \text{ mol } H_2}{1 \text{ mol CO}} \times \frac{2.0 \text{ g } H_2}{\text{mol}} = 4.29 \text{ g } H_2$$

4.29 g H_2 will react with 30.0 g CO

10.0 g H_2 - 4.29 g H_2 = 5.71 g H_2 remain unreacted

21. Percentage yield calculation

$Fe(s) + CuSO_4(aq) \rightarrow Cu(s) + FeSO_4(aq)$

$$400 \text{ g } CuSO_4 \times \frac{1 \text{ mol } CuSO_4}{159.6 \text{ g } CuSO_4} \times \frac{1 \text{ mol Cu}}{1 \text{ mol } CuSO_4} \times \frac{63.5 \text{ g Cu}}{1 \text{ mol Cu}}$$

= 159 g Cu (theoretical yield)

$$\% \text{ Yield} = \frac{\text{actual yield}}{\text{theoretical yield}} \times 100\% = \frac{151 \text{ g Cu}}{159 \text{ g Cu}} \times 100\%$$

= 95.0% yield of Cu

22. Reaction: $C_6H_{12}O_6 \rightarrow 2\ C_2H_5OH + 2\ CO_2$

(a) $$750 \text{ g } C_6H_{12}O_6 \times \frac{1 \text{ mol } C_6H_{12}O_6}{180.0 \text{ g } C_6H_{12}O_6} \times \frac{2 \text{ mol } C_2H_5OH}{1 \text{ mol } C_6H_{12}O_6}$$

$$\times \frac{46.0 \text{ g } C_2H_5OH}{1 \text{ mol } C_2H_5OH} \times 0.846 = 324 \text{ g } C_2H_5OH$$

(The factor 0.846 is used because only an 84.6% yield is obtained.)

(b) $$475 \text{ g } C_2H_5OH \times \frac{1 \text{ mol } C_2H_5OH}{46.0 \text{ g } C_2H_5OH} \times \frac{1 \text{ mol } C_6H_{12}O_6}{2 \text{ mol } C_2H_5OH} \times \frac{180.0 \text{ g } C_6H_{12}O_6}{\text{mol } C_6H_{12}O_6}$$

$$\times \frac{1}{0.846} = 1.10 \times 10^3 \text{ g } C_6H_{12}O_6$$

(The factor $\frac{1}{0.846}$ takes into account that 475 g C_2H_5OH is only 84.6% of the theoretical yield. Thus, a greater amount of $C_6H_{12}O_6$ is needed.)

23. Reaction: $3\ C + 2\ SO_2 \rightarrow CS_2 + 2\ CO_2$

$$950 \text{ g } CS_2 \times \frac{1 \text{ mol } CS_2}{76.2 \text{ g } CS_2} \times \frac{3 \text{ mol C}}{1 \text{ mol } CS_2} \times \frac{12.0 \text{ g C}}{\text{mol C}} \times \frac{1}{0.860} = 522 \text{ g C}$$

(The factor $\frac{1}{0.860}$ takes into account that 950 g CS_2 is only 86.0% of the theoretical yield. Thus, a greater amount of C is needed.)

24. Reaction: $CaC_2 + 2\ H_2O \rightarrow C_2H_2 + Ca(OH)_2$

Convert 0.540 mol C_2H_2 to grams of CaC_2 in the sample.

$$0.540 \text{ mol } C_2H_2 \times \frac{1 \text{ mol } CaC_2}{1 \text{ mol } C_2H_2} \times \frac{64.1 \text{ g } CaC_2}{\text{mol } CaC_2} = 34.6 \text{ g of } CaC_2 \text{ in impure sample}$$

The 34.6 g of CaC_2 in the sample is a part of or a percent of the original sample (44.5 g). Calculate the percent.

$$\frac{34.6 \text{ g } CaC_2}{44.5 \text{ g Sample}} \times 100\% = 77.8\%\ CaC_2$$

25. Which salt will produce the most AgCl when equal weights of $CaCl_2$ and $MgCl_2$ react with $AgNO_3$?

Reactions: $CaCl_2 + 2\ AgNO_3 \rightarrow Ca(NO_3)_2 + 2\ AgCl$

$MgCl_2 + 2\ AgNO_3 \rightarrow Mg(NO_3)_2 + 2\ AgCl$

1 mole of each salt will produce the same amount (2 moles) of AgCl. Formula weights: $MgCl_2$, 95.3; $CaCl_2$, 111.1. $MgCl_2$ has a higher percentage of Cl than $CaCl_2$ because Mg has a lower atomic weight than Ca. Therefore, on an equal weight basis, $MgCl_2$ will produce more AgCl than $CaCl_2$.

Calculations show that 1.00 g $MgCl_2$ produces 3.01 g AgCl, and 1.00 g $CaCl_2$ produces 2.56 g AgCl.

26. Reaction: $Li_2O + H_2O \rightarrow 2\ LiOH$

$$\frac{2500 \text{ g } H_2O}{\text{astronaut-day}} \times 30 \text{ days} \times 3 \text{ astronauts} \times \frac{1 \text{ mol } H_2O}{18.0 \text{ g } H_2O} \times \frac{1 \text{ mol } Li_2O}{1 \text{ mol } H_2O}$$

$$\times \frac{29.8 \text{ g } Li_2O}{1 \text{ mol } Li_2O} \times \frac{1 \text{ kg } Li_2O}{1000 \text{ g } Li_2O} = 3.73 \times 10^2 \text{ kg } Li_2O$$

27. Preparation of HCl by the reaction of H_2SO_4 and NaCl

$H_2SO_4 + 2\ NaCl \rightarrow Na_2SO_4 + 2\ HCl$

$$25.0 \text{ L HCl sol'n} \times \frac{1000 \text{ mL HCl sol'n}}{\text{L HCl sol'n}} \times \frac{1.20 \text{ g HCl sol'n}}{1.00 \text{ mL HCl sol'n}}$$

$$\times \frac{0.42 \text{ g HCl}}{1.00 \text{ g sol'n}} \times \frac{1 \text{ mol HCl}}{36.5 \text{ g HCl}} \times \frac{1 \text{ mol } H_2SO_4}{2 \text{ mol HCl}} \times \frac{98.1 \text{ g } H_2SO_4}{1 \text{ mol } H_2SO_4}$$

$$\times \frac{1.00 \text{ g } H_2SO_4 \text{ sol'n}}{0.96 \text{ g } H_2SO_4} \times \frac{1 \text{ kg}}{1000 \text{ g}} = 18 \text{ kg concentrated } H_2SO_4$$

28. Tablets of $Al(OH)_3$ needed to neutralize HCl in gastric juice.

$Al(OH)_3(s) + 3\ HCl(aq) \rightarrow AlCl_3(aq) + 3\ H_2O(l)$

$$2.5 \text{ L} \times \frac{3.0 \text{ g HCl}}{\text{L}} \times \frac{1 \text{ mol HCl}}{36.5 \text{ g HCl}} \times \frac{1 \text{ mol } Al(OH)_3}{3 \text{ mol HCl}} \times \frac{78.0 \text{ g } Al(OH)_3}{1 \text{ mol } Al(OH)_3}$$

$$\times \frac{1000 \text{ mg}}{\text{g}} \times \frac{1 \text{ tablet}}{400 \text{ mg}} = 13 \text{ tablets}$$

29. Percentage of $KClO_3$ in a mixture of $KClO_3$ and NaCl

$2\ KClO_3 \rightarrow 2\ KCl + 3\ O_2$

12.82 g mixture - 9.45 g residue = 3.37 g O_2 lost by heating

Because the O_2 lost came only from $KClO_3$, we can use it (O_2) to calculate the amount of $KClO_3$ in the mixture.

$$3.37 \text{ g } O_2 \times \frac{1 \text{ mol } O_2}{32.0 \text{ g } O_2} \times \frac{2 \text{ mol } KClO_3}{3 \text{ mol } O_2} \times \frac{122.6 \text{ g } KClO_3}{1 \text{ mol } KClO_3} = 8.61 \text{ g } KClO_3$$

in the mixture

8.61 g $KClO_3$ represents a part of or a percent of the original sample (12.82 g).

$$\frac{8.61 \text{ g } KClO_3}{12.82 \text{ g sample}} \times 100\% = 67.2\% \ KClO_3$$

30. Based on the equation for the preparation of phosphine from calcium phosphide

$Ca_3P_2 + 6\ H_2O \rightarrow 3\ Ca(OH)_2 + 2\ PH_3$

Which of the following statements are correct?

(a) Correct

$$1 \text{ mol } Ca_3P_2 \times \frac{2 \text{ mol } PH_3}{1 \text{ mol } Ca_3P_2} = 2 \text{ mol } PH_3$$

(b) Incorrect. 1 g of Ca_3P_2 would produce 0.37 g PH_3

$$1 \text{ g } Ca_3P_2 \times \frac{1 \text{ mol } Ca_3P_2}{182.3 \text{ g } Ca_3P_2} \times \frac{2 \text{ mol } PH_3}{1 \text{ mol } Ca_3P_2} \times \frac{34.0 \text{ g } PH_3}{1 \text{ mol } PH_3} = 0.37 \text{ g } PH_3$$

(c) Correct. (See equation)

(d) Correct. (See equation)

(e) Incorrect. Two mol of Ca_3P_2 requires 12 mol of H_2O to produce 4.0 mol of PH_3

$$2 \text{ mol } Ca_3P_2 \times \frac{6 \text{ mol } H_2O}{1 \text{ mol } Ca_3P_2} = 12 \text{ mol } H_2O$$

(f) Correct. 2 moles of Ca_3P_2 will react with 12 moles of H_2O (3 moles of H_2O are present in excess) and 6 moles of $Ca(OH)_2$ will be formed.

(g) Incorrect.

$$200 \text{ g } Ca_3P_2 \times \frac{1 \text{ mol } Ca_3P_2}{182.3 \text{ g } Ca_3P_2} \times \frac{6 \text{ mol } H_2O}{1 \text{ mol } Ca_3P_2} \times \frac{18.0 \text{ g } H_2O}{1 \text{ mol } H_2O}$$

$= 118$ g H_2O; 118 g H_2O react with 200 g Ca_3P_2

The amount of water present (110 g) is less than needed to react with 200 g Ca_3P_2. The water in the limiting reactant.

(h) Incorrect. Water is the limiting reactant.

$$100 \text{ g } H_2O \times \frac{1 \text{ mol } H_2O}{18.0 \text{ g } H_2O} \times \frac{2 \text{ mol } PH_3}{6 \text{ mol } H_2O} \times \frac{34.0 \text{ g } PH_3}{1 \text{ mol } PH_3} = 63.0 \text{ g } PH_3$$

31. Based on the equation for the preparation of hydrogen cyanide

$$2\ CH_4 + 3\ O_2 + 2\ NH_3 \rightarrow 2\ HCN + 6\ H_2O$$

which of the following statements are correct?

(a) Correct.

(b) Incorrect.

$$16 \text{ mol } O_2 \times \frac{2 \text{ mol HCN}}{3 \text{ mol } O_2} = 10.7 \text{ mol } O_2 \text{ (not 12 mol } O_2)$$

(c) Correct

(d) Incorrect.

$$12 \text{ mol HCN} \times \frac{6 \text{ mol } H_2O}{2 \text{ mol HCN}} = 36 \text{ mol } H_2O \text{ (not 4 mol } H_2O)$$

(e) Correct

(f) Incorrect. The O_2 is the limiting reactant.

$$3 \text{ mol } O_2 \times \frac{2 \text{ mol HCN}}{3 \text{ mol } O_2} = 2 \text{ mol HCN (not 3 mol of HCN)}$$

B. 1. In Figure 12.1, color is the evidence of diffusion, for bromine is colored, air is colorless. If hydrogen and oxygen had been the two gases, this would not work because both gases are colorless. Two ways could be used to show the diffusion - change of density would be one, for before diffusion, the hydrogen bulb would be much less dense; after diffusion the densities would be equal. A second method would require introduction of spark gaps into both flasks. Before diffusion, neither gas would show reaction when sparked. After diffusion, the gases in both flasks would explode because of the explosive mixture of the two gases.

2. The air pressure inside the balloon is greater than the air pressure outside the balloon. The pressure inside must equal the sum of the outside air pressure plus the pressure exerted by the stretched rubber in the balloon.

3. The two major constituents of dry air are nitrogen and oxygen.

4. The pressure represented by 1 torr is equal in magnitude to the pressure exerted by 1 mm Hg pressure.

5. The molecules of hydrogen gas at the higher temperature will be moving faster. This must be so because the pressure is greater at the higher temperature and the masses and volume are the same.

6. The pressure corresponding to a volume of 4 liters is one atmosphere.

7. Any point on the graph represents a particular volume and pressure and the product of the pressure and volume for any point on the graph is a constant, i.e., 4 liter-atmospheres. This agrees with Boyle's Law, $PV = k$.

8. In Figure 12.9 if T_2 had been lower than T_1, then V_2 would have been lower than V_1.

9. The pressure of the gases in the bottle is less than the pressures of gases outside of the bottle. The fact that the water level inside the bottle is higher than the water level outside the bottle indicates that the atmospheric pressure outside the bottle is greater than the pressure exerted by the gases inside the bottle.

10. Air, being approximately 80% nitrogen and 20% oxygen, has a density intermediate between that of nitrogen and oxygen. Hence, any gas having a molecular weight greater than oxygen will have a density greater than air. This would include H_2S and all gases listed below it in table 12.3. (This assumes all density comparisons are made at the same temperature.)

C. 1. The basic assumptions of the Kinetic-molecular Theory are:

(a) Gases consist of tiny molecules.

(b) The distance between molecules is large compared to the size of the molecules.

(c) Gas molecules have no attraction for each other.

(d) Gas molecules move in straight lines, colliding with other molecules and with the container walls.

(e) Collisions with other molecules or with container walls are elastic, involving no loss of energy.

(f) The average kinetic energy for molecules is the same for all gases at the same temperature, and its value increases with increased temperature.

2. The order of increasing molecular velocities is the order of decreasing molecular weights: Rn, F_2, N_2, CH_4, He, H_2. At the same temperature the kinetic energies of the gases are the same and equal to $1/2\ mv^2$. For the kinetic energies to be the same the velocities must increase as molecular weights decrease.

3. The average kinetic energies of the molecules in question 2 are the same because they are all at the same temperature.

4. The four parameters used to describe the behavior of gases are: pressure; volume; temperature; and number of moles.

5. An ideal gas is one whose behavior is described exactly by the Kinetic Molecular Theory and by the gas laws for all possible values of P, V, and T.

6. At low temperature the behavior of a gas is least likely to behave ideally. This is so because at low temperatures the velocities of the molecules decrease and attractive forces between molecules can become operative (ideally there are no attractive forces between molecules of a gas).

7. At high pressures gases are least likely to behave ideally. As the pressure increases, gas molecules are compressed closer together. At sufficiently high pressures the actual volume of the molecules becomes an appreciable fraction of the total volume occupied by the gas. (Ideally the volume of the molecules is negligible compared to the containing volume.) With sufficiently high pressure, some gases liquefy.

8. Comparison, at the same temperature and pressure, equal volumes of H_2 and O_2 as to:

(a) Number of molecules. Equal. Avogadro's law.

(b) Mass. Number of molecules equal; therefore, mass of O_2 16 times the mass of H_2.

(c) Number of moles. Number of molecules equal; therefore, moles of O_2 = moles of H_2.

(d) Average kinetic energies of the molecules. Temperatures the same; therefore, kinetic energies are equal.

(e) Rate of effusion. Rate H_2 = 4 × rate O_2

$$\frac{\text{Rate } H_2}{\text{Rate } O_2} = \sqrt{\frac{\text{molecular weight } O_2}{\text{molecular weight } H_2}} = \sqrt{\frac{32}{2}} = \frac{4}{1}$$

(f) Density. Density O_2 = 16 × density H_2

$$D_{O_2} = \frac{\text{mass } O_2}{\text{volume } O_2}\ ; \quad D_{H_2} = \frac{\text{mass } H_2}{\text{volume } H_2}\ ;$$

volume O_2 = volume H_2; mass O_2 = 16 × mass H_2

$$\frac{D_{O_2}}{D_{H_2}} = \frac{\text{mass } O_2/\text{volume } O_2}{\text{mass } H_2/\text{volume } H_2} = \frac{\text{mass } O_2}{\text{mass } H_2} = \frac{16}{1}$$

9. Behavior of gases as described by the kinetic-molecular theory.

(a) Boyle's law. Boyle's law states the volume of a gas, at constant temperature, is inversely proportional to the pressure. The kinetic-molecular theory assumes the volume occupied by gases is mostly empty space. Decreasing the volume of a gas by compressing it increases the concentration of gas molecules resulting in more collisions of the molecules with the container wall and increased pressure.

(b) Charles' law. Charles' law states the volume of a gas at constant pressure is directly proportional to the absolute temperature. According to the kinetic-molecular theory the kinetic energies of gas molecules are proportional to the absolute temperature. Increasing the temperature of a gas sample causes the molecules to move faster and in order for the pressure not to increase the volume of the gas must increase.

(c) Dalton's law. Dalton's law states the pressure of a mixture of gases is the sum of the pressures exerted by the individual gases. According to the kinetic molecular theory there are no attractive forces between gas molecules; therefore, in a mixture of gases each gas acts independently and the total pressure exerted will be the sum of the pressures exerted by the individual gases.

10. $N_2(g) + O_2(g) \rightarrow 2\ NO(g)$

1 vol. 1 vol. 2 vol.

According to Avogadro's law, equal volumes of nitrogen and oxygen at the same temperature and pressure contain the same number of molecules. In the reaction, nitrogen and oxygen molecules react in a 1:1 volume ratio. Since two volumes of nitrogen monoxide are produced, one molecule of nitrogen and one molecule of oxygen must produce two molecules of nitrogen monoxide. Therefore, each nitrogen and oxygen molecule must be made up of two atoms.

11. We refer gases to STP because some reference point is needed to relate volume and moles. A temperature and pressure must be specified to determine the moles of gas in a given volume, and 0°C, 760 torr are convenient reference points.

12. Heating a mole of N_2 gas at constant pressure effects the following in the ways indicated:

    (a) Density will decrease. Heating the gas at constant pressure will increase its volume. The mass does not change so the increased volume results in a lower density.

    (b) Specific gravity decreases. Since the density decreases (see preceding question) the specific gravity will also decrease because specific gravity is the ratio of the density of the gas divided by a reference density usually air.

    (c) Mass does not change. Heating a substance does not change its mass.

    (d) Average kinetic energy of the molecules increases. This is a basic assumption of the kinetic-molecular theory.

    (e) Average velocity of the molecules will increase. Increasing the temperature increases the average kinetic energies of the molecules; hence, the average velocity of the molecules will increase also.

    (f) Number of N_2 molecules remains unchanged. Heating does not alter the number of molecules present, except if extremely high temperatures were attained. Then the N_2 molecules might dissociate into N atoms resulting in fewer N_2 molecules.

13. The following statements are correct: b,d,f1,f4,h,i,n,o,p,q, and t.

D. Review problems.

1. The barometer reads 715 mm Hg. Calculate the corresponding pressure in:

    (a) atmospheres

    $$715 \text{ mm} \times \frac{1 \text{ atm}}{760 \text{ mm}} = 0.941 \text{ atm}$$

    (b) inches of Hg

    $$715 \text{ mm} \times \frac{1 \text{ cm}}{10 \text{ mm}} \times \frac{1 \text{ in. Hg}}{2.54 \text{ cm}} = 28.1 \text{ in. Hg}$$

(c) lb/in.^{2}

$$715 \text{ mm} \times \frac{1 \text{ atm}}{760 \text{ mm}} \times \frac{14.7 \text{ lb/in.}^2}{\text{atm}} = 13.8 \text{ lb/in.}^2$$

(d) torrs

$$715 \text{ mm} \times \frac{1 \text{ torr}}{1 \text{ mm}} = 715 \text{ torr}$$

(e) millibars

$$715 \text{ mm} \times \frac{1 \text{ atm}}{760 \text{ mm}} \times \frac{1013 \text{ mbar}}{1 \text{ atm}} = 953 \text{ mbar}$$

(f) $715 \text{ mm} \times \frac{1 \text{ atm}}{760 \text{ mm}} \times \frac{101.3 \text{ kPa}}{1 \text{ atm}} = 95.3 \text{ kPa}$

2. Express the following pressure in atmospheres

(a) $28 \text{ mm Hg} \times \frac{1 \text{ atm}}{760 \text{ mm}} = 0.037 \text{ atm}$

(b) $6000 \text{ cm Hg} \times \frac{1 \text{ atm}}{76 \text{ cm}} = 79 \text{ atm}$

(c) $795 \text{ torr} \times \frac{1 \text{ atm}}{760 \text{ torr}} = 1.05 \text{ atm}$

(d) $5.00 \text{ kPa} \times \frac{1 \text{ atm}}{101.3 \text{ kPa}} = 4.94 \times 10^{-2} \text{ atm}$

3. A gas occupies a volume of 400 mL at 500 mm Hg pressure. What will be its volume if the pressure is changed to:

(a) 760 mm Hg. Increase in pressure causes a decrease in volume.

$$400 \text{ mL} \times \frac{500 \text{ mm Hg}}{760 \text{ mm Hg}} = 263 \text{ mL}$$

(b) 250 torr. Decrease in pressure causes an increase in volume.
500 mm Hg = 500 torr

$$400 \text{ mL} \times \frac{500 \text{ torr}}{250 \text{ torr}} = 800 \text{ mL}$$

(c) 2.00 atm. 1.00 atm = 760 mm Hg

$$400 \text{ mL} \times \frac{500 \text{ mm Hg}}{2 \times 760 \text{ mm Hg}} = 132 \text{ mL}$$

4. Pressure is inversely proportional to volume at constant temperature

(a) $640 \text{ mm Hg} \times \frac{500 \text{ mL}}{855 \text{ mL}} = 374 \text{ mm Hg}$

(b) $640 \text{ mm Hg} \times \frac{500 \text{ mL}}{450 \text{ mL}} = 711 \text{ mm Hg}$

5. $1.0 \text{ atm} \times \frac{2500 \text{ L}}{25 \text{ L}} = 1.0 \times 10^2 \text{ atm}$

6. Given 6.00 L of N_2 gas at -25°C (273 - 25 = 248 K) calculate the volume at:

Remember to use absolute temperature in calculations.

(a) 0.0°C (0.0°C = 273 K); $6.00 \text{ L} \times \frac{273 \text{ K}}{248 \text{ K}} = 6.60 \text{ L}$

(b) 0.0°F; $(0.0°\text{F} - 32°\text{F}) \times \frac{1°\text{C}}{1.8°\text{F}} = -18°\text{C}$

-18°C = (273 - 18) K = 255 K

$6.00 \text{ L} \times \frac{255 \text{ K}}{248 \text{ K}} = 6.17 \text{ L}$

(c) 100 K; $6.00 \text{ L} \times \frac{100 \text{ K}}{248 \text{ K}} = 2.42 \text{ L}$

(d) 345 K; $6.00 \text{ L} \times \frac{345 \text{ K}}{248 \text{ K}} = 8.35 \text{ L}$

7. To double the volume of a gas at constant pressure the temperature (K) must be doubled.

27°C = 300 K; V = initial volume; 2 V = final volume.

$300 \text{ K} \times \frac{2V}{V} = 600 \text{ K}$ (or 327°C)

8. V = volume at 22°C and 740 torr

2 V = volume after temperature increase, pressure still at 740 torr.

V = final volume after increase in pressure, temperature constant

$740 \text{ torr} \times \frac{2 \text{ V}}{\text{V}} = 1.48 \times 10^3$ torr, the pressure required to reduce the volume from 2 V to V.

9. Starting and ending volume is 250 mL. The pressure change will increase the volume, so the temperature change must decrease the volume.

22°C = 295 K

$250 \text{ mL} = 250 \text{ mL} \times \frac{700 \text{ torr}}{500 \text{ torr}} \times \frac{X}{295 \text{ K}}$

$X = 295 \text{ K} \times \frac{500 \text{ torr}}{700 \text{ torr}} \times \frac{250 \text{ mL}}{250 \text{ mL}} = 211 \text{ K}$; 211 K - 273 = -62°C

The temperature must be changed to -62°C.

10. Lowering the temperature will lower the pressure at constant volume.

25°C = 298 K; -196°C = 77 K

$P = 252 \text{ atm} \times \frac{77 \text{ K}}{298 \text{ K}} = 65 \text{ atm}$

11. Increasing the temperature increases the pressure at constant volume (Assume the volume of a tire is constant.)

$$°C = \frac{71°F - 32}{1.8} = 22°C = 295\ K;\quad T = 295\ K \times \frac{44\ psi}{30\ psi} = 433\ K$$

$$433\ K - 273 = 160°C;\quad 160°C \times \frac{1.8°F}{1.0°C} + 32 = 320°F$$

The tire will burst when the temperature reaches 320°F.

12. Given 410 mL of gas at 27°C (300 K) and 740 mm Hg pressure. Calculate volume at:

(a) STP

Temperature change, 300 K to 273 K (decreases volume)
Pressure change, 740 mm Hg to 760 mm Hg (decreases volume)

$$V \text{ at STP} = 410\ mL \times \frac{273\ K}{300\ K} \times \frac{740\ mm\ Hg}{760\ mm\ Hg} = 363\ mL$$

(b) 250°C (523 K) and 680 mm Hg

Temperature change, 300 K to 523 K (increases volume)
Pressure change, 740 mm Hg to 680 mm Hg (increases volume)

$$V = 410\ mL \times \frac{523\ K}{300\ K} \times \frac{740\ mm\ Hg}{680\ mm\ Hg} = 778\ mL$$

13. Given 5.30 L of H_2 at STP, calculate volume at 70°C (343 K) and 830 torr pressure?

Temperature change, 273 K to 343 K (increases volume)
Pressure change, 760 torr to 830 torr (decreases volume)

$$V = 5.30\ L \times \frac{343\ K}{273\ K} \times \frac{760\ torr}{830\ torr} = 6.10\ L$$

14. Given 800 mL of gas at STP, calculate the pressure when the volume is 250 mL at 30°C (303 K)

Volume change, 800 mL to 250 mL (increase in pressure)
Temperature change, 273 K to 303 K (increase in pressure)

$$P = 1.00\ atm \times \frac{800\ mL}{250\ mL} \times \frac{303\ K}{273\ K} = 3.55\ atm$$

15. Given 1400 L He at 0.950 atm and 18°C (291 K)

$$4.0\ torr \times \frac{1.0\ atm}{760\ torr} = 0.0053\ atm$$

	1	2
P	0.950 atm	0.0053 atm
V	1400 L	V_2
T	18°C (291 K)	2.0°C (275 K)

$$V_2 = 1400\ L \times \frac{275\ K}{291\ K} \times \frac{0.950\ atm}{0.0053\ atm} = 2.4 \times 10^5\ L$$

16.

	1	2
P	2.50 atm	1.50 atm
V	22.4 L	V_2
T	27°C (300 K)	-5.00°C (268 K)

$$V_2 = 22.4 \text{ L} \times \frac{268 \text{ K}}{300 \text{ K}} \times \frac{2.50 \text{ atm}}{1.50 \text{ atm}} = 33.4 \text{ L}$$

17. Calculate the volume at STP

	1	2
P	400 torr	760 torr
V	600 mL	V_2
T	40°C (313 K)	273 K

$$V_2 = 600 \text{ mL } N_2O \times \frac{400 \text{ torr}}{760 \text{ torr}} \times \frac{273 \text{ K}}{313 \text{ K}} = 275 \text{ mL at STP}$$

Calculate the number of molecules in 275 mL N_2O

$$275 \text{ mL} \times \frac{1 \text{ L}}{1000 \text{ mL}} \times \frac{1 \text{ mol}}{22.4 \text{ L}} \times \frac{6.022 \times 10^{23} \text{ molecules}}{\text{mol}}$$

$= 7.39 \times 10^{21}$ molecules N_2O

Calculate the number of atoms in 7.39×10^{21} molecules of N_2O.

$$7.39 \times 10^{21} \text{ molecules } N_2O \times \frac{3 \text{ atoms}}{\text{molecule}} = 2.22 \times 10^{22} \text{ atoms}$$

18. $P_{total} = P_{O_2} + P_{H_2O \text{ vapor}} = 720 \text{ torr}$

P_{H_2O} at 20°C = 17.5 torr; P_{O_2} = 720 torr - 17.5 torr = 703 torr

19. $P_{total} = P_{H_2} + P_{N_2} + P_{O_2}$ (Dalton's Law)

P_{total} = 700 torr + 200 torr + 300 torr = 1200 torr

20. Volume of dry methane, CH_4.
Volume of wet methane is 2.50 L at 25°C and 720 torr.
Vapor pressure of H_2O at 25°C = 23.8 torr
720 torr - 23.8 torr = 696 torr (pressure of CH_4)
After drying (removal of H_2O vapor), the pressure of CH_4 changes from 696 torr to 720 torr, resulting from a decrease in volume.

$$V = 2.50 \text{ L} \times \frac{696 \text{ torr}}{720 \text{ torr}} = 2.42 \text{ L (dry } CH_4)$$

21. Calculate the pressure of each gas when put into the 10.0 L container (Boyle's law). Then add the two pressures together (Dalton's law). Assume that the temperature is constant.

CO_2: $P = 500 \text{ torr} \times \frac{5.00 \text{ L}}{10.0 \text{ L}} = 250 \text{ torr}$

CH_4: $P = 400 \text{ torr} \times \frac{3.00 \text{ L}}{10.0 \text{ L}} = 120 \text{ torr}$

$P_{total} = 250 \text{ torr} + 120 \text{ torr} = 370 \text{ torr}$

22. One mole of a gas occupies 22.4 L at STP. Volume of 2.5 moles of CO_2 at STP;

$2.5 \text{ mol} \times \frac{22.4 \text{ L}}{\text{mol}} = 56 \text{ L}$

23. The number of moles of a gas is proportional to the pressure (temperature and volume remaining constant).

(a) $60 \text{ mol} \times \frac{850 \text{ lbs/in}^2}{1500 \text{ lbs/in}^2} = 34 \text{ mol}$

(b) $60 \text{ mol } H_2 \times \frac{2.0 \text{ g } H_2}{\text{mol}} = 1.2 \times 10^2 \text{ g } H_2$

24. Grams of CO_2 in 2500 mL of CO_2 at STP. 1 mol gas = 22.4 L at STP

$2500 \text{ mL } CO_2 \times \frac{1 \text{ L}}{1000 \text{ mL}} \times \frac{1 \text{ mol}}{22.4 \text{ L}} \times \frac{44.0 \text{ g}}{\text{mol}} = 4.91 \text{ g } CO_2$

25. Molecular weight of a gas. 560 mL at STP has a mass of 1.08 g.

$\frac{1.08 \text{ g}}{560 \text{ mL}} \times \frac{22.4 \times 10^3 \text{ mL}}{\text{mol}} = 43.2 \text{ g/mol (mol. wt.)}$

26. Volume of NO_2 at STP. 1 mol gas = 22.4 L at STP

(a) $1.0 \text{ mol } NO_2 \times \frac{22.4 \text{ L}}{\text{mol}} = 22.4 \text{ L}$

(b) $17.05 \text{ g } NO_2 \times \frac{1 \text{ mol}}{46.0 \text{ g}} \times \frac{22.4 \text{ L}}{\text{mol}} = 8.30 \text{ L}$

(c) $1.20 \times 10^{24} \text{ molecules } NO_2 \times \frac{1 \text{ mol}}{6.022 \times 10^{23} \text{ molecules}}$

$\times \frac{22.4 \text{ L}}{\text{mol}} = 44.6 \text{ L}$

27. Molecules of NH_3 in 1.00 L at STP.

$1.00 \text{ L } NH_3 \times \frac{1 \text{ mol}}{22.4 \text{ L}} \times \frac{6.022 \times 10^{23} \text{ molecules}}{\text{mol}} = 2.69 \times 10^{22} \text{ molecules}$

28. Moles of Cl_2 in 1.00 m^3 at STP.

$$1.00\ m^3 \times \left(\frac{100\ cm}{m}\right)^3 \times \frac{1\ mL}{1\ cm^3} \times \frac{1\ L}{1000\ mL} \times \frac{1\ mol}{22.4\ L} = 44.6\ mol\ Cl_2$$

29. Molecular weight of a gas from a known density.

$$\frac{1.78\ g}{L} \times \frac{22.4\ L}{mol} = 39.9\ g/mol\ (molecular\ weight)$$

30. Density calculations at STP. 1 mol gas = 22.4 L at STP.

(a) Kr, formula weight = 83.8 g/mol

$$d = \frac{83.8\ g\ Kr}{mol} \times \frac{1\ mol}{22.4\ L} = 3.74\ g/L$$

(b) He, formula weight = 4.0 g/mol

$$d = \frac{4.0\ g\ He}{mol} \times \frac{1\ mol}{22.4\ L} = 0.18\ g/L$$

(c) SO_3, molecular weight = 80.1 g/mol

$$d = \frac{80.1\ g\ SO_3}{mol} \times \frac{1\ mol}{22.4\ L} = 3.58\ g/L$$

(d) C_4H_8, molecular weight = 56.0 g/mol

$$d = \frac{56.0\ g\ C_4H_8}{mol} \times \frac{1\ mol}{22.4\ L} = 2.50\ g/L$$

31. (a) Density of F_2 at STP, molecular weight = 38.0 g/mol

$$d = \frac{38.0\ g\ F_2}{mol} \times \frac{1\ mol}{22.4\ L} = 1.70\ g/L$$

(b) Density of F_2 at 27°C (300 K) and 1 atm pressure. Assume 1 mol of F_2 and use the ideal gas equation to calculate the volume it occupies at 27°C and 1 atm pressure.

$$PV = nRT \qquad V = \frac{nRT}{P}$$

$$V = \frac{1.00\ mol \times 300\ K \times 0.0821\ L\ atm}{1.00\ atm \times mol\ K} = 24.6\ L$$

One mole of F_2 (38.0 g) occupies 24.6 L at 27°C and 1 atm pressure.

$$d = \frac{38.0\ g}{mol} \times \frac{1\ mol}{24.6\ L} = 1.54\ g/L$$

32. Temperature where the density of methane is 1.0 g/L at 1 atm pressure. At STP, 22.4 L of CH_4 has a mass of 16.0 g.

$$\frac{22.4\ L}{16.0\ g} = \frac{1.40\ L}{1.00\ g} \text{ at STP}$$

Volume must change from 1.40 L to 1.00 L by decreasing temperature.

$$T = 273\ K \times \frac{1.00\ L}{1.40\ L} = 195\ K; \quad 195\ K - 273 = -78°C$$

33. Calculations using the ideal gas equation, PV = nRT

(a) Volume of H_2; n = 0.510 mol; T = 47°C (320 K); P = 1.6 atm

$$V = \frac{nRT}{P}; \quad V = \frac{0.510\ mol \times 320\ K \times 0.0821\ L\ atm}{1.6\ atm \times mol\ K} = 8.4\ L\ H_2$$

(b) Grams of CH_4

$$n = \frac{grams}{molecular\ weight} = \frac{g}{M}; \quad PV = \frac{gRT}{M}; \quad g = \frac{PVM}{RT}$$

$$P = \frac{600\ torr \times 1\ atm}{760\ torr} = 0.789\ atm; \quad V = 16.0\ L; \quad M = \frac{16.0\ g}{mol};$$

T = 27°C (300 K)

$$g = \frac{0.789\ atm \times 16.0\ L \times 16.0\ g/mol}{300\ K \times 0.0821\ L\ atm/mol\ K} = 8.20\ g\ CH_4$$

(c) Density of CO_2

$$PV = \frac{gRT}{M}; \quad \frac{g}{V} = \frac{PM}{RT} = density$$

P = 4.00 atm; T = -20°C (253 K); M = 44.0 g/mol

$$Density = \frac{4.00\ atm \times 44.0\ g/mol}{253\ K \times 0.0821\ L\ atm/mol\ K} = 8.47\ g/L$$

(d) Molecular weight of a gas.

$$PV = \frac{gRT}{M}; \quad M = \frac{gRT}{PV}; \quad density = \frac{g}{V} = \frac{2.58\ g}{L}; \ T = 27°C\ (300\ K);$$

P = 1 atm

$$M = \frac{2.58\ g/L \times 300\ K \times 0.0821\ L\ atm}{1\ atm \times mol\ K} = 63.5\ g/mol$$

34. Molecular weight (M) of a gas.

$$n = \frac{g}{M}; \quad PV = \frac{gRT}{M}; \quad M = \frac{gRT}{PV}$$

1.15 g; 30°C (303 K); 0.813 atm; 0.215 L

$$M = \frac{1.15\ g \times 303\ K \times 0.0821\ L\ atm}{0.813\ atm \times 0.215\ L \times mol\ K} = 164\ g/mol \text{ (molecular weight)}$$

35. Calculate the volume of 2.3 mol of Ne at 27°C (300 K) and 750 torr. Using the ideal gas equation, PV = nRT

$$n = 2.3\ mol; \quad P = \frac{750}{760}\ atm = 0.987\ atm; \quad T = 300\ K; \quad R = \frac{0.0821\ L\ atm}{mol\ K}$$

$$V = \frac{nRT}{P} = \frac{2.3 \text{ mol} \times 300 \text{ K} \times 0.0821 \text{ L atm}}{0.987 \text{ atm} \times \text{mol K}} = 57 \text{ L}$$

36. Calculate the volume of 5.00 mol H_2 and 0.500 mol CO_2 at STP

Total moles = 5.50 mol; $5.50 \text{ mol} \times \frac{22.4 \text{ L}}{\text{mol}} = 123 \text{ L}$

37. Calculate the temperature of 4.5 mol of a gas that occupies 0.250 L at 4.15 atm pressure.

$PV = nRT \qquad T = \frac{PV}{nR}$

$$T = \frac{4.15 \text{ atm} \times 0.250 \text{ L}}{4.50 \text{ mol} \times 0.0821 \text{ L atm/mol K}} = 2.81 \text{ K } (-270°\text{C})$$

38. Moles of N_2 that occupy 5.20 L at 250 K and 0.500 atm.

$PV = nRT \qquad n = \frac{PV}{RT}$

$$n = \frac{0.500 \text{ atm} \times 5.20 \text{ L}}{250 \text{ K} \times 0.0821 \text{ L atm/mol K}} = 0.127 \text{ mol } N_2$$

39. Volume of H_2 at STP produced by reacting 8.30 mol of Al with H_2SO_4

$2 \text{ Al(s)} + 3 \text{ H}_2\text{SO}_4\text{(aq)} \rightarrow \text{Al}_2(\text{SO}_4)_3\text{(aq)} + 3 \text{ H}_2\text{(g)}$

$$8.3 \text{ mol Al} \times \frac{3 \text{ mol H}_2}{2 \text{ mol Al}} \times \frac{22.4 \text{ L}}{\text{mol}} = 279 \text{ L H}_2$$

40. Given the equation $4 \text{ NH}_3\text{(g)} + 5 \text{ O}_2\text{(g)} \rightarrow 4 \text{ NO(g)} + 6 \text{ H}_2\text{O(g)}$

(a) Calculate the moles of NH_3 required to produce 5.5 mol NO

$$5.5 \text{ mol NO} \times \frac{4 \text{ mol NH}_3}{4 \text{ mol NO}} = 5.5 \text{ mol NH}_3$$

(b) Calculate the moles of NH_3 that will react with 7.0 mol of O_2.

$$7.0 \text{ mol O}_2 \times \frac{4 \text{ mol NH}_3}{5 \text{ mol O}_2} = 5.6 \text{ mol NH}_3$$

(c) Calculate the liters of NO that can be made from 12 mol of O_2 and 10 mol of NH_3 at STP

First calculate the liters of NO from each reactant

$$12.0 \text{ mol O}_2 \times \frac{4 \text{ mol NO}}{5 \text{ mol O}_2} \times \frac{22.4 \text{ L}}{\text{mol}} = 215 \text{ L NO}$$

$$10.0 \text{ mol NH}_3 \times \frac{4 \text{ mol NO}}{4 \text{ mol NH}_3} \times \frac{22.4 \text{ L}}{\text{mol}} = 224 \text{ L NO}$$

O_2 is the limiting reactant. 215 L of NO will be produced.

(d) Calculate the liters of NO produced by reaction of 800 mL of O_2. Volume to volume calculation.

$$800 \text{ mL } O_2 \times \frac{4 \text{ L NO}}{5 \text{ L } O_2} = 640 \text{ mL NO} = 0.640 \text{ L NO}$$

(e) Calculate the liters of NO that can be made from 3.0 L of NH_3 and 3.0 L of O_2.

The 3.0 L of O_2 is the limiting reactant (from equation: 4 L of NH_3 react with 5 L of O_2).

$$3.0 \text{ L } O_2 \times \frac{4 \text{ L NO}}{5 \text{ L } O_2} = 2.4 \text{ L NO}$$

(f) Calculate the grams of O_2 needed to produce 60 L of NO at STP

L NO $\rightarrow$ mol NO $\rightarrow$ mol O_2 $\rightarrow$ g O_2

$$60 \text{ L NO} \times \frac{1 \text{ mol NO}}{22.4 \text{ L NO}} \times \frac{5 \text{ mol } O_2}{4 \text{ mol NO}} \times \frac{32.0 \text{ g } O_2}{\text{mol } O_2} = 1.1 \times 10^2 \text{ g } O_2$$

(g) Calculate the grams of NH_3 needed to produce 32 L of products, NO plus H_2O, at STP.

4 L of NH_3 will produce 4 L of NO + 6 L of H_2O (10 L of product).

$$32 \text{ L } (NO + H_2O) \times \frac{4 \text{ L } NH_3}{10 \text{ L } (NO + H_2O)} \times \frac{1 \text{ mol } NH_3}{22.4 \text{ L}} \times \frac{17.0 \text{ g } NH_3}{1 \text{ mol } NH_3}$$

$$= 9.7 \text{ g } NH_3$$

41. Given the equation $4\ FeS_2 + 11\ O_2 \xrightarrow{\Delta} 2\ Fe_2O_3 + 8\ SO_2$

(a) How many liters of O_2, at STP, will react with 0.600 kg of FeS_2?

$$0.600 \text{ kg } FeS_2 \times \frac{1000 \text{ g } FeS_2}{\text{kg } FeS_2} \times \frac{1 \text{ mol } FeS_2}{120.0 \text{ g } FeS_2} \times \frac{11 \text{ mol } O_2}{4 \text{ mol } FeS_2}$$

$$\times \frac{22.4 \text{ L } O_2}{\text{mol } O_2} = 308 \text{ L } O_2$$

(b) How many liters of SO_2, at STP, will be produced from 0.600 kg of FeS_2? Because the starting amount of FeS_2 is the same as in part (a), it is simpler to calculate the volume of SO_2 from the volume of O_2 that reacts with 0.600 kg FeS_2

$$308 \text{ L } O_2 \times \frac{8 \text{ L } SO_2}{11 \text{ L } O_2} = 224 \text{ L } SO_2$$

42. $C_2H_2(g) + 2\ HF(g) \rightarrow C_2H_4F_2(g)$

$1.0\ \text{mol}\ C_2H_2 \rightarrow 1.0\ \text{mol}\ C_2H_4F_2$

$$5.0\ \text{mol HF} \times \frac{1\ \text{mol}\ C_2H_4F_2}{2\ \text{mol HF}} = 2.5\ \text{mol}\ C_2H_4F_2$$

C_2H_2 is the limiting reactant.

1.0 mol C_2H_2 reacts with 2.0 mol HF to produce 1.0 mol $C_2H_4F_2$ and leave 3.0 mol HF unreacted.

At the end of the reaction, the mixture will consist of 1.0 mol $C_2H_4F_2$ plus 3.0 mol HF (4.0 mol of gases) in a 10 L flask.

Use the ideal gas equation to calculate the pressure.

PV = nRT; T = 0°C (273 K)

$$P = \frac{nRT}{V} = \frac{4.0\ \text{mol} \times 273\ \text{K} \times 0.0821\ \text{L atm}}{10\ \text{L} \times \text{mol K}} = 9.0\ \text{atm}$$

43. The relative rate of effusion of two gases are inversely proportional to their formula weights.

$$\frac{\text{rate of effusion of He}}{\text{rate of effusion of } N_2} = \sqrt{\frac{\text{formula weight } N_2}{\text{atomic weight He}}} = \sqrt{\frac{28.0}{4.0}} = \frac{2.6}{1}$$

He effuses 2.6 times as fast as N_2.

44. Effusion

(a) $\frac{\text{Rate He}}{\text{Rate } CH_4} = \sqrt{\frac{16.0}{4.0}} = \frac{2}{1}$; He effuses twice as fast as CH_4

(b) Let y = distance (cm) CH_4 travels

100 cm - y = distance traveled by He.

The He, having twice the speed of the CH_4 travels twice the distance of the CH_4. Therefore,

2 y = 100 cm - y; 3 y = 100 cm; y = 33.3 cm

He travel 2 y cm or 67 cm

45. Calculation of molecular formula.

% composition: 85.7% C, 14.3% H.

$$85.7\ \text{g C} \times \frac{1\ \text{mol C}}{12.0\ \text{g C}} = 7.14\ \text{mol C};\ 14.3\ \text{g H} \times \frac{1\ \text{mol H}}{1.0\ \text{g H}} = 14.3\ \text{mol H}$$

$C_{\frac{7.14}{7.14}}H_{\frac{14.3}{7.14}} = C_{1.00}H_{2.00}$; simplest formula = CH_2

From the density: $\frac{2.50\text{ g}}{\text{L}} \times \frac{22.4\text{ L}}{\text{mol}} = 56.0\text{ g/mol}$ = formula weight

Weight of simplest formula = 14.0

$\frac{56.0}{14.0} = 4$; $(CH_2)_4 = C_4H_8$ (molecular formula)

46. Given the reaction $2\ CO + O_2 \rightarrow 2\ CO_2$

Calculate the limiting reactant

$$10\text{ mol CO} \times \frac{2\text{ mol }CO_2}{2\text{ mol CO}} = 10\text{ mol }CO_2$$

$$8\text{ mol }O_2 \times \frac{2\text{ mol }CO_2}{1\text{ mol }O_2} = 16\text{ mol }CO_2$$

(a) CO is the limiting reactant. 10 mol CO react with 5 mol O_2 to produce 10 mol CO_2. Therefore, 3 mol of O_2 will remain unreacted. At the end of the reaction, 10 mol of CO_2 and 3 mol O_2 will be present. No CO will be present.

(b) 13 mol of gas (10 mol CO_2 + 3 mol O_2) is present in the 10 L vessel at 0°C (273 K).

$$P = \frac{nRT}{V} = \frac{13\text{ mol} \times 273\text{ K} \times 0.0821\text{ L atm}}{10\text{ L} \times \text{mol K}} = 29\text{ atm}$$

47. Only the $KClO_3$ in the mixture will yield O_2.

Calculate the grams of $KClO_3$ in the 1.20 g mixture of $KClO_3$ and KCl that produces 250 mL of O_2 (STP) when heated.

$2\ KClO_3(s) \rightarrow 2\ KCl(s) + 3\ O_2(g)$ First calculate the grams of $KClO_3$ that will yield 250 mL O_2. 250 mL = 0.250 L

$$0.250\text{ L }O_2 \times \frac{1\text{ mol }O_2}{22.4\text{ L }O_2} \times \frac{2\text{ mol }KClO_3}{3\text{ mol }O_2} \times \frac{122.6\text{ g }KClO_3}{1\text{ mol }KClO_3}$$

= 0.912 g $KClO_3$ in the 1.20 g mixture

Now calculate the percentage by weight of $KClO_3$ in the mixture.

$$\frac{0.912\text{ g }KClO_3}{1.20\text{ g mixture}} \times 100\% = 76.0\%\ KClO_3$$

B. 1. The potential energy is greater in the liquid water than in the ice. The heat necessary to melt the ice increases the potential energy, giving the molecules greater freedom of motion.

2. At 0°C, all three substances, H_2S, H_2Se, and H_2Te, are gases, because they all have boiling points below 0°C.

3. The pressure of the atmosphere must be 1.00 atmosphere, otherwise the water would be boiling at some other temperature.

4. 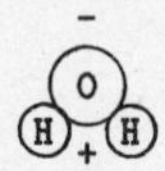

5. Melting point, boiling point, heat of fusion, heat of vaporization, density, and crystalline structure in the solid state are some of the physical properties of water that would be different if the molecules were linear (and nonpolar) instead of bent (and highly polar).

6. Prefixes preceding the word hydrate are used in naming hydrates to indicate the number of molecules of water present in the formulas. The prefixes used are mono, di, tri, tetra, penta, hexa, and octa to represent, respectively, 1, 2, 3, 4, 5, 6, 7, and 8 molecules of water.

7. The distillation setup in Figure 13.9 would be satisfactory for separating salt and water, but not for separating ethyl alcohol and water. In the first case, the water is easily vaporized, the salt is not, so the water boils off and condenses in the condenser. In the second case, both substances are easily vaporized, so both would vaporize (though not to an equal degree) and the condensed liquid would contain both substances.

8. The ethyl alcohol would be at 70°C. The liquid is boiling, which means its vapor pressure equals the confining pressure, and from Table 13.2 we find that ethyl alcohol has a vapor pressure of 543 torr at 70°C.

9. The water in the containers would both have the same vapor pressure, for it is a function of the temperature of the liquid. The pressure exerted by water vapor molecules would be less in the open container, because the opportunity to escape prevents the establishment of equilibrium.

10. In Figure 13.5, it would be case (b) in which the atmosphere would reach saturation.

11. If ethyl ether and ethyl alcohol were both placed in the closed container, (a) both substances would be present in the vapor, for both are volatile liquids; (b) ethyl ether would have more molecules in the vapor because it has a higher vapor pressure.

12. The equilibrium vapor pressure observed in (c) would remain unchanged. The presence of more water in (b) does not change the magnitude of the equilibrium vapor pressure of the water. The temperature of the water determines the magnitude of the equilibrium vapor pressure.

13. At 30 torr, H_2O would boil at approximately 29°C, ethyl alcohol at 14°C, and ethyl ether and ethyl chloride at some temperature below 0°C.

14. (a) At a pressure of 500 torr, water boils at 88°C.
    (b) The normal boiling point of ethyl alcohol is 78°C.
    (c) At a pressure of 0.50 atm (380 torr) ethyl ether boils at 16°C.

15. Based on Figure 13.8:
    (a) Line BC is horizontal because the temperature remains constant during the entire process of melting.
    (b) During BC, both solid and liquid phases will be present.
    (c) The line DE represents the change from liquid to vapor, that is, vaporization. The horizontal line is at the boiling temperature of the substance.

C. Review questions.

1. Physical properties of water:

   (a) melting point, 0°C; (b) boiling point, 100°C; (c) colorless; (d) odorless; (e) tasteless; (f) heat of fusion of 335 J (80 cal) per gram; (g) heat of vaporization of 2.26 kJ (540 cal) per gram; (h) density 1.0 g/mL; (i) specific heat = 4.184 J/g °C.

2. For water to have its maximum density, the temperature must be 4°C, and the pressure sufficient to keep it liquid. d = 1.00 g/mL.

3. If you apply heat to an ice-water mixture, the heat goes to melt ice, rather than warm the water, so the temperature remains constant until all the ice is melted.

4. Ice at 0°C contains less heat than water at 0°C. Heat must be added to convert ice to water, so the water will contain that much additional heat.

5. Ice floats in water because it is less dense than water. The density of ice at 0°C is 0.915 g/mL, vs. 1.00 g.mL for water. Ice would sink in ethyl alcohol, which has a density of 0.79 g/mL.

6. The heat of vaporization of water would be lower if water molecules were linear instead of bent. If linear the molecules of water would be nonpolar. The relatively high heat of vaporization of water is, in good part, a result of the molecule being highly polar.

7. Ethyl alcohol exhibits hydrogen bonding; ethyl ether does not. This is indicated by the high heat of vaporization of ethyl alcohol even though its molecular weight is much less than the molecular weight of ethyl ether.

8. A linear water molecule, being nonpolar, would exhibit less hydrogen bonding than the highly polar, bent, water molecule. The polar molecule has a greater attractive force than a nonpolar molecule.

9. Ammonia exhibits hydrogen bonding; methane does not. The ammonia molecule is polar; the methane molecule is not. The nitrogen atom in ammonia is quite electronegative; the carbon atom in methane is much less electronegative.

10. Water at 80°C will have fewer hydrogen bonds than water at 40°C. At the higher temperature the molecules of water are moving faster than at the lower temperature. This results in less hydrogen bonding at the higher temperature.

11. $H_2NCH_2CH_2NH_2$ has two polar NH_2 groups. It should therefore, show more hydrogen bonding and therefore, a higher boiling point (49°C).

12. (a) Order of increasing rate of evaporation: Mercury, acetic acid, water, toluene, benzene, carbon tetrachloride, methyl alcohol, bromine.

    (b) Highest boiling point: Mercury; lowest boiling point: bromine.

13. Rubbing alcohol feels cold when applied to the skin because the evaporation of the alcohol absorbs heat from the skin. The alcohol has a fairly high vapor pressure and, consequently, evaporates quite rapidly. This produces the cooling effect.

14. Water boils when its vapor pressure equals the prevailing atmospheric pressure over the water. In order for water to boil at 50°C the pressure over the water would need to be reduced to a point equal to the vapor pressure of the water (92.5 torr).

15. Two conditions must be met for water molecules to evaporate. (1) The evaporating molecules must be on the surface of the water, and (2) the molecules must have sufficient kinetic energy to overcome attractive forces of other water molecules. Consequently, higher energy molecules escape; lower energy molecules are left behind. Evaporation, thus cools the remaining liquid.

16. In a pressure cooker the temperature at which the water boils increases above its normal boiling point because the water vapor (steam) formed by boiling cannot escape, resulting in an increased pressure over the water and consequently in an increased boiling temperature.

17. Vapor pressure varies with temperature. The temperature at which the vapor pressure of a liquid equals the prevailing pressure is the boiling point.

18. As temperature increases, molecular velocities increase, so at greater velocities, it becomes easier for molecules to break away from the attractive forces in the liquid.

19. Water has a relatively high boiling point because there is high attraction between molecules due to hydrogen bonding.

20. Ammonia would have a higher vapor pressure than SO_2 at -40°C.

21. As the temperature of a liquid increases the kinetic energy of the molecules and the vapor pressure of the liquid increases. When the vapor pressure of the liquid equals the external pressure boiling begins with many of the molecules having enough energy to escape from the liquid. Bubbles of vapor are formed throughout the liquid and these bubbles rise to the surface and escape as boiling continues.

22. HF has a higher boiling point than HCl because of the strong hydrogen bonding in HF. Neither F_2 nor Cl_2 will have hydrogen bonding, so the compound, F_2, with the smaller mass, has the lower boiling point.

23. Yes, ice can be cooled below 0°C, the same as any other solid.

24. The boiling liquid remains at constant temperature because the added heat energy is being used to convert the liquid to a gas, i.e., to supply the heat of vaporization for the liquid at its boiling point.

25. 2336°C, the boiling point of copper.

26. A lake freezes from the top down because as the temperature drops to freezing or below, the water on the surface would tend to cool faster than the water deeper down. As the surface water freezes the ice formed floats because the ice is less dense than the liquid water below it.

27. If the lake is in an area where the temperature is below freezing for part of the year, the expected temperature would be 4°C at the bottom of the lake. This is because the surface water would cool to 4°C (maximum density) and sink.

28. Reactions of metals with water:

$2\ Al + 3\ H_2O(steam) \rightarrow 3\ H_2(g) + Al_2O_3$ requires steam

$Ca + 2\ H_2O \rightarrow H_2(g) + Ca(OH)_2$ slowly at room temperature

$3\ Fe + 4\ H_2O(steam) \rightarrow 4\ H_2(g) + Fe_3O_4$ requires steam

$2\ Na + 2\ H_2O \rightarrow H_2(g) + 2\ NaOH$ vigorously at room temperature

$Zn + H_2O(steam) \rightarrow H_2(g) + ZnO$ requires steam

29. The formation of hydrogen and oxygen from water is an endothermic reaction. We know from the following evidence:

    (a) We must continually put in energy (electrical) to make the reaction proceed. It stops as soon as the energy input stops.

    (b) The reverse reaction, the burning of hydrogen in oxygen gives off energy (heat).

30. (a) The word anhydride comes from the Greek, anhydros, meaning waterless. An anhydride is a substance that will react (combine) with water to form an acid or a base.

    (b) An acid anhydride will be an oxide of a nonmetal.

    (c) A basic anhydride will be an oxide of a metal.

31. (a) Acid anhydrides: H_2SO_3, SO_2; H_2SO_4, SO_3; HNO_3, N_2O_5; $HClO_4$, Cl_2O_7; H_2CO_3, CO_2; H_3PO_4, P_2O_5.

    (b) Basic anhydrides: $NaOH$ - Na_2O; KOH - K_2O; $Ba(OH)_2$ - BaO; $Ca(OH)_2$ - CaO; $Mg(OH)_2$ - MgO.

32. The balanced equations are:

    (a) $Ba(OH)_2 \xrightarrow{\Delta} BaO + H_2O$

    (b) $2\ CH_3OH + 3\ O_2 \rightarrow 2\ CO_2 + 4\ H_2O$

    (c) $2\ Rb + 2\ H_2O \rightarrow 2\ RbOH + H_2$

    (d) $SnCl_2 \cdot 2\ H_2O \xrightarrow{\Delta} SnCl_2 + 2\ H_2O$

    (e) $HNO_3 + NaOH \rightarrow NaNO_3 + H_2O$

    (f) $Li_2O + H_2O \rightarrow 2\ LiOH$

    (g) $2\ KOH \xrightarrow{\Delta} K_2O + H_2O$

    (h) $Ba + 2\ H_2O \rightarrow Ba(OH)_2 + H_2$

    (i) $Cl_2 + H_2O \rightarrow HCl + HClO$

    (j) $SO_3 + H_2O \rightarrow H_2SO_4$

    (k) $H_2SO_3 + 2\ KOH \rightarrow K_2SO_3 + 2\ H_2O$

    (l) $CO_2 + H_2O \rightarrow H_2CO_3$

33. The conversion of oxygen to ozone is an endothermic reaction. Evidence of this is:

(a) It usually requires an electrical discharge to form ozone from oxygen (not proof by itself).

(b) The equation is shown requiring 286 kJ for the reaction.

(c) If the reaction were exothermic, it would, once started, sustain itself. This does not occur.

34. Oxygen atom, O $:\dot{\underset{..}{O}}\cdot$ Oxygen molecule, O_2 $:\ddot{O}::\ddot{O}:$

Ozone molecule, O_3 $\cdot\underset{..}{O}\cdot\cdot\ddot{O}\cdot\cdot\underset{.}{O}\cdot$ Peroxide ion, O_2^{2-} $:\underset{..}{\ddot{O}}:\underset{..}{\ddot{O}}:^{2-}$

The Lewis dot formula for the O_2 molecule is really not correct but is the best that can be done at this point in the course.

The peroxide ion has 18 electrons, 14 of which are valence electrons.

The oxygen molecule has 16 electrons, 12 of which are valence electrons

35. Ozone in the atmosphere absorbs ultraviolet radiation by the reaction:

$$O_3 \xrightarrow[\text{radiation}]{\text{ultraviolet}} O_2 + O + \text{Heat}$$

Therefore the amount of ultraviolet radiation reaching the earth is greatly reduced.

36. Names of hydrates:

(a) $BaBr_2 \cdot 2\ H_2O$ barium bromide dihydrate

(b) $AlCl_3 \cdot 6\ H_2O$ aluminum chloride hexahydrate

(c) $FePO_4 \cdot 4\ H_2O$ iron(III) phosphate tetrahydrate

(d) $MgNH_4PO_4 \cdot 6\ H_2O$ magnesium ammonium phosphate hexahydrate

(e) $FeSO_4 \cdot 7\ H_2O$ iron(II) sulfate heptahydrate

(f) $SnCl_4 \cdot 5\ H_2O$ tin(IV) chloride pentahydrate

37. When $CuSO_4$, which is gray-white, is exposed to moisture, it hydrates to $CuSO_4 \cdot 5\ H_2O$, which is blue. Thus the appearance of a blue color indicates that the anhydrous $CuSO_4$ has picked up moisture.

38. In metal dioxides, all the bonds are ionic; metal ions ionically bonded to oxide ions. Oxygen atoms are not bonded to each other in metal dioxides. In metal peroxides, covalent O-O bonds exist within the peroxide ions. However, the peroxide ions are ionically bonded to metal ions.

39. Deionized water is water from which the ions have been removed.

(a) Hard water is water containing dissolved calcium and magnesium salts.

(b) Soft water is free of ions that cause hardness.

(c) Distilled water is water that has been vaporized by boiling and recondensed. It is free of nonvolatile impurities but may still contain any volatile impurities that were initially present in the water.

40. Soap can soften hard water by forming a precipitate with, and thus removing, the calcium and magnesium ions. This precipitate is a greasy scum and is very objectional, so this is a poor way to soften water.

41. Chlorine is commonly used to destroy bacteria in water. Ozone is also used in some places.

42. Ozone.

43. When organic pollutants in water are oxidized by dissolved oxygen (molecular), there may not be sufficient dissolved oxygen to sustain marine life in the water, such as fish. Most marine life depends on dissolved oxygen for respiration.

44. You should not drink liquids that are stored in ceramic containers because they are likely to have dissolved some of the lead from the ceramic. If the ceramic is glazed, the liquid is less apt to dissolve substances from the ceramic.

45. $Na_2Zeolite(s) + Mg^{2+}(aq) \rightarrow MgZeolite(s) + 2\ Na^+(aq)$

46. Softening of hard water using sodium carbonate:

$CaCl_2(aq) + Na_2CO_3(aq) \rightarrow CaCO_3(s) + 2\ NaCl(aq)$

Calcium ions are precipitated as $CaCO_3$.

47. The following statements are correct: a,b,c,f,h,l,m,o,p,s,t,u,w,y.

D. Review problems.

1. Moles

(a) $100\ g\ CoCl_2 \cdot 6H_2O \times \frac{1\ mol}{237.9\ g} = 0.420\ mol\ CoCl_2 \cdot 6H_2O$

(b) $100\ g\ FeI_2 \cdot 4H_2O \times \frac{1\ mol}{381.6\ g} = 0.262\ mol\ FeI_2 \cdot 4H_2O$

2. Moles of water

(a) $100 \text{ g } CoCl_2 \cdot 6H_2O \times \frac{1 \text{ mol } CoCl_2 \cdot 6H_2O}{237.9 \text{ g } CoCl_2 \cdot 6H_2O} \times \frac{6 \text{ mol } H_2O}{1 \text{ mol } CoCl_2 \cdot 6H_2O}$

$= 2.52 \text{ mol } H_2O$

(b) $100 \text{ g } FeI_2 \cdot 4H_2O \times \frac{1 \text{ mol } FeI_2 \cdot 4H_2O}{381.6 \text{ g } FeI_2 \cdot 4H_2O} \times \frac{4 \text{ mol } H_2O}{1 \text{ mol } FeI_2 \cdot 4H_2O}$

$= 1.05 \text{ mol } H_2O$

3. Epsom salts.

Epsom salt contains 7 mol H_2O per mol $ZnSO_4 \cdot 7H_2O$.

$7 \times 18.0 \text{ g } H_2O = 126 \text{ g } H_2O$

$\frac{126 \text{ g } H_2O}{246.4 \text{ g } MgSO_4 \cdot 7H_2O} \times 100\% = 51.1\% \ H_2O$

4. 1 mol $Al_2(SO_4)_3 \cdot 18H_2O$ contains 18 mol H_2O

$18.0 \text{ g } H_2O \times 18 = 324 \text{ g } H_2O$

$\frac{324 \text{ g } H_2O}{666.3 \text{ g } Al_2(SO_4)_3 \cdot 18H_2O} \times 100\% = 48.6\% \ H_2O$

5. The compound contains 14.2% H_2O and 85.8% $Pb(C_2H_3O_2)_2$

$14.2 \text{ g } H_2O \times \frac{1 \text{ mol}}{18.0 \text{ g}} = 0.789 \text{ mol } H_2O$

$85.8 \text{ g } Pb(C_2H_3O_2)_2 \times \frac{1 \text{ mol}}{325.2 \text{ g}} = 0.264 \text{ mol } Pb(C_2H_3O_2)_2$

Mole ratios: $\frac{0.789}{0.264} = 2.99 \text{ mol } H_2O$; $\frac{0.264}{0.264} = 1.00 \text{ mol } Pb(C_2H_3O_2)_2$

The hydrate formula is $Pb(C_2H_3O_2)_2 \cdot 3H_2O$

6. Formula of a hydrate:

$25.0 \text{ g } FePO_4 \cdot XH_2O$

$16.9 \text{ g } FePO_4$

$8.1 \text{ g } H_2O$

$8.1 \text{ g } H_2O \times \frac{1 \text{ mol}}{18.0 \text{ g}} = 0.45 \text{ mol } H_2O$

$16.9 \text{ g } FePO_4 \times \frac{1 \text{ mol}}{150.8 \text{ g}} = 0.112 \text{ mol } FePO_4$

$(FePO_4)_{\frac{0.112}{0.112}}(H_2O)_{\frac{0.45}{0.112}} = (FePO_4)_{1.0}(H_2O)_{4.0}$

The hydrate formula is $FePO_4 \cdot 4H_2O$

7. Total joules needed = joules to heat the water + joules to vaporize the water

$$\text{joules to heat} = 120\ \text{g} \times (100 - 20)\,°\text{C} \times \frac{4.184\ \text{J}}{\text{g}\ °\text{C}} = 4.0 \times 10^4\ \text{J}$$

$$\text{joules to vaporize} = 120\ \text{g} \times \frac{2.26 \times 10^3\ \text{J}}{\text{g}} = 2.71 \times 10^5\ \text{J}$$

$$\text{Total joules} = 0.40 \times 10^5\ \text{J} + 2.71 \times 10^5\ \text{J} = 3.11 \times 10^5\ \text{J}$$

8. Total joules removed = joules removed to cool the water to °C + joules removed to freeze the water

$$126\ \text{g}\ H_2O \times (24 - 0)\,°\text{C} \times \frac{4.184\ \text{J}}{\text{g}\ H_2O\ °\text{C}} + 126\ \text{g}\ H_2O \times \frac{335\ \text{J}}{\text{g}}$$

(joules to cool to 0°C) (joules to freeze)

$$= 1.3 \times 10^4\ \text{J} + 4.22 \times 10^4\ \text{J} = 5.5 \times 10^4\ \text{J}$$

9. Total calories needed = calories to melt the ice (0°C) + calories to heat the water from 0°C to 100°C + calories to vaporize the water

$$225\ \text{g ice} \times \frac{80\ \text{cal}}{\text{g ice}} + 225\ \text{g} \times 100\,°\text{C} \times \frac{1\ \text{cal}}{\text{g}\ °\text{C}} + 225\ \text{g} \times \frac{540\ \text{cal}}{\text{g}}$$

(cal to melt) + (cal to heat) + (cal to vaporize)

$$= 1.6 \times 10^5\ \text{cal}$$

10. $\frac{2.26\ \text{kJ}}{1\ \text{g}\ H_2O} \times \frac{18.0\ \text{g}\ H_2O}{\text{mol}\ H_2O} = 40.7\ \text{kJ/mol}\ H_2O(4.07 \times 10^4\ \text{J/mol}\ H_2O)$

11. The water must be cooled 25°C

$100\ \text{g} \times \frac{80\ \text{cal}}{\text{g}} = 8000$ cal required to melt the ice

$300\ \text{g} \times 25\,°\text{C} \times \frac{1\ \text{cal}}{\text{g}\ °\text{C}} = 7500$ cal must be lost by H_2O to cool the H_2O to 0°C. Yes, sufficient ice is used to cool the water to 0°C.

12. $300\ \text{g}\ H_2O \times (100 - 25)\,°\text{C} \times \frac{1\ \text{cal}}{\text{g}\ °\text{C}} = 2.25 \times 10^4$ cal required to heat the H_2O to 100°C

$35.0\ \text{g steam} \times \frac{540\ \text{cal}}{\text{g}} = 1.89 \times 10^4$ cal available from the steam.

No, there is not sufficient steam. 2.25×10^4 calories are required and 1.89×10^4 calories are available from the steam.

13. $50.0\ \text{mol steam} \times \frac{18.0\ \text{g}}{\text{mol}} \times \frac{2.26\ \text{kJ}}{\text{g}} = 2.03 \times 10^3\ \text{kJ}\ (2.03 \times 10^6\ \text{J})$

liberated by condensing the steam at 100°C to liquid water at 100°C.

$$50.0 \text{ mol } H_2O \times \frac{18.0 \text{ g}}{\text{mol}} \times \frac{4.184 \text{ J}}{\text{g°C}} \times (100 - 30)\text{°C} = 2.6 \times 10^5 \text{ J}$$

liberated in cooling the water from 100°C to 30°C.

2.03×10^6 J + 0.26×10^6 J = 2.29×10^6 J (total energy liberated)

14. $100 \text{ g ice} \times \frac{2.01 \text{ J}}{\text{g°C}} \times 10.0\text{°C} = 2.01 \times 10^3$ J (2.01 kJ) to convert ice at −10.0°C to ice at 0°C.

$100 \text{ g ice} \times \frac{335 \text{ J}}{\text{g}} = 3.35 \times 10^4$ J (33.5 kJ) to convert ice at 0°C to water at 0°C.

$100 \text{ g } H_2O \times \frac{4.184 \text{ J}}{\text{g°C}} \times 20.0\text{°C} = 8.37 \times 10^3$ J (8.37 kJ) to heat water at 0°C to water at 20.0°C.

2.01 kJ + 33.5 kJ + 8.37 kJ = 43.9 kJ (total energy needed)

15. $2\ H_2O(l) \longrightarrow 2\ H_2(g) + O_2(g)$

$$25.0 \text{ L } O_2 \times \frac{1 \text{ mol } O_2}{22.4 \text{ L } O_2} \times \frac{2 \text{ mol } H_2O}{1 \text{ mol } O_2} \times \frac{18.0 \text{ g } H_2O}{1 \text{ mol } H_2O} = 40.2 \text{ g } H_2O$$

16. $1 \text{ mol } H_2O(l) \times \frac{18.0 \text{ g}}{\text{mol}} \times \frac{1.0 \text{ mL}}{\text{g}} = 18.0 \text{ mL liquid } H_2O$

$1 \text{ mol } H_2O(g) \times \frac{22.4 \text{ L}}{\text{mol}} = 22.4 \text{ L } H_2O \text{ gas}$

18.0 mL (liquid) vs. 22.4 L (gas)

17. (a) $2\ H_2O_2 \longrightarrow 2\ H_2O_2 + O_2$

$$12.0 \text{ mol } H_2O_2 \times \frac{1 \text{ mol } O_2}{2 \text{ mol } H_2O_2} = 6.00 \text{ mol } O_2$$

(b) $PV = nRT \qquad V = \frac{nRT}{P}$

$T = 35 + 273\text{°C} = 308 \text{ K} \qquad P = 710 \text{ torr} \times \frac{1 \text{ atm}}{760 \text{ torr}} = 0.934 \text{ atm}$

$$V = \frac{6.00 \text{ mol} \times 308 \text{ K} \times 0.0821 \text{ L-atm}}{0.934 \text{ atm} \times \text{mol K}} = 162 \text{ L } O_2$$

18. $2\ H_2O_2 \longrightarrow 2\ H_2O_2 + O_2$

$$625 \text{ g } H_2O_2 \text{ sol'n} \times \frac{3.0 \text{ g } H_2O_2}{100 \text{ g sol'n}} \times \frac{1 \text{ mol } H_2O_2}{34.0 \text{ g } H_2O_2} \times \frac{1 \text{ mol } O_2}{2 \text{ mol } H_2O_2}$$

$$\times \frac{22.4 \text{ L } O_2}{\text{mol } O_2} = 6.2 \text{ L } O_2$$

19\. (a) $2\ K + 2\ H_2O \longrightarrow 2\ KOH + H_2$

$$1.00\text{ mol K} \times \frac{2\text{ mol }H_2O}{2\text{ mol K}} \times \frac{18.0\text{ g }H_2O}{1\text{ mol }H_2O} = 18.0\text{ g }H_2O$$

(b) $Ca + 2\ H_2O \longrightarrow Ca(OH)_2 + H_2$

$$1.00\text{ mol Ca} \times \frac{2\text{ mol }H_2O}{1\text{ mol Ca}} \times \frac{18.0\text{ g }H_2O}{1\text{ mol }H_2O} = 36.0\text{ g }H_2O$$

(c) $2\ Na + 2\ H_2O \longrightarrow 2\ NaOH + H_2$

$$1.00\text{ g Na} \times \frac{1\text{ mol Na}}{23.0\text{ g Na}} \times \frac{2\text{ mol }H_2O}{2\text{ mol Na}} \times \frac{18.0\text{ g }H_2O}{1\text{ mol }H_2O} = 0.783\text{ g }H_2O$$

(d) $SO_3 + H_2O \rightarrow H_2SO_4$

$$1.00\text{ mol }SO_3 \times \frac{1\text{ mol }H_2O}{1\text{ mol }SO_3} \times \frac{18.0\text{ g }H_2O}{1\text{ mol }H_2O} = 18.0\text{ g }H_2O$$

(e) $MgO + H_2O \rightarrow Mg(OH)_2$

$$1.00\text{ g MgO} \times \frac{1\text{ mol MgO}}{40.3\text{ g MgO}} \times \frac{1\text{ mol }H_2O}{1\text{ mol MgO}} \times \frac{18.0\text{ g }H_2O}{1\text{ mol }H_2O} = 0.447\text{ g }H_2O$$

(f) $N_2O_5 + H_2O \rightarrow 2\ HNO_3$

$$1.00\text{ g }N_2O_5 \times \frac{1\text{ mol }N_2O_5}{108\text{ g }N_2O_5} \times \frac{1\text{ mol }H_2O}{1\text{ mol }N_2O_5} \times \frac{18.0\text{ g }H_2O}{1\text{ mol }H_2O} = 0.167\text{ g }H_2O$$

20\. $$\frac{1.00\text{ mol }H_2O}{1.00\text{ day}} \times \frac{6.022 \times 10^{23}\text{ molecules }H_2O}{\text{mol }H_2O} \times \frac{1.00\text{ day}}{24\text{ hrs}} \times \frac{1\text{ hr}}{60\text{ min}}$$

$$\times \frac{1\text{ min}}{60\text{ sec}} = 6.97 \times 10^{18}\text{ molecules/sec}$$

21\. g sol'n - g H_2O = g H_2SO_4

$$122\text{ mL sol'n} \times \frac{1.26\text{ g sol'n}}{\text{mL sol'n}} - 100\text{ mL} \times \frac{1.00\text{ g }H_2O}{\text{mL }H_2O} = 54\text{ g }H_2SO_4\text{ added}$$

22\. $2\ H_2 + O_2 \rightarrow 2\ H_2O$

$$80.0\text{ mL }H_2 \times \frac{1\text{ mL }O_2}{2\text{ mL }H_2} = 40.0\text{ mL }O_2$$

(a) Yes, O_2 remains unreacted

(b) 60.0 mL O_2 - 40.0 mL O_2 = 20.0 mL O_2 unreacted

B. 1. Note that Na^+ is smaller than Cl^-

(a)

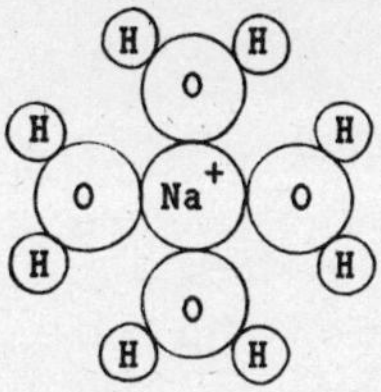

(b)

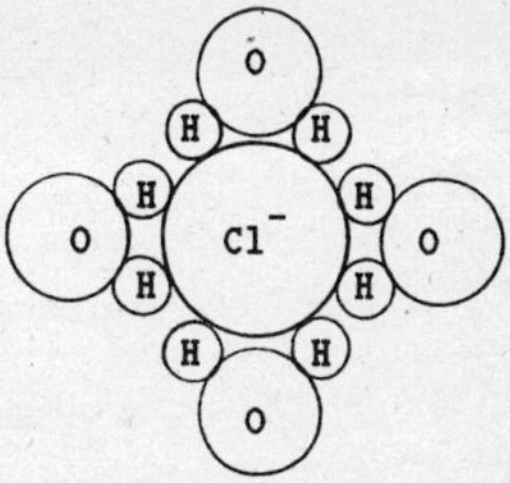

These diagrams are intended to show the orientation of the water molecules about the ions and not the number of water molecules.

2. Reasonably soluble: a. KOH b. $NiCl_2$ d. $AgC_2H_3O_2$

e. Na_2CrO_4 h. $CaCl_2$ i. $Fe(NO_3)_3$

Insoluble: c. ZnS f. PbI_2 g. $MgCO_3$ j. $BaSO_4$

3. From Table 14.2, approximately 4.5 grams of NaF would be soluble in 100 g of water at 50°C. Data such as Figure 14.3 is desirable, but often we must use data that is less complete.

4. From Table 14.3, solubilities in water at 25°C are:

a. KCl, 35 g/100 g water; b. $KClO_3$, 9 g/100 g water;

c. KNO_3, 39 g/100 g water

Though all are read from the same graph, the KNO_3 is the most difficult to estimate because of the steep slope.

5. Potassium fluoride has relatively high solubility when compared to lithium or sodium fluoride. For lithium and sodium halides the order of solubility is, in order of increasing solubility, F^-, Cl^-, Br^-, I^-; but for potassium halides the order of increasing solubility is Cl^-, Br^-, F^-, I^-.

6. (a) $KClO_3$ at 60°C, 25 g; (b) HCl at 20°C, 72 g;

(c) Li_2SO_4 at 80°C, 31 g; (d) KNO_3 at 0°C, 14 g.

7. KNO_3

8. A 1 molal solution of camphor will show a greater freezing point depression than a 2 molal solution of benzene.

$$\frac{1 \text{ mol solute}}{\text{kg camphor}} \times \frac{40°\text{C kg camphor}}{\text{mol solute}} = 40°\text{C (freezing point lowering)}$$

$$\frac{2 \text{ mol solute}}{\text{kg benzene}} \times \frac{5.1°\text{C kg benzene}}{\text{mol solute}} = 10.2°\text{C (freezing point lowering)}$$

9. Volume of 1 cm cube = 1 cm^3. Volume of 0.01 cm cube = 1×10^{-6} cm^3. Therefore there would be 10^6 of these cubes made from the larger cube. Area of one face of small cube = $(0.01 \text{ cm})^2 = 1 \times 10^{-4}$ cm^2. Total surface area of all 0.01 cm cubes cut from 1 cm cube,
1×10^6 cubes $\times 6 \times 10^{-4}$ cm^2/cube = 6×10^2 cm^2

10. 63 g NH_4Cl/150 g H_2O = 42 g NH_4Cl/100 g H_2O. From Figure 14.3, the solubility of NH_4Cl in water is approximately 42 g/100 g H_2O at 30°C, 45 g/100 g H_2O at 40°C. Therefore the solution of 63 g/150 g of water would be saturated at 10°, 20°, and 30°C; unsaturated at 40° and 50°C.

11. Because the dissolving process involves solvent molecules attaching to the solute ions or molecules, this rate decreases as more of the solvent molecules are already attached to solute molecules. As the solution gets more and more saturated, the number of unused solvent molecules is much less. Also the rate of recrystallization increases as the concentration of dissolved solute increases.

12. Yes. A volumetric flask is satisfactory for preparing normal solutions. You determine the equivalents of solute required for the volume of solution being prepared; weigh out the corresponding number of grams of solute; place the solute in the volumetric flask; add water to dissolve the solute; dilute to the mark with water, and mix.

13. Because the concentration of water is greater in the thistle tube, the water will flow through the membrane from the thistle tube to the urea solution in the beaker. Therefore the solution level in the thistle tube will fall.

C. Review Questions

1. The two components of a solution are the solute and the solvent. The solute is the component that is dissolved or the least abundant component. The solvent is the dissolving agent or the most abundant component.

2. It is not always apparent in a solution which component is the solute. For example, in a solution made from equal volumes of two liquids the designation of solute and solvent would be simply a matter of preference on the part of the person making the designations.

3. The ions or molecules of a dissolved solute do not settle out because the individual particles are so small that the force of molecular collisions is large compared to the force of gravity.

4. Yes. It is possible to have one solid dissolved in another solid. Metal alloys are of this type. Atoms of one metal are dissolved among atoms of another metal.

5. Orange. The three reference solutions, KCl, $KMnO_4$, and $K_2Cr_2O_7$, are the same in that all contain K^+ ions in solution. The different colors must result from the different anions dissolved in the solutions: MnO_4^-, purple, and $Cr_2O_7^{2-}$, orange. Therefore, we might expect the $Cr_2O_7^{2-}$ ion present in an aqueous solution of $Na_2Cr_2O_7$ to impart an orange color to the solution.

6. Carbon tetrachloride and benzene are both nonpolar molecules, so there are no strong intermolecular forces between molecules of either substance, or with each other, so they are miscible with each other. Sodium chloride consists of ions strongly attracted to each other by electrical attractions. The carbon tetrachloride molecules, being nonpolar, have no strong forces to pull the ions apart, so sodium chloride is insoluble in carbon tetrachloride.

7. Coca Cola has two main features, taste and carbonation (fizz). The carbonation is due to a dissolved gas, carbon dioxide. Since dissolved gases become less soluble as temperature is increased, hot Coca Cola would be totally flat, with no carbonation. It would therefore be unappealing to most people.

8. Air is considered to be a solution because it is a homogeneous mixture of several substances and does not have a fixed composition.

9. A teaspoon of sugar would definitely dissolve more rapidly in 200 mL of hot coffee than in 200 mL of iced tea. The much greater thermal agitation of the hot coffee will help break the sugar molecules away from the undissolved solid and disperse them throughout the solution. Other solutes in coffee or tea would have no significant effect; the temperature difference is the key feature.

10. The solubility of gases in liquids is greatly affected by the pressure of the gas above the liquid. The greater the pressure, the more soluble the gas. There is very little effect of pressure on the dissolving of solids in liquids.

11. For a given mass of solute the smaller the particles the faster the dissolution of the solute. This is so because the smaller particles have a greater surface area exposed to the dissolving action of the solvent.

12. In a saturated solution, the net rate of dissolving is zero, that is, there is no further increase in the amount of dissolved solute,

even though undissolved solute is continuously dissolving. This is so because dissolved solute is continuously coming out of solution at a rate equal to the rate of dissolving.

13. When crystals of $AgNO_3$ and NaCl are mixed, the contact is not intimate enough for the double displacement reaction to occur. When solutions of the two chemicals are mixed, the ions are free to move and come into intimate contact with each other so reaction occurs easily. The AgCl formed is insoluble.

14. A 16 molar solution of nitric acid is a solution where the ratio of moles of solute (HNO_3) to liters of solution is 16 to 1.

15. The two solutions contain the same number of chloride ions. One liter of 1 M NaCl contains 1 mole of NaCl; therefore the solution contains 1 mole of chloride ion. 0.5 liter of 1 M $MgCl_2$ contains 0.5 mole of $MgCl_2$; therefore the solution contains 1 mole of chloride ion.

$$0.5\ \text{L} \times \frac{1\ \text{mol MgCl}_2}{\text{L}} \times \frac{2\ \text{mol Cl}^-}{1\ \text{mol MgCl}_2} = 1\ \text{mol Cl}^-$$

16. "Dilute" and "concentrated" are rather general terms which are not very useful because they are so indefinite. A solution that is termed dilute in one situation might well be considered concentrated under different circumstances.

17. A solution of H_2SO_4 that is 18 molar contains 18 moles of H_2SO_4 per liter of solution. Since there can be 2 equivalent weights of H_2SO_4 per mole of H_2SO_4, a solution of H_2SO_4 that is 18 molar would contain 36 equivalent weights of H_2SO_4 per liter of solution. Such a solution would be 36 normal.

18. The number of grams of NaCl in 750 mL of 5 molar solution is:

$$0.750\ \text{L} \times \frac{5.00\ \text{mol NaCl}}{\text{L}} \times \frac{58.1\ \text{g NaCl}}{1\ \text{mol NaCl}} = 218\ \text{g NaCl.}$$

Dissolve the 218 g of NaCl in a minimum amount of water then dilute the resulting solution to a final volume of 750 mL.

19. Ranking of the specified bases in descending order of the volume of each required to react with 1 liter of 1 M HCl. The volume of each required to yield 1 mole of OH^- ion is shown.

(a)	1 M NaOH	1 liter
(d)	0.6 M $Ba(OH)_2$	0.83 liter
(c)	2 M KOH	0.50 liter
(b)	1.5 M $Ca(OH)_2$	0.33 liter

20. The boiling point of a liquid (or solution) is the temperature at which the vapor pressure of the liquid equals the pressure of the atmosphere. Since a solution containing a nonvolatile solute has a lower vapor pressure than the pure solvent, the boiling point of the solution must be at a higher temperature than for the pure solvent in order for the vapor pressure of the solution to equal the atmospheric pressure.

21. The freezing point is the temperature at which a liquid changes to a solid. The vapor pressure of a solution is lower than that of a pure solute. Therefore, the vapor pressure curve of the solution intersects the vapor pressure curve of the pure solid at a temperature lower than the freezing point of the pure solvent. At this point of intersection, the vapor pressure of the solution equals the vapor pressure of the pure solid solvent.

22. A glass of Seven-Up and crushed ice would be colder than a glass of water and crushed ice. The ice will keep the solution at its freezing point, and the Seven-Up will have a lower freezing point because of the dissolved solutes.

23. Water and ice are different phases of the same substance in equilibrium at the freezing point of water, 0°C. The presence of the methanol lowers the freezing point of the water. If the ratio of the alcohol to water is high, the freezing point can be lowered as much as 10°C or more.

24. Effectiveness in lowering the freezing point of 500 g of water:

(a) 100 g of ethyl alcohol (2.17 moles) is more effective than 100 g of sucrose (0.292 mole).

(b) 20 g of ethyl alcohol (0.435 mole) is more effective than 100 g of sucrose (0.292 mole).

(c) 20 g of methyl alcohol (0.625 mole) is more effective than 20 g of ethyl alcohol (0.435 mole).

25. 5 molal NaCl = 5 moles NaCl/kg H_2O; 5 molar NaCl = 5 moles NaCl/ liter solution. The volume of the 5 molar solution will be larger than 1 liter (1 L water + 5 moles NaCl). The volume of the 5 molar solution is exactly 1 L (5 moles NaCl + sufficient water to give 1 L of solution). The 5 molal solution is less concentrated (lower molarity) than the 5 molar solution.

26. The normality of H_2SO_4 equals twice the molarity, N = 2M.

27. A semipermeable membrane will allow water molecules to pass through in both directions. If it has pure water on one side and 10% sugar solution on the other, there is a higher concentration of water molecules on the pure water side, more water molecules are passing from the pure water to the sugar solution than vice versa.

28. The urea solution will have the greater osmotic pressure because it has 1.67 moles of solute/kg of water whereas the glucose solution has only 0.83 moles of solute/kg of water.

29. A lettuce leaf immersed in salad dressing containing salt and vinegar will soon become limp and wilted due to osmosis, as the solvent inside the lettuce leaf, water, flows into the dressing where the solute concentration is higher. With loss of fluids, the lettuce becomes limp. In water, the effect of osmosis will be in the opposite direction, though of small magnitude because the difference in concentration is not large. Water will flow into the lettuce, maintaining a high liquid content inside the lettuce.

30. The concentration os solutes is higher in seawater than in body fluids. Hence the survivors who drank seawater suffered additional dehydration because of the transfer of water by osmosis from the body tissues into the intestinal tract.

31. The following statements are correct: a,b,f,h,j,k,l,n,p,r,s,t,v,x, y,and z.

D. Review Problems

1. Weight percent calculations.

(a) 25.0 g NaBr + 100 g H_2O = 125 g solution

$$\frac{25.0 \text{ g NaBr}}{125 \text{ g solution}} \times 100\% = 20.0\% \text{ NaBr}$$

(b) 1.20 g K_2SO_4 + 10.0 g H_2O = 11.2 g solution

$$\frac{1.20 \text{ g } K_2SO_4}{11.2 \text{ g solution}} \times 100\% = 10.7\% \ K_2SO_4$$

(c) 40.0 g $Mg(NO_3)_2$ + 500 g H_2O = 540 g solution

$$\frac{40.0 \text{ g } Mg(NO_3)_2}{540 \text{ g solution}} \times 100\% = 7.41\% \ Mg(NO_3)_2$$

2. A 12.5% $AgNO_3$ solution contains 12.5 g $AgNO_3$ per 100 g solution

(a) $$30.0 \text{ g } AgNO_3 \times \frac{100 \text{ g solution}}{12.5 \text{ g } AgNO_3} = 240 \text{ g solution}$$

(b) $$0.400 \text{ mol } AgNO_3 \times \frac{169.9 \text{ g } AgNO_3}{\text{mol } AgNO_3} \times \frac{100 \text{ g sol'n}}{12.5 \text{ g } AgNO_3} = 544 \text{ g sol'n}$$

3. Weight percent calculations.

(a) 60.0 g NaCl + 200.0 g H_2O = 260.0 g solution

$$\frac{60.0 \text{ g NaCl}}{260.0 \text{ g sol'n}} \times 100\% = 23.1\% \text{ NaCl}$$

(b) $0.25 \text{ mol } HC_2H_3O_2 \times \frac{60.0 \text{ g}}{\text{mol}} = 15 \text{ g } HC_2H_3O_2$

$3.0 \text{ mol } H_2O \times \frac{18.0 \text{ g}}{\text{mol}} = 54 \text{ g } H_2O$

$\frac{15 \text{ g } HC_2H_3O_2}{69 \text{ g solution}} \times 100\% = 22\% \; HC_2H_3O_2$

(c) $1.0 \text{ molal solution of } C_6H_{12}O_6 \text{ in } H_2O = \frac{1 \text{ mol } C_6H_{12}O_6}{1000 \text{ g } H_2O}$

$1.0 \text{ mol } C_6H_{12}O_6 \times \frac{180 \text{ g } C_6H_{12}O_6}{\text{mol}} = 180 \text{ g } C_6H_{12}O_6$

$1000 \text{ g } H_2O + 180 \text{ g } C_6H_{12}O_6 = 1180 \text{ g solution}$

$\frac{180 \text{ g } C_6H_{12}O_6}{1180 \text{ g solution}} \times 100\% = 15.3\% \; C_6H_{12}O_6$

4. (a) $65 \text{ g solution} \times \frac{5.0 \text{ g KCl}}{100 \text{ g solution}} = 3.3 \text{ g KCl}$

(b) $250 \text{ g solution} \times \frac{15.0 \text{ g } K_2CrO_4}{100 \text{ g solution}} = 37.5 \text{ g } K_2CrO_4$

(c) $100 \text{ g } H_2O \times \frac{6.0 \text{ g } NaHCO_3}{94.0 \text{ g } H_2O} = 6.4 \text{ g } NaHCO_3$

5. $\text{g solution} = 25.0 \text{ g KCl} \times \frac{100 \text{ g solution}}{5.50 \text{ g KCl}} = 455 \text{ g solution}$

Alternate: $\frac{25.0 \text{ g KCl}}{\text{g solution}} \times 100\% = 5.50\% \text{ KCl}$

$\text{g solution} = \frac{25.0 \text{ g KCl} \times 100\%}{5.50\%} = 455 \text{ g solution}$

6. (a) $500 \text{ g solution} \times \frac{0.90 \text{ g NaCl}}{100 \text{ g solution}} = 4.5 \text{ g NaCl}$

(b) $4.5 \text{ g NaCl} \times \frac{91 \text{ g } H_2O}{9.0 \text{ g NaCl}} = 46 \text{ g } H_2O$ needed to give a solution that is 9.0% NaCl.

$500 \text{ g solution} - 4.5 \text{ g NaCl} = 495 \text{ g } H_2O$ in 500 g solution

$495 \text{ g } H_2O - 46 \text{ g } H_2O = 449 \text{ g } H_2O$ that must be evaporated.

7. From Figure 14.3 the solubility of KNO_3 in water at 20°C is 32 g per 100 g H_2O

$50.0\ g\ KNO_3 \times \frac{100\ g\ H_2O}{32\ g\ KNO_3} = 156\ g\ H_2O$ needed to make a saturated solution

$175\ g\ H_2O - 156\ g\ H_2O = 19\ g\ H_2O$ that need to be evaporated.

8. Weight/volume percent

(a) $\frac{22.0\ g\ CH_3OH}{100\ mL\ solution} \times 100\% = 22.0\%\ CH_3OH$

(b) $\frac{4.20\ g\ NaCl}{12.5\ mL\ solution} \times 100\% = 33.6\%\ NaCl$

9. Volume/volume percent

(a) $\frac{10.0\ mL\ CH_3OH}{40.0\ mL\ solution} \times 100\% = 25.0\%\ CH_3OH$

(b) $\frac{2.0\ mL\ CCl_4}{9.0\ mL\ solution} \times 100\% = 22\%\ CCl_4$

10. $\frac{150\ mL\ alcohol}{X\ mL\ solution} \times 100\% = 70\%$ alcohol solution

$X\ mL\ solution = \frac{150\ mL \times 100\%}{70\%} = 214\ mL\ solution$

11. (a) $1.00\ L\ HNO_3\ sol'n \times \frac{1000\ mL}{L} \times \frac{1.21\ g\ HNO_3\ sol'n}{mL\ sol'n}$

$\times \frac{35.0\ g\ HNO_3}{100\ g\ HNO_3\ sol'n} = 424\ g\ HNO_3$

(b) $500\ g\ HNO_3 \times \frac{1000\ mL}{424\ g\ HNO_3} \times \frac{1.00\ L}{1000\ mL} = 1.18\ L$

12. (a) $\frac{0.10\ mol}{250\ mL} \times \frac{1000\ mL}{L} = 0.40\ M$

(b) $\frac{2.5\ mol\ NaCl}{0.650\ L} = 3.8\ M\ NaCl$

(c) $\frac{0.025\ mol\ HCl}{10\ mL} \times \frac{1000\ mL}{L} = 2.5\ M\ HCl$

(d) $\frac{0.35\ mol\ BaCl_2 \cdot 2H_2O}{593\ mL} \times \frac{1000\ mL}{1\ L} = 0.59\ M\ BaCl_2 \cdot 2H_2O$

13. (a) $\frac{53.0\ g\ Na_2CrO_4}{1.00\ L} \times \frac{1\ mol\ Na_2CrO_4}{162.0\ g} = 0.327\ M\ Na_2CrO_4$

(b) $\frac{260\ g\ C_6H_{12}O_6}{800\ mL} \times \frac{1000\ mL}{L} \times \frac{1\ mol}{180\ g} = 1.81\ M\ C_6H_{12}O_6$

(c) $\frac{1.50 \text{ g } Al_2(SO_4)_3}{2.00 \text{ L}} \times \frac{1 \text{ mol}}{342.3 \text{ g}} = 2.19 \times 10^{-3} \text{ M } Al_2(SO_4)_3$

(d) $\frac{0.0282 \text{ g } Ca(NO_3)_2}{1.00 \text{ mL}} \times \frac{1000 \text{ mL}}{\text{L}} \times \frac{1 \text{ mol}}{164.1 \text{ g}} = 0.172 \text{ M } Ca(NO_3)_2$

14. Molarity = $\frac{\text{moles solute}}{\text{liter of solution}}$

(a) $40.0 \text{ L} \times \frac{1.0 \text{ mol LiCl}}{\text{L}} = 40 \text{ mol LiCl}$

(b) $25.0 \text{ mL} \times \frac{1 \text{ L}}{1000 \text{ mL}} \times \frac{3.0 \text{ mol } H_2SO_4}{1 \text{ L}} = 0.075 \text{ mol } H_2SO_4$

(c) $349 \text{ mL} \times \frac{1 \text{ L}}{1000 \text{ mL}} \times \frac{0.0010 \text{ mol NaOH}}{1 \text{ L}} = 3.5 \times 10^{-4} \text{ mol NaOH}$

(d) $5000 \text{ mL} \times \frac{1 \text{ L}}{1000 \text{ mL}} \times \frac{3.1 \text{ mol } CoCl_2}{1 \text{ L}} = 16 \text{ mol } CoCl_2$

15. (a) $150 \text{ L} \times \frac{1.0 \text{ mol NaCl}}{\text{L}} \times \frac{58.5 \text{ g NaCl}}{1 \text{ mol NaCl}} = 8.8 \times 10^3 \text{ g NaCl}$

(b) $0.035 \text{ L} \times \frac{10.0 \text{ mol HCl}}{\text{L}} \times \frac{36.5 \text{ g HCl}}{1 \text{ mol HCl}} = 13 \text{ g HCl}$

(c) $260 \text{ mL} \times \frac{18 \text{ mol } H_2SO_4}{1000 \text{ mL}} \times \frac{98.1 \text{ g } H_2SO_4}{1 \text{ mol } H_2SO_4} = 4.6 \times 10^2 \text{ g } H_2SO_4$

(d) $8.00 \text{ mL} \times \frac{1 \text{ L}}{1000 \text{ mL}} \times \frac{8.00 \text{ mol } Na_2C_2O_4}{1 \text{ L}} \times \frac{134 \text{ g } Na_2C_2O_4}{\text{mol } Na_2C_2O_4}$

$= 8.58 \text{ g } Na_2C_2O_4$

16. $\frac{\text{Moles solute}}{\text{M}}$ = liters of solution; 1 L = 1000 mL

(a) $0.430 \text{ mol KCl} \times \frac{1 \text{ L}}{0.256 \text{ mol KCl}} \times \frac{1000 \text{ mL}}{\text{L}} = 1.68 \times 10^3 \text{ mL}$

(b) $10.0 \text{ mol KCl} \times \frac{1 \text{ L}}{0.256 \text{ mol KCl}} \times \frac{1000 \text{ mL}}{\text{L}} = 3.91 \times 10^4 \text{ mL}$

(c) $20.0 \text{ g KCl} \times \frac{1 \text{ mol KCl}}{74.6 \text{ g KCl}} \times \frac{1 \text{ L}}{0.256 \text{ mol KCl}} \times \frac{1000 \text{ mL}}{\text{L}}$

$= 1.05 \times 10^3 \text{ mL}$

(d) $71.0 \text{ g } Cl^- \times \frac{1 \text{ mol } Cl^-}{35.5 \text{ g } Cl^-} \times \frac{1 \text{ mol KCl}}{1 \text{ mol } Cl^-} \times \frac{1 \text{ L}}{0.256 \text{ mol KCl}}$

$\times \frac{1000 \text{ mL}}{\text{L}} = 7.81 \times 10^3 \text{ mL}$

17. Assume 1.00 L of solution

$$\frac{1000 \text{ mL solution}}{\text{L}} \times \frac{1.21 \text{ g } HNO_3 \text{ sol'n}}{\text{mL sol'n}} \times \frac{35.0 \text{ g } HNO_3}{100 \text{ g } HNO_3 \text{ sol'n}}$$

$$\times \frac{1 \text{ mol } HNO_3}{63.0 \text{ g } HNO_3} = 6.72 \text{ M } HNO_3$$

18. Dilution problems $V_1M_1 = V_2M_2$

(a) 200 mL 12 M HCl + 200 mL H_2O

V_1 = 200 mL; M_1 = 12 M; V_2 = 400 mL; M_2 = M_2

200 mL × 12 M = 400 mL × M_2

$$M = \frac{200 \text{ mL} \times 12 \text{ M}}{400 \text{ mL}} = 6.0 \text{ M HCl}$$

(b) 60.0 mL 0.60 M $ZnSO_4$ + 500 mL H_2O

V_1 = 60.0 mL; M_1 = 0.60 M; V_2 = 560 mL; M_2 = M_2

60.0 mL × 0.60 M = 560 mL × M_2

$$M_2 = \frac{60.0 \text{ mL} \times 0.60 \text{ M}}{560 \text{ mL}} = 0.064 \text{ M } ZnSO_4$$

(c) First calculate the moles of HCl in each solution. Then calculate the molarity.

$$100 \text{ mL} \times \frac{1 \text{ L}}{1000 \text{ mL}} \times \frac{1.0 \text{ mol HCl}}{\text{L}} = 0.10 \text{ mol HCl}$$

$$150 \text{ mL} \times \frac{1 \text{ L}}{1000 \text{ mL}} \times \frac{2.0 \text{ mol HCl}}{\text{L}} = 0.30 \text{ mol HCl}$$

Total moles = 0.40 mol HCl

Total volume = 100 mL + 150 mL = 250 mL (0.250 L)

$$\frac{0.40 \text{ mol HCl}}{0.250 \text{ L}} = 1.6 \text{ M HCl}$$

19. Dilution problems. $V_1M_1 = V_2M_2$

(a) V_1 × 12 M = 400 mL × 6.0 M

$$V_1 = \frac{400 \text{ mL} \times 6.0 \text{ M}}{12 \text{ M}} = 200 \text{ mL 12 M HCl}$$

(b) V_1 × 15 M = 50 mL × 6.0 M

$$V_1 = \frac{50 \text{ mL} \times 6.0 \text{ M}}{15 \text{ M}} = 20 \text{ mL 15 M } NH_3$$

(c) V_1 × 16 M HNO_3 = 100 mL × 2.5 M

$$V_1 = \frac{100 \text{ mL} \times 2.5 \text{ M}}{16 \text{ M}} = 16 \text{ mL } 16 \text{ M } HNO_3$$

(d) $$10.0 \text{ N } H_2SO_4 = \frac{10.0 \text{ equivalents } H_2SO_4}{\text{L}} \times \frac{1 \text{ mol } H_2SO_4}{2 \text{ equivalents } H_2SO_4}$$

$= 5.00 \text{ M } H_2SO_4$; $V_1 \times 18 \text{ M} = 250 \text{ mL} \times 5.00 \text{ M}$

$$V_1 = \frac{250 \text{ mL} \times 5.00 \text{ M}}{18 \text{ M}} = 69 \text{ mL } 18 \text{ M } H_2SO_4$$

20. First calculate the molarity of the solution

$$\frac{80.0 \text{ g } H_2SO_4}{500 \text{ mL}} \times \frac{1000 \text{ mL}}{\text{L}} \times \frac{1 \text{ mol } H_2SO_4}{98.1 \text{ g } H_2SO_4} = 1.63 \text{ M } H_2SO_4$$

$M_1V_1 = M_2V_2$; $1.63 \text{ M } H_2SO_4 \times 500 \text{ mL} = 0.10 \text{ M } H_2SO_4 \times V_2$

$$V_2 = \frac{1.63 \text{ M } H_2SO_4 \times 500 \text{ mL}}{0.10 \text{ M } H_2SO_4} = 8.2 \times 10^3 \text{ mL} = 8.2 \text{ L}$$

21. 250 mL = 0.250 L

$$0.250 \text{ L} \times \frac{0.75 \text{ mol } H_2SO_4}{\text{L}} = 0.19 \text{ mol } H_2SO_4$$

(a) Final volume after mixing = 250 mL + 150 mL = 400 mL = 0.400 L

$$\frac{0.19 \text{ mol } H_2SO_4}{0.400 \text{ L}} = 0.48 \text{ M } H_2SO_4$$

(b) $$250 \text{ mL} \times \frac{1 \text{ L}}{1000 \text{ mL}} \times \frac{0.70 \text{ mol } H_2SO_4}{\text{L}} = 0.18 \text{ mol } H_2SO_4$$

Total moles H_2SO_4 = 0.19 mol + 0.18 mol = 0.37 mol H_2SO_4

Final volume = 250 mL + 250 mL = 500 mL = 0.500 L

$$\frac{0.37 \text{ mol } H_2SO_4}{0.500 \text{ L}} = 0.74 \text{ M } H_2SO_4$$

(c) $$400 \text{ mL} \times \frac{1 \text{ L}}{1000 \text{ mL}} \times \frac{2.50 \text{ mol } H_2SO_4}{\text{L}} = 1.00 \text{ mol } H_2SO_4$$

Total mol H_2SO_4 = 0.19 + 1.00 = 1.19 mol H_2SO_4

Final volume = 250 mL + 400 mL = 650 mL = 0.650 L

$$\frac{1.19 \text{ mol } H_2SO_4}{0.650 \text{ L}} = 1.83 \text{ M } H_2SO_4$$

22. Dilution problem. Note that the problem asks for the mL of water to be added, not the final volume.

$300 \text{ mL} \times 1.40 \text{ M} = V_2 \times 0.500 \text{ M}$

$$V_2 = \frac{300\ \text{mL} \times 1.40\ \text{M}}{0.500\ \text{M}} = 840\ \text{mL (final volume)}$$

840 mL - 300 mL = 540 mL of water to be added

23. Dilution problem

$10.0\ \text{mL} \times 16\ \text{M} = 500\ \text{mL} \times M_2$

$$M_2 = \frac{10.0\ \text{mL} \times 16\ \text{M}}{500\ \text{mL}} = 0.32\ \text{M}\ HNO_3$$

24. Dilution problem

$V_1 \times 5.00\ \text{M KOH} = 250\ \text{mL} \times 0.625\ \text{M}$

$$V_1 = \frac{250\ \text{mL} \times 0.625\ \text{M}}{5.00\ \text{M}} = 31.3\ \text{mL}\ 5.00\ \text{M KOH}$$

To make 250 mL of 0.625 M KOH, take 31.3 mL of 5.00 M KOH and dilute with water to a volume of 250 mL.

25. Stoichiometry problem. $BaCl_2 + K_2CrO_4 \longrightarrow BaCrO_4 + 2\ KCl$

mL $BaCl_2 \longrightarrow$ mol $BaCl_2 \longrightarrow$ mol $BaCrO_4 \longrightarrow$ g $BaCrO_4$

(a) $$100\ \text{ml} \times \frac{0.300\ \text{mol}\ BaCl_2}{1000\ \text{mL}} \times \frac{1\ \text{mol}\ BaCrO_4}{1\ \text{mol}\ BaCl_2} \times \frac{253.3\ \text{g}\ BaCrO_4}{\text{mol}\ BaCrO_4}$$

$= 7.60\ \text{g}\ BaCrO_4$

(b) $$50.0\ \text{mL} \times \frac{0.300\ \text{mol}\ K_2CrO_4}{1000\ \text{mL}} \times \frac{1\ \text{mol}\ BaCl_2}{1\ \text{mol}\ K_2CrO_4} \times \frac{1000\ \text{mL}\ BaCl_2}{1.0\ \text{mol}\ BaCl_2}$$

$= 15\ \text{mL of}\ 1.0\ \text{M}\ BaCl_2$

26. Stoichiometry problem. $3\ MgCl_2 + 2\ Na_3PO_4 \rightarrow Mg_3(PO_4)_2 + 6\ NaCl$

(a) mL $MgCl_2 \rightarrow$ mol $MgCl_2 \rightarrow$ mol $Na_3PO_4 \rightarrow$ mL Na_3PO_4

$$50.0\ \text{mL} \times \frac{0.250\ \text{mol}\ MgCl_2}{1000\ \text{mL}\ MgCl_2} \times \frac{2\ \text{mol}\ Na_3PO_4}{3\ \text{mol}\ MgCl_2} \times \frac{1000\ \text{mL}\ Na_3PO_4}{0.250\ \text{mol}\ Na_3PO_4}$$

$= 33.3\ \text{mL}\ 0.250\ \text{M}\ Na_3PO_4$

(b) mL $MgCl_2 \rightarrow$ mol $MgCl_2 \rightarrow$ mol $Mg_3(PO_4)_2 \rightarrow$ g $Mg_3(PO_4)_2$

$$50.0\ \text{mL} \times \frac{0.250\ \text{mol}\ MgCl_2}{1000\ \text{mL}\ MgCl_2} \times \frac{1\ \text{mol}\ Mg_3(PO_4)_2}{3\ \text{mol}\ MgCl_2} \times \frac{262.9\ \text{g}\ Mg_3(PO_4)_2}{\text{mol}\ Mg_3(PO_4)_2}$$

$= 1.10\ \text{g}\ Mg_3(PO_4)_2$

27. Stoichiometry problem. $Mg + 2\ HCl \rightarrow MgCl_2 + H_2$

(a) mL HCl $\rightarrow$ mol HCl $\rightarrow$ mol H_2

$$200 \text{ mL} \times \frac{3.00 \text{ mol HCl}}{1000 \text{ mL}} \times \frac{1 \text{ mol } H_2}{2 \text{ mol HCl}} = 0.300 \text{ mol } H_2$$

(b) $PV = nRT$; $P = \frac{720 \text{ torr} \times 1 \text{ atm}}{760 \text{ torr}} = 0.947$ atm;

$T = 27°C = 27 + 273 = 300$ K; $n = 0.300$ mol

$$V = \frac{nRT}{P} = \frac{0.300 \text{ mol} \times 300 \text{ K} \times 0.0821 \text{ L atm}}{0.947 \text{ atm} \times \text{mol K}} = 7.80 \text{ L } H_2$$

28. Stoichiometry problem. $Mg + 2\ HCl \rightarrow MgCl_2 + H_2$

150 mL = 0.150 L; L H_2 → mol H_2 → mol HCl → M HCl

$$3.50 \text{ L } H_2 \times \frac{1 \text{ mol } H_2}{22.4 \text{ L } H_2} \times \frac{2 \text{ mol HCl}}{1 \text{ mol } H_2} \times \frac{1}{0.150 \text{ L HCl}} = \frac{2.08 \text{ mol HCl}}{\text{L}}$$

= 2.08 M HCl

29. The balanced equation is

$$6\ FeCl_2 + K_2Cr_2O_7 + 14\ HCl \rightarrow 6\ FeCl_3 + 2\ CrCl_3 + 2\ KCl + 7\ H_2O$$

(a) $2.0 \text{ mol } FeCl_2 \times \frac{2 \text{ mol KCl}}{6 \text{ mol } FeCl_2} = 0.67 \text{ mol KCl}$

(b) $1.0 \text{ mol } FeCl_2 \times \frac{2 \text{ mol } CrCl_3}{6 \text{ mol } FeCl_2} = 0.33 \text{ mol } CrCl_3$

(c) $0.050 \text{ mol } K_2Cr_2O_7 \times \frac{6 \text{ mol } FeCl_2}{1 \text{ mol } K_2Cr_2O_7} = 0.30 \text{ mol } FeCl_2$

(d) $0.025 \text{ mol } FeCl_2 \times \frac{1 \text{ mol } K_2Cr_2O_7}{6 \text{ mol } FeCl_2} \times \frac{1000 \text{ mL } K_2Cr_2O_7}{0.060 \text{ mol } K_2Cr_2O_7}$

= 69 mL of 0.060 M $K_2Cr_2O_7$

(e) $15.0 \text{ mL } FeCl_2 \times \frac{6.0 \text{ mol } FeCl_2}{1000 \text{ mL } FeCl_2} \times \frac{14 \text{ mol HCl}}{6 \text{ mol } FeCl_2} \times \frac{1000 \text{ mL HCl}}{6 \text{ mol HCl}}$

= 35 mL of 6.0 M HCl

30. $2\ KMnO_4 + 16\ HCl \rightarrow 2\ MnCl_2 + 5\ Cl_2 + 8\ H_2O + 2\ KCl$

(a) $0.50 \text{ mol } KMnO_4 \times \frac{5 \text{ mol } Cl_2}{2 \text{ mol } KMnO_4} = 1.3 \text{ mol } Cl_2$

(b) $1.0 \text{ L} \times \frac{2.0 \text{ mol } KMnO_4}{1 \text{ L}} \times \frac{16 \text{ mol HCl}}{2 \text{ mol } KMnO_4} = 16 \text{ mol HCl}$

(c) $200\ \text{mL} \times \dfrac{0.50\ \text{mol KMnO}_4}{1000\ \text{mL}} \times \dfrac{16\ \text{mol HCl}}{2\ \text{mol KMnO}_4} \times \dfrac{1000\ \text{mL HCl}}{6.0\ \text{mol HCl}}$

$= 1.3 \times 10^2\ \text{mL } 6.0\ M\ \text{HCl}$

(d) $75.0\ \text{mL} \times \dfrac{6.0\ \text{mol HCl}}{1000\ \text{mL}} \times \dfrac{5\ \text{mol Cl}_2}{16\ \text{mol HCl}} \times \dfrac{22.4\ \text{L Cl}_2}{\text{mol Cl}_2} = 3.2\ \text{L Cl}_2$

31. For an acid the number of hydrogen ions reacting per formula unit of acid equals the number of equivalent weights per mole of acid. For a base the number of hydroxide ions reacting per formula unit of base equals the number of equivalent weights per mole of base.

(a) HCl; $\dfrac{1\ \text{mol HCl}}{1\ \text{eq wt}} \times \dfrac{36.5\ \text{g HCl}}{1\ \text{mol HCl}} = \dfrac{36.5\ \text{g HCl}}{\text{eq wt}}$

NaOH; $\dfrac{1\ \text{mol NaOH}}{\text{eq wt}} \times \dfrac{40.0\ \text{g NaOH}}{1\ \text{mol NaOH}} = \dfrac{40.0\ \text{g NaOH}}{\text{eq wt}}$

(b) HCl; $\dfrac{1\ \text{mol HCl}}{\text{eq wt}} \times \dfrac{36.5\ \text{g HCl}}{1\ \text{mol HCl}} = 36.5\ \text{g HCl/eq wt}$

$Ba(OH)_2$; $\dfrac{1\ \text{mol Ba(OH)}_2}{2\ \text{eq wt}} \times \dfrac{171.3\ \text{g Ba(OH)}_2}{\text{mol Ba(OH)}_2} = 85.7\ \text{g Ba(OH)}_2\text{/eq wt}$

(c) H_2SO_4; $\dfrac{1\ \text{mol H}_2\text{SO}_4}{2\ \text{eq wt}} \times \dfrac{98.1\ \text{g H}_2\text{SO}_4}{1\ \text{mol H}_2\text{SO}_4} = \dfrac{49.1\ \text{g H}_2\text{SO}_4}{\text{eq wt}}$

$Ca(OH)_2$; $\dfrac{1\ \text{mol Ca(OH)}_2}{2\ \text{eq wt}} \times \dfrac{74.1\ \text{g Ca(OH)}_2}{1\ \text{mol Ca(OH)}_2} = \dfrac{37.1\ \text{g Ca(OH)}_2}{\text{eq wt}}$

(d) H_2SO_4; $\dfrac{1\ \text{mol H}_2\text{SO}_4}{1\ \text{eq wt}} \times \dfrac{98.1\ \text{g H}_2\text{SO}_4}{1\ \text{mol H}_2\text{SO}_4} = \dfrac{98.1\ \text{g H}_2\text{SO}_4}{\text{eq wt}}$

KOH; $\dfrac{1\ \text{mol KOH}}{1\ \text{eq wt}} \times \dfrac{56.1\ \text{g KOH}}{1\ \text{mol KOH}} = \dfrac{56.1\ \text{g KOH}}{\text{eq wt}}$

(e) H_3PO_4; $\dfrac{1\ \text{mol H}_3\text{PO}_4}{2\ \text{eq wt}} \times \dfrac{98.0\ \text{g H}_3\text{PO}_4}{1\ \text{mol H}_3\text{PO}_4} = \dfrac{49.0\ \text{g H}_3\text{PO}_4}{\text{eq wt}}$

LiOH; $\dfrac{1\ \text{mol LiOH}}{1\ \text{eq wt}} \times \dfrac{23.9\ \text{g LiOH}}{1\ \text{mol LiOH}} = \dfrac{23.9\ \text{g LiOH}}{\text{eq wt}}$

32. Normality

(a) 4.0 M HCl: $\dfrac{4.0\ \text{mol}}{\text{L}} \times \dfrac{1\ \text{eq}}{\text{L}} = \dfrac{4.0\ \text{eq}}{\text{L}} = 4.0\ N\ \text{HCl}$

(b) 0.243 M HNO_3: $\dfrac{0.243\ \text{mol}}{\text{L}} \times \dfrac{1\ \text{eq}}{\text{mol}} = \dfrac{0.243\ \text{eq}}{\text{mol}} = 0.243\ N\ \text{HNO}_3$

(c) 3.0 M H_2SO_4: $\dfrac{3.0\ \text{mol}}{\text{L}} \times \dfrac{2\ \text{eq}}{\text{mol}} = \dfrac{6.0\ \text{eq}}{\text{L}} = 6.0\ N\ \text{H}_2\text{SO}_4$

(d) 1.85 M H_3PO_4: $\dfrac{1.85\ \text{mol}}{\text{L}} \times \dfrac{3\ \text{eq}}{\text{mol}} = \dfrac{5.55\ \text{eq}}{\text{L}} = 5.55\ N\ \text{H}_3\text{PO}_4$

(e) $0.250\ M\ HC_2H_3O_2$: $\dfrac{0.250\ mol}{L} \times \dfrac{1\ eq}{mol} = \dfrac{0.250\ eq}{L} = 0.250\ N\ HC_2H_3O_2$

33. Eq acid = Eq base   Eq wt. of $Ca(OH)_2 = \dfrac{74.1\ g}{mol} \times \dfrac{1\ mol}{2\ eq} = \dfrac{37.1\ g}{eq}$

$N_A \times 0.03626\ L = 2.50\ g\ Ca(OH)_2 \times \dfrac{1\ eq}{37.1\ g}$

$N_A = \dfrac{2.50\ g \times 1\ eq}{0.03626\ L \times 37.1\ g} = \dfrac{1.86\ eq}{L} = 1.86\ N\ H_2SO_4$

34. Calculate the number of equivalents of each base.

$12.0\ g\ Mg(OH)_2 \times \dfrac{1\ eq\ base}{29.2\ g\ Mg(OH)_2} = 0.411$ equiv. base, $Mg(OH)_2$

$10.0\ g\ Al(OH)_3 \times \dfrac{1\ eq\ base}{26.0\ g\ Al(OH)_3} = 0.385$ equiv. base, $Al(OH)_3$

1 equivalent of base will neutralize 1 equivalent of acid. Therefore, 12.0 g $Mg(OH)_2$ will neutralize more stomach acid than 10.0 g $Al(OH)_3$.

35. $N_AV_A = N_BV_B$

(a) $0.1254\ N \times 20.22\ mL = 0.2550\ N \times V_B$

$V_B = \dfrac{0.1254 \times 20.22\ mL}{0.2550\ N} = 9.943\ mL\ NaOH$

(b) $0.1246\ N \times 14.86\ mL = 0.2550\ N \times V_B$

$V_B = \dfrac{0.1246\ N \times 14.86\ mL}{0.2550\ N} = 7.261\ mL\ NaOH$

(c) $0.1430\ M\ H_2SO_4 = 0.2860\ N\ H_2SO_4$ (there are 2 equivalents of H in 1 mol of H_2SO_4).

$0.2860\ N \times 18.00\ mL = 0.2550\ N \times V_B$

$V_B = \dfrac{0.2860\ N \times 18.00\ mL}{0.2550\ N} = 20.19\ mL\ NaOH$

36. Molality = m = $\dfrac{mol\ solute}{kg\ solvent}$

(a) $\dfrac{g\ CH_3OH}{100\ g\ H_2O} \longrightarrow \dfrac{g\ CH_3OH}{kg\ H_2O} \longrightarrow \dfrac{mol\ CH_3OH}{kg\ H_2O} = m$

$\dfrac{14.0\ g\ CH_3OH}{100\ g\ H_2O} \times \dfrac{1000\ g\ H_2O}{kg\ H_2O} \times \dfrac{1\ mol\ CH_3OH}{32.0\ g\ CH_3OH} = \dfrac{4.38\ mol\ CH_3OH}{kg\ H_2O}$

$= 4.38\ m\ CH_3OH$

(b) $\frac{\text{mol } C_6H_6}{250 \text{ g } CCl_4} \longrightarrow \frac{\text{mol } C_6H_6}{\text{kg } CCl_4} = \text{m}$

$$\frac{2.50 \text{ mol } C_6H_6}{250 \text{ g } CCl_4} \times \frac{1000 \text{ g } CCl_4}{\text{kg } CCl_4} = \frac{10.0 \text{ mol } C_6H_6}{\text{kg } CCl_4} = 10.0 \text{ m } C_6H_6$$

(c) $\frac{\text{g } C_6H_{12}O_6}{\text{g } H_2O} \longrightarrow \frac{\text{g } C_6H_{12}O_6}{\text{kg } H_2O} \longrightarrow \frac{\text{mol } C_6H_{12}O_6}{\text{kg } H_2O} = \text{m}$

$$\frac{1.0 \text{ g } C_6H_{12}O_6}{1.0 \text{ g } H_2O} \times \frac{1000 \text{ g } H_2O}{\text{kg } H_2O} \times \frac{1 \text{ mol } C_6H_{12}O_6}{180 \text{ g } C_6H_{12}O_6} = \frac{5.6 \text{ mol } C_6H_{12}O_6}{\text{kg } H_2O}$$

$= 5.6 \text{ m } C_6H_{12}O_6$

37. (a) With equal masses of CH_3OH and C_2H_5OH the substance with the lower formula weight will represent more moles of solute in solution. Therefore, the CH_3OH will be more effective than C_2H_5OH as an antifreeze.

(b) Equal molal solutions will lower the freezing point of the solution by the same amount.

38. Assume 1000 mL of solution to calculate the amounts of H_2SO_4 and H_2O in the solution. Then calculate molality and molarity.

$1000 \text{ mL sol'n} \times \frac{1.29 \text{ g}}{\text{mL}} = 1.29 \times 10^3 \text{ g}$ (weight of 1000 mL sol'n)

$1.29 \times 10^3 \text{ g sol'n} \times \frac{38 \text{ g } H_2SO_4}{100 \text{ g sol'n}} = 4.9 \times 10^2 \text{ g } H_2SO_4$ in 1000 mL sol'n

$1.29 \times 10^3 \text{ g} - 4.9 \times 10^2 \text{ g} = 8.0 \times 10^2 \text{ g } H_2O$ in 1000 mL sol'n.

$$\text{m} = \frac{4.9 \times 10^2 \text{ g } H_2SO_4}{8.0 \times 10^2 \text{ g } H_2O} \times \frac{1000 \text{ g } H_2O}{\text{kg } H_2O} \times \frac{1 \text{ mol } H_2SO_4}{98.1 \text{ g } H_2SO_4} = 6.2 \text{ m } H_2SO_4$$

$$\text{M} = \frac{4.9 \times 10^2 \text{ g } H_2SO_4}{\text{L}} \times \frac{1 \text{ mol } H_2SO_4}{98.1 \text{ g } H_2SO_4} = 5.0 \text{ M } H_2SO_4$$

39. 1.00 lb = 454 g sugar

$4.00 \text{ lb } H_2O \times \frac{454 \text{ g}}{\text{lb}} = 1.82 \times 10^3 \text{ g } H_2O$ (1.82 kg H_2O)

$$454 \text{ g } C_{12}H_{22}O_{11} \times \frac{1 \text{ mol } C_{12}H_{22}O_{11}}{342 \text{ g}} = 1.33 \text{ mol } C_{12}H_{22}O_{11}$$

Calculate the change in freezing point, Δt_f, using the data in the following equation. K_f for H_2O = 1.86°C kg solvent/mol solute.

$$\Delta t_f = mK_f = \frac{\text{mol solute}}{\text{kg solvent}} \times K_f$$

$$\Delta t_f = \frac{1.33 \text{ mol } C_{12}H_{22}O_{11}}{1.82 \text{ kg } H_2O} \times \frac{1.86°C \text{ kg } H_2O}{\text{mol solute}} = 1.36°C$$

Freezing point of solution = 0°C - 1.36°C = -1.36°C

$$-1.36°C \times \frac{1.8°F}{1°C} + 32 = 29.6°F \text{ (freezing point of solution)}$$

If this sugar solution is put outside where it is 20°F the solution will freeze.

40. Freezing point, -5.4°C. Freezing point depression is 5.4°C

$$\Delta t_f = mK_f;\ m = \frac{\Delta t_f}{K} = \frac{5.4°C}{1.86°C \text{ kg solvent/mol solute}} = 2.9 \text{ m}$$

$$K_b \text{ for } H_2O = \frac{0.512°C \text{ kg solvent}}{\text{mol solute}} = \frac{0.512°C}{m}$$

$$\Delta t_b = mK_b = 2.9 \text{ m} \times \frac{0.512°C}{m} = 1.5°C$$

Boiling point = 100.0°C + 1.5°C = 101.5°C

41. $m = \frac{\text{mol solute}}{\text{kg solvent}}$

(a) $\frac{100 \text{ g } C_2H_6O_2}{150 \text{ g } H_2O} \times \frac{1 \text{ mol } C_2H_6O_2}{62.0 \text{ g}} \times \frac{1000 \text{ g } H_2O}{\text{kg}} = 10.8 \text{ m}$

(b) $\Delta t_b = mK_b = 10.8 \text{ m} \times \frac{0.512°C}{m} = 5.53°C$

Boiling point = 100.00°C + 5.53°C = 105.53°C

(c) $\Delta t_f = mK_f = 10.8 \text{ m} \times \frac{1.86°C}{m} = 20.1°C$

Freezing point = 0.00°C - 20.1°C = -20.1°C

42. (a) $\frac{2.68 \text{ g } C_{10}H_8}{38.4 \text{ g } C_6H_6} \times \frac{1 \text{ mol } C_{10}H_8}{128.0 \text{ g } C_{10}H_8} \times \frac{1000 \text{ g } C_6H_6}{\text{kg } C_6H_6} = 0.545 \text{ m}$

(b) K_f for benzene is 5.1°C/m; freezing point of benzene is 5.5°C

$\Delta t_f = 0.545 \text{ m} \times \frac{5.1°C}{m} = 2.8°C$

Freezing point of solution = 5.5°C - 2.8°C = 2.7°C

(c) K_b for benzene is 2.53°C/m; boiling point of benzene is 80.1°C

$\Delta t_b = 0.545 \text{ m} \times \frac{2.53°C}{m} = 1.38°C$

Boiling point of solution = 80.1°C + 1.38°C = 81.5°C

43. Freezing point of acetic acid is 16.6°C; $K_f = 3.90°C/m$

$\Delta t_f = 16.6°C - 13.2°C = 3.4°C$

$\Delta t_f = mK_f;\ m = \frac{3.4°C}{3.90°C/m} = 0.87\ m$

Convert 8.00 g unknown/60.0 g $HC_2H_3O_2$ to g/mol

$$\frac{8.00\ \text{g unknown}}{60.0\ \text{g}\ HC_2H_3O_2} \times \frac{1000\ \text{g}\ HC_2H_3O_2}{\text{kg}\ HC_2H_3O_2} \times \frac{1\ \text{kg}\ HC_2H_3O_2}{0.87\ \text{mol unknown}} = 153\ \text{g/mol}$$

The molecular weight of the unknown is 153 g/mol

44. $\Delta t_f = 2.50°C$ (H_2O solvent); $\Delta t_f = mK_f;\ K_f = \frac{1.86°C}{m}$

$$m = 2.50°C \times \frac{1\ m}{1.86°C} = 1.34\ m = \frac{1.34\ \text{mol unknown}}{\text{kg}\ H_2O}$$

Convert 4.80 g unknown/22.0 g H_2O to g/mol

$$\frac{4.80\ \text{g unknown}}{22.0\ \text{g}\ H_2O} \times \frac{1000\ \text{g}\ H_2O}{\text{kg}\ H_2O} \times \frac{1\ \text{kg}\ H_2O}{1.34\ \text{mol unknown}} = 163\ \text{g/mol}$$

Molecular weight of the unknown is 163 g/mol

45. Freezing point lowering = 0.372°C

$\Delta t_f = mK_f;\ K_f = \frac{1.86°C}{m}$

$$0.372°C \times \frac{1\ m}{1.86°C} = 0.200\ m = \frac{0.200\ \text{mol}\ C_2H_6O_2}{\text{kg}\ H_2O}$$

$$6.2\ \text{g}\ C_2H_6O_2 \times \frac{1\ \text{mol}\ C_2H_6O_2}{62\ \text{g}\ C_2H_6O_2} = 0.10\ \text{mol}\ C_2H_6O_2$$

$$0.10\ \text{mol}\ C_2H_6O_2 \times \frac{1\ \text{kg}\ H_2O}{0.200\ \text{mol}\ C_2H_6O_2} \times \frac{1000\text{g}\ H_2O}{\text{kg}\ H_2O} = 500\ \text{g}\ H_2O$$

46. (a) Freezing point lowering = 20.0°C; 12.0 L H_2O = 12000 g H_2O

$$20.0°C \times \frac{1\ m}{1.86°C} = 10.8\ m\ \text{(to protect to -20°C)}$$

$$12000\ \text{g}\ H_2O \times \frac{10.8\ \text{mol}\ C_2H_6O_2}{1000\ \text{g}\ H_2O} \times \frac{62.0\ \text{g}\ C_2H_6O_2}{1\ \text{mol}\ C_2H_6O_2}$$

$$= 8.04 \times 10^3\ \text{g}\ C_2H_6O_2$$

(b) $8.04 \times 10^3\ \text{g}\ C_2H_6O_2 \times \frac{1.00\ \text{mL}}{1.11\ \text{g}} = 7.24 \times 10^3\ \text{mL}\ C_2H_6O_2$

(c) $-20.0°C \times \frac{1.8°F}{1.0°C} + 32°F = -4.0°F$

47. $NaOH + HCl \longrightarrow NaCl + H_2O$; 150 mL = 0.150 L

$$0.150 \text{ L HCl} \times \frac{1.0 \text{ mol HCl}}{\text{L HCl}} \times \frac{1 \text{ mol NaOH}}{1 \text{ mol HCl}} \times \frac{40.0 \text{ g NaOH}}{1 \text{ mol NaOH}}$$

$$\times \frac{100 \text{ g NaOH sol'n}}{10.0 \text{ g NaOH}} = 60 \text{ g } 10\% \text{ NaOH sol'n}$$

48. $NaOH + HCl \longrightarrow NaCl + H_2O$

$$1.0 \text{ molal HCl} = \frac{1 \text{ mol HCl}}{1 \text{ kg } H_2O} = \frac{36.5 \text{ g HCl}}{1000 \text{ g } H_2O} = \frac{1 \text{ mol HCl}}{1036.5 \text{ g sol'n}}$$

$$250 \text{ g HCl sol'n} \times \frac{1 \text{ mol HCl}}{1036.5 \text{ g sol'n}} \times \frac{1 \text{ mol NaOH}}{1 \text{ mol HCl}} \times \frac{40.0 \text{ g NaOH}}{1 \text{ mol NaOH}}$$

$$\times \frac{100 \text{ g NaOH sol'n}}{10.0 \text{ g NaOH}} = 96.5 \text{ g NaOH solution}$$

49. (a) $1.0 \text{ L syrup} \times \frac{1000 \text{ mL}}{\text{L}} \times \frac{1.06 \text{ g syrup}}{\text{mL syrup}} \times \frac{15.0 \text{ g sugar}}{100 \text{ g syrup}}$

$= 1.6 \times 10^2$ g sugar

(b) $\frac{1.6 \times 10^2 \text{ g } C_{12}H_{22}O_{11}}{\text{L}} \times \frac{1 \text{ mol } C_{12}H_{22}O_{11}}{342 \text{ g } C_{12}H_{22}O_{11}} = 0.47 \text{ M}$

(c) $m = \frac{\text{mol}}{\text{kg}}$; $\frac{15.0 \text{ g } C_{12}H_{22}O_{11}}{85.0 \text{ g } H_2O} \times \frac{1000 \text{ g } H_2O}{\text{kg } H_2O} \times \frac{1 \text{ mol } C_{12}H_{22}O_{11}}{342 \text{ g } C_{12}H_{22}O_{11}}$

$= 0.516$ molal

50. $K_f = 5.1°C/m$; $\Delta t_f = 0.614°C$; $\Delta t_f = mK_f$

$$\frac{3.84 \text{ g } C_4H_2N}{250 \text{ g } C_6H_6} \times \frac{1000 \text{ g } C_6H_6}{\text{kg } C_6H_6} = 15.4 \text{ g } C_4H_2N/\text{kg } C_6H_6$$

$$m = 0.614°C \times \frac{1 \text{ m}}{5.1°C} = 0.12 \text{ m} = \frac{0.12 \text{ mol } C_4H_2N}{\text{kg } C_6H_6}$$

$$\frac{15.4 \text{ g } C_4H_2N}{\text{kg } C_6H_6} \times \frac{1 \text{ kg } C_6H_6}{0.12 \text{ mol } C_4H_2N} = \frac{128 \text{ g}}{\text{mol}} \text{ (mol. wt.)}$$

Empirical formula weight = 64.0 g

Molecular weight = 128 g

2 × empirical formula = molecular formula

Molecular formula = $(C_4H_2N)_2 = C_8H_4N_2$

B. 1. The Arrhenius definition of an acid is restricted to aqueous solution. The Brønsted-Lowry definition of an acid is not restricted to aqueous solutions.

2. An electrolyte must be in solution in order for the light to glow.

3. The following classes of compounds are electrolytes: acids, bases, and salts.

4. The orientation of the polar water molecule about the Na^+ and Cl^- ions is different. The positive end of the polar water molecule is directed towards the Cl^- while the negative end of the water molecule is directed towards the Na^+. Also, more water molecules can cluster around the larger Cl^- ion than around the smaller Na^+ ion.

5. A hydrogen ion concentration of 0.003 M will have a pH between 2 and 3.

6. Tomatoes with a pH lower than the pH of blood is more acidic than blood.

C. Review questions.

1. By the Arrhenius theory, and acid is a substance that produces hydrogen ions, and a base is a substance that produces hydroxide ions in aqueous solution. By the Brønsted-Lowry theory, an acid is a proton donor, and a base is a proton acceptor. Since a proton is a hydrogen ion, then the two theories are very similar for acids, but not for bases. A chloride ion can accept a proton (to become an HCl molecule) so it is a Brønsted-Lowry base, but would not be a base by the Arrhenius theory, since it does not produce hydroxide ions.

   By the Lewis theory, an acid is an electron pair acceptor, and a base is an electron pair donor. Many individual substances would be similarly classified as bases by Brønsted-Lowry or Lewis theories, since a substance with an electron pair to donate can accept a proton, but the Lewis definition is almost exclusively applied to reactions where the acid and base combine into a single molecule. The Brønsted-Lowry definition is usually applied to reactions that involve a transfer of a proton from the acid to the base. The Arrhenius definition is most often applied to individual substances, not to reactions. According to the Arrhenius theory neutralization involves the reaction between a hydrogen ion and a hydroxide ion to form water. Neutralization according to the Brønsted-Lowry theory involves the transfer of a proton to a negative ion. The formation of a coordinate-covalent bond constitutes a Lewis neutralization.

2. Neutralization reactions.

Arrhenius: $HCl + NaOH \longrightarrow NaCl + H_2O$ ($H^+ + OH^- \longrightarrow H_2O$)

Brønsted-Lowry: $HCl + KCN \longrightarrow HCN + KCl$ ($H^+ + CN^- \longrightarrow HCN$)

Lewis: $AlCl_3 + NaCl \longrightarrow AlCl_4^- + Na^+$

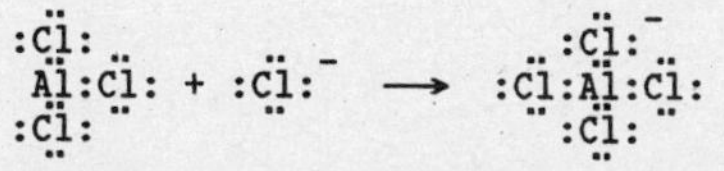

3. Conjugate acid-base pairs (conjugate acid listed first for each pair).

(a) $HCl - Cl^-$; $NH_4^+ - NH_3$

(b) $HCO_3^- - CO_3^{2-}$; $H_2O - OH^-$

(c) $H_3O^- - H_2O$; $H_2CO_3 - HCO_3^-$

(d) $HC_2H_3O_2 - C_2H_3O_2^-$; $H_3O^+ - H_2O$

(e) $H_2SO_4 - HSO_4^-$; $H_2C_2H_3O_2^+ - HC_2H_3O_2$

(f) step 1: $H_2SO_4^- - HSO_4^-$; $H_3O^+ - H_2O$

step 2: $HSO_4^- - SO_4^{2-}$; $H_3O^+ - H_2O$

(g) $HClO_4 - ClO_4^-$; $H_3O^+ - H_2O$

(h) $H_3O^+ - H_2O$; $CH_3OH - CH_3O^-$

4. (a) $:\ddot{B}r:^-$ (b) $:\ddot{O}:H^-$ (c) $:C\vdots\vdots N:^-$

These ions are considered to be bases according to the Brønsted-Lowry theory because they can accept a proton at any of their unshared pairs of electrons. They are considered to be bases according to the Lewis acid-base theory because they can donate an electron pair.

5. Balancing equations.

(a) $Mg + 2\ HCl \longrightarrow MgCl_2 + H_2$

(b) $BaO + 2\ HBr \longrightarrow BaBr_2 + H_2O$

(c) $2\ Al + 3\ H_2SO_4 \longrightarrow Al_2(SO_4)_3 + 3\ H_2$

(d) $Na_2CO_3 + 2\ HCl \longrightarrow 2\ NaCl + H_2O + CO_2$

(e) $Fe_2O_3 + 6\ HBr \longrightarrow 2\ FeBr_3 + 3\ H_2O$

(f) $Ca(OH)_2 + H_2CO_3 \longrightarrow CaCO_3 + 2\ H_2O$

(g) $NaOH + HBr \longrightarrow NaBr + H_2O$

(h) $KOH + HCl \longrightarrow KCl + H_2O$

(i) $Ca(OH)_2 + 2\ HI \longrightarrow CaI_2 + 2\ H_2O$

(j) $Al(OH)_3 + 3\ HBr \longrightarrow AlBr_3 + 3\ H_2O$

(k) $Na_2O + 2\ HClO_4 \longrightarrow 2\ NaClO_4 + H_2O$

(l) $3\ LiOH + FeCl_3 \longrightarrow Fe(OH)_3 + 3\ LiCl$

(m) $2\ NH_4OH + FeCl_2 \longrightarrow Fe(OH)_2 + 2\ NH_4Cl$

6. The classes of compounds into which electrolytes generally fall are acids, bases, and salts.

7. The following compounds are electrolytes: (a) HCl, acid; (b) CO_2, acid anhydride; (c) $CaCl_2$, salt; (g) $NaHCO_3$, salt; (i) $AgNO_3$, salt; (j) HCOOH, acid; (k) RbOH, base; (l) K_2CrO_4, salt.

8. Names of the compounds in Table 15.3

H_2SO_4	sulfuric acid	H_2CO_3	carbonic acid
HCl	hydrochloric acid	H_2SO_3	sulfurous acid
$HClO_4$	perchloric acid	$H_2C_2O_4$	oxalic acid
KOH	potassium hydroxide	HClO	hypochlorous acid
$Ba(OH)_2$	barium hydroxide	HF	hydrofluoric acid
HNO_3	nitric acid	$HC_2H_3O_2$	acetic acid
HBr	hydrobromic acid	HNO_2	nitrous acid
NaOH	sodium hydroxide	H_2S	hydrosulfuric acid
$Ca(OH)_2$	calcium hydroxide	H_3BO_3	boric acid
		NH_4OH	ammonium hydroxide

9. Hydrogen chloride dissolved in water conducts an electric current. HCl reacts with polar water molecules to produce H_3O^+ and Cl^- ions, which conduct the electric current. Benzene is a nonpolar solvent, so cannot pull the HCl molecules apart. Since there are no ions in the benzene solution, it does not conduct an electric current.

10. In their crystalline structure, salts exist as positive and negative ions in definite geometric arrangement to each other, held together by the attraction of the opposite charges. When dissolved in water, the salt dissociates as the ions are pulled away from each other by the polar water molecules.

11. Testing the electrical conductivity of the solutions shows that CH_3OH is a nonelectrolyte, while NaOH is an electrolyte. This tells us that the OH group in CH_3OH must be covalently bonded to the CH_3 group, rather than ionically bonded.

12. Molten NaCl conducts electricity because, when melted, the ions are free to move and thus conduct electricity; whereas in the solid state they are immobile and do not conduct electricity.

13. Dissociation is the breaking free of alreading existing ions. Ionization is the formation of ions from molecules. The dissolving of NaCl is a dissociation, for the ions already existed. The dissolving of HCl in water is a ionization process, because ions are formed from the molecules.

14. Strong electrolytes are those which are essentially 100% ionized or dissociated in water. Weak electrolytes are those which are only slightly ionized in water.

15. Ions are hydrated in solution because there is electrical attraction between the charged ions and the polar water molecules.

16. Simplified equations showing the dissociation or ionization of compounds in water:

(a) $Cu(NO_3)_2 \longrightarrow Cu^{2+} + 2\ NO_3^-$

(b) $HC_2H_3O_2 \longrightarrow H^+ + C_2H_3O_2^-$

(c) $HNO_2 \longrightarrow H^+ + NO_2^-$

(d) $LiOH \longrightarrow Li^+ + OH^-$

(e) $NH_4Br \longrightarrow NH_4^+ + Br^-$

(f) $K_2SO_4 \longrightarrow 2\ K^+ + SO_4^{2-}$

(g) $NaClO_3 \longrightarrow Na^+ + ClO_3^-$

(h) $K_3PO_4 \longrightarrow 3\ K^+ + PO_4^{3-}$

17. The main distinction between water solutions of strong and weak electrolytes is the degree of ionization of the electrolyte. A solution of an electrolyte contains many more ions than a solution of a nonelectrolyte. Strong electrolytes are essentially 100% ionized. Weak electrolytes are only slightly ionized in water.

18. (a) In a neutral solution the concentrations of H^+ and OH^- are equal.

(b) In an acid solution the concentration of H^+ is greater than the concentration of OH^-.

(c) In a basic solution the concentration of OH^- is greater than the concentration of H^+.

19. The net ionic equation for an acid-base reaction in an aqueous solution is $H^+ + OH^- \longrightarrow H_2O$.

20. The HCl molecule is a polar molecule and, consequently, is much more soluble in the polar solvent, water, than in the nonpolar solvent, benzene.

21. Pure water is neutral because the concentrations of acid and base ions are equal.

22. Balanced net ionic equations.

(a) $Ba^{2+} + SO_4^{2-} \longrightarrow BaSO_4(s)$

(b) $2\ H^+ + CO_3^{2-} \longrightarrow H_2O(l) + CO_2(g)$

(c) $Mg(s) + 2\ HC_2H_3O_2(aq) \longrightarrow Mg^{2+} + 2\ C_2H_3O_2^- + H_2(g)$

(d) $H_2S(g) + Cd^{2+} \longrightarrow CdS(s) + 2\ H^+$

(e) $Zn(s) + 2\ H^+(aq) \longrightarrow Zn^{2+}(aq) + H_2(g)$

(f) $Al^{3+} + PO_4^{3-} \longrightarrow AlPO_4(s)$

23. The more acidic solution in each pair is listed followed by the reason why.

(a) One molar H_2SO_4. The concentration of H^+ from 1 M H_2SO_4 will exceed 1 M because there are two ionizable hydrogens per formula unit of H_2SO_4. The concentration of H^+ from 1 M HCl cannot exceed 1 M because there is only one ionizable hydrogen per formula unit of HCl.

(b) One molar HCl. HCl is a strong electrolyte while $HC_2H_3O_2$ is a weak electrolyte.

(c) Two molar HCl. 2 M HCl will give a 2 M H^+ concentration. 1 M HCl will give a 1 M H^+ concentration.

(d) One molar H_2SO_4. 1 M H_2SO_4 is the same as 2 N H_2SO_4 and as such will give a greater concentration of H^+ than 1 N H_2SO_4.

24. A hydronium ion is a hydrated hydrogen ion.

$$\underset{\text{(hydrogen ion)}}{H^+} + H_2O \longrightarrow \underset{\text{(hydronium ion)}}{H_3O^+}$$

25. The magnitude of the freezing point depression depends on the concentration of particles in the solution. Beginning with the solution having the highest freezing point the order is: $C_{12}H_{22}O_{11}$ (1 mole of particles per mole of $C_{12}H_{22}O_{11}$), $HC_2H_3O_2$ (slightly more than 1 mole of particles per mole of $HC_2H_3O_2$), HCl (2 moles of particles per mole of HCl), $CaCl_2$ (3 moles of particles per mole of $CaCl_2$).

26. (a) pH = -log H^+; at 25°C $H^+ = 1 \times 10^{-7}$. pH = -log 10^{-7} = 7

    At 100°C $H^+ = 1 \times 10^{-6}$. pH = -log 10^{-6} = 6

    pH greatest at 25°C

    (b) H^+ concentration greatest at 100°C.

    (c) The water is neutral at both temperatures, because the concentration of H^+ equals the concentration of OH^- at both temperatures.

27. A pH difference of 1 unit means a 10-fold difference in the concentration of H^+ in solution. For example, the pH of 0.10 M HCl is 1.0; the pH of 0.010 M HCl is 2.0.

28. The acetic acid solution freezes at a lower temperature than the alcohol solution. The acetic acid ionizes slightly while the alcohol does not. The ionization of the acetic acid increases its particle concentration above that of the alcohol solution, resulting in a lower freezing point for the acetic acid solution.

29. CH_3OH would be more economical. On a mole to mole basis both are equally effective. Because a mole of CH_3OH weighs less than a mole of C_2H_5OH there are more moles of CH_3OH than C_2H_5OH per pound. Hence, the CH_3OH would be more economical.

30. A colloidal dispersion of gas in gas cannot exist because the size of gas molecules is always smaller than colloidal.

31. The two general methods for preparing colloidal dispersions:

    (a) dispersion. Large particles are physically broken down to colloidal size and then dispersed;

    (b) condensation. Ion or molecules in solution are caused to condense into colloidal-size particles.

32. Colloidal particles tend to remain dispersed for two reasons:

    (a) Continual collisions with molecules of the dispersing medium (Brownian movement);

(b) Like charges on the colloidal particles prevent them from coalescing to larger particles which would settle out.

33. In homogenized milk the fat globules have been reduced to colloidal size so they remain dispersed indefinitely, whereas in nonhomogenized milk the large fat globules coalesce, forming a definite cream layer.

34. A Cottrell precipitator would be of no value in reducing ozone concentration because ozone is not a charged particle.

35. Activated charcoal would be more effective in adsorbing hydrogen sulfide at 0°C than 100°C. There are attractive forces between the charcoal and the adsorbed gas. As temperature increases, the thermal agitation increases, breaking the adsorbed gas molecules loose from the surface of the charcoal.

36. The following statements are correct: a,c,f,i,j,k,l,n,p,r,t and u.

D. Review Problems.

1. Calculation of molarity of ions.

(a) $$0.015\ M\ NaCl \times \frac{1\ mol\ Na^+}{1\ mol\ NaCl} = 0.015\ M\ Na^+$$

$$0.015\ M\ NaCl \times \frac{1\ mol\ Cl^-}{1\ mol\ NaCl} = 0.015\ M\ Cl^-$$

(b) $$4.25\ M\ NaKSO_4 \times \frac{1\ mol\ Na^+}{1\ mol\ NaKSO_4} = 4.25\ M\ Na^+$$

$$4.25\ M\ NaKSO_4 \times \frac{1\ mol\ K^+}{1\ mol\ NaKSO_4} = 4.25\ M\ K^+$$

$$4.25\ M\ NaKSO_4 \times \frac{1\ mol\ SO_4^{2-}}{1\ mol\ NaKSO_4} = 4.25\ M\ SO_4^{2-}$$

(c) $$0.75\ M\ ZnBr_2 \times \frac{1\ mol\ Zn^{2+}}{1\ mol\ ZnBr_2} = 0.75\ M\ Zn^{2+}$$

$$0.75\ M\ ZnBr_2 \times \frac{2\ mol\ Br^-}{1\ mol\ ZnBr_2} = 1.5\ M\ Br^-$$

(d) $$1.65\ M\ Al_2(SO_4)_3 \times \frac{3\ mol\ SO_4^{2-}}{1\ mol\ Al_2(SO_4)_3} = 4.95\ M\ SO_4^{2-}$$

$$1.65\ M\ Al_2(SO_4)_3 \times \frac{2\ mol\ Al^{3+}}{1\ mol\ Al_2(SO_4)_3} = 3.30\ M\ Al^{3+}$$

(e) $0.20\ M\ CaCl_2 \times \frac{1\ mol\ Ca^{2+}}{1\ mol\ CaCl_2} = 0.20\ M\ Ca^{2+}$

$0.20\ M\ CaCl_2 \times \frac{2\ mol\ Cl^-}{1\ mol\ CaCl_2} = 0.40\ M\ Cl^-$

(f) $\frac{22.0\ g\ KI}{500\ mL} \times \frac{1\ mol\ KI}{166.0\ g} \times \frac{1000\ mL}{L} = 0.265\ M\ KI$

$0.265\ M\ KI \times \frac{1\ mol\ K^+}{1\ mol\ KI} = 0.265\ M\ K^+$

$0.265\ M\ KI \times \frac{1\ mol\ I^-}{1\ mol\ KI} = 0.265\ M\ I^-$

(g) $\frac{900\ g\ (NH_4)_2SO_4}{20.0\ L} \times \frac{1\ mol}{132.1\ g} = 0.341\ M\ (NH_4)_2SO_4$

$0.341\ M\ (NH_4)_2SO_4 \times \frac{2\ mol\ NH_4^+}{mol\ (NH_4)_2SO_4} = 0.682\ M\ NH_4^+$

$0.341\ M\ (NH_4)_2SO_4 \times \frac{1\ mol\ SO_4^{2-}}{mol\ (NH_4)_2SO_4} = 0.341\ M\ SO_4^{2-}$

(h) $\frac{0.0120\ g\ Mg(ClO_3)_2}{0.00100\ L} \times \frac{1\ mol}{191.3\ g} = 0.0627\ M\ Mg(ClO_3)_2$

$0.0627\ M\ Mg(ClO_3)_2 \times \frac{1\ mol\ Mg^{2+}}{mol\ Mg(ClO_3)_2} = 0.0627\ M\ Mg^{2+}$

$0.0627\ M\ Mg(ClO_3)_2 \times \frac{2\ mol\ ClO_3^-}{mol\ Mg(ClO_3)_2} = 0.125\ M\ ClO_3^-$

2. The molarities of each ion as calculated in question one will be used to calculate the grams of each ion present in 100 mL of solution (100 mL = 0.100 L).

(a) $0.100\ L \times \frac{0.015\ mol\ Na^+}{L} \times \frac{23.0\ g\ Na^+}{mol\ Na^+} = 0.035\ g\ Na^+$

$0.100\ L \times \frac{0.015\ mol\ Cl^-}{L} \times \frac{35.5\ g\ Cl^-}{mol\ Cl^-} = 0.053\ g\ Cl^-$

(b) $0.100\ L \times \frac{4.25\ mol\ Na^+}{L} \times \frac{23.0\ g\ Na^+}{mol\ Na^+} = 9.78\ g\ Na^+$

$0.100\ L \times \frac{4.25\ mol\ K^+}{L} \times \frac{39.1\ g\ K^+}{mol\ K^+} = 16.6\ g\ K^+$

$$0.100\ L \times \frac{4.25\ mol\ SO_4^{2-}}{L} \times \frac{96.1\ g\ SO_4^{2-}}{mol\ SO_4^{2-}} = 40.8\ g\ SO_4^{2-}$$

(c) $$0.100\ L \times \frac{0.75\ M\ Zn^{2+}}{L} \times \frac{65.4\ g\ Zn^{2+}}{mol\ Zn^{2+}} = 4.9\ g\ Zn^{2+}$$

$$0.100\ L \times \frac{1.5\ mol\ Br^{-}}{L} \times \frac{79.9\ g\ Br^{-}}{mol\ Br^{-}} = 12\ g\ Br^{-}$$

(d) $$0.100\ L \times \frac{3.30\ mol\ Al^{3+}}{L} \times \frac{27.0\ g\ Al^{3+}}{mol\ Al^{3+}} = 8.91\ g\ Al^{3+}$$

$$0.100\ L \times \frac{4.95\ mol\ SO_4^{2-}}{L} \times \frac{96.1\ g\ SO_4^{2-}}{mol\ SO_4^{2-}} = 47.6\ g\ SO_4^{2-}$$

(e) $$0.100\ L \times \frac{0.20\ mol\ Ca^{2+}}{L} \times \frac{40.1\ g\ Ca^{2+}}{mol\ Ca^{2+}} = 0.80\ g\ Ca^{2+}$$

$$0.100\ L \times \frac{0.40\ mol\ Cl^{-}}{L} \times \frac{35.5\ g\ Cl^{-}}{mol\ Cl^{-}} = 1.4\ g\ Cl^{-}$$

(f) $$0.100\ L \times \frac{0.265\ mol\ K^{+}}{L} \times \frac{39.1\ g\ K^{+}}{mol\ K^{+}} = 1.04\ g\ K^{+}$$

$$0.100\ L \times \frac{0.265\ mol\ I^{-}}{L} \times \frac{126.9\ g\ I^{-}}{mol\ I^{-}} = 3.36\ g\ I^{-}$$

(g) $$0.100\ L \times \frac{0.682\ mol\ NH_4^{+}}{L} \times \frac{18.0\ g\ NH_4^{+}}{mol\ NH_4^{+}} = 1.23\ g\ NH_4^{+}$$

$$0.100\ L \times \frac{0.341\ mol\ SO_4^{2-}}{L} \times \frac{96.1\ g\ SO_4^{2-}}{mol\ SO_4^{2-}} = 3.28\ g\ SO_4^{2-}$$

(h) $$0.100\ L \times \frac{0.0627\ mol\ Mg^{2+}}{L} \times \frac{24.3\ g\ Mg^{2+}}{mol\ Mg^{2+}} = 0.152\ g\ Mg^{2+}$$

$$0.100\ L \times \frac{0.125\ mol\ ClO_3^{-}}{L} \times \frac{83.5\ g\ ClO_3^{-}}{mol\ ClO_3^{-}} = 1.04\ g\ ClO_3^{-}$$

3. $CaI \rightarrow Ca^{2+} + 2\ I^{-}$; $0.520\ M\ I^{-}$

$$\frac{0.520 \text{ mol } I^-}{L} \times \frac{1 \text{ mol } Ca^{2+}}{2 \text{ mol } I^-} = \frac{0.260 \text{ mol } Ca^{2+}}{L} = 0.260 \text{ M } Ca^{2+}$$

4. (a) If you mix two different volumes of 1.0 M NaCl the concentration of the resulting solution will remain at 1.0 M NaCl.

$$1.0 \text{ M NaCl} \times \frac{1 \text{ mol } Na^+}{\text{mol NaCl}} = 1.0 \text{ M } Na^+$$

$$1.0 \text{ M NaCl} \times \frac{1 \text{ mol } Cl^-}{\text{mol NaCl}} = 1.0 \text{ M } Cl^-$$

(b) $HCl + NaOH \rightarrow NaCl + H_2O$

$$30.0 \text{ mL HCl} \times \frac{1 \text{ L}}{1000 \text{ mL}} \times \frac{1.0 \text{ mol HCl}}{L} = 0.030 \text{ mol HCl}$$

$$30.0 \text{ mL NaOH} \times \frac{1 \text{ L}}{1000 \text{ mL}} \times \frac{1.0 \text{ mol NaOH}}{\text{L NaOH}} = 0.030 \text{ mol NaOH}$$

0.030 mole of HCl reacts with 0.030 mole of NaOH and produces 0.030 mole NaCl. Final volume, 0.060 L.

$$\frac{0.030 \text{ mol NaCl}}{0.060 \text{ L}} = 0.50 \text{ M NaCl}$$

Since there is one mole each of sodium and chloride ions per mole of NaCl, the molar concentration of Na^+ and Cl^- will be: 0.50 M Na^+, and 0.50 M Cl^-.

(c) 100.0 mL of 2.0 M KCl and 100.0 mL of 1.0 M $CaCl_2$ are mixed giving a final volume of 200.0 mL and concentrations of 1.0 M KCl and 0.5 M CaCl . The concentration of K^+ will be 1.0 M and and the concentration of Ca^{2+} will be 0.5 M. The chloride ion concentration will be 2.0 M (1.0 M from the KCl + 2 × 0.5 M from the $CaCl_2$).

(d) $KOH + HCl \rightarrow KCl + H_2O$

$$100.0 \text{ mL} \times \frac{1 \text{ L}}{1000 \text{ mL}} \times \frac{0.40 \text{ mol KOH}}{L} = 0.040 \text{ mol KOH}$$

$$100.0 \text{ mL} \times \frac{1 \text{ L}}{1000 \text{ mL}} \times \frac{0.80 \text{ mol HCl}}{L} = 0.080 \text{ mol HCl}$$

0.040 mol KOH reacts with 0.080 mol HCl. 0.040 mol of HCl remains and 0.040 mol of KCl is produced. Final volume is 200.0 mL and contains 0.040 mol of HCl and 0.040 mol of KCl. Moles of ions are: 0.040 mol H^+, 0.040 mol K^+, and 0.080 mol Cl^-.

Concentrations of ions are:

$$\frac{0.040 \text{ mol } H^+}{0.200 \text{ L}} = 0.20 \text{ M } H^+; \quad \text{M } K^+ = \text{M } H^+ = 0.20 \text{ M}$$

$$\frac{0.080 \text{ mol Cl}^-}{0.200 \text{ L}} = 0.40 \text{ M Cl}^-$$

(e) $$35.0 \text{ mL} \times \frac{1 \text{ L}}{1000 \text{ mL}} \times \frac{0.20 \text{ mol Ba(OH)}_2}{\text{L}} = 0.0070 \text{ mol Ba(OH)}_2$$

$$35.0 \text{ mL} \times \frac{1 \text{ L}}{1000 \text{ mL}} \times \frac{0.20 \text{ mol H}_2\text{SO}_4}{\text{L}} = 0.0070 \text{ mol H}_2\text{SO}_4$$

Final volume = 35.0 mL + 35.0 mL = 70.0 mL

$$\begin{array}{lllll} H_2SO_4 & + \; Ba(OH)_2 & \rightarrow & BaSO_4(s) & + \; 2\,H_2O \\ 0.0070 \text{ mol} & 0.0070 \text{ mol} & & 0.0070 \text{ mol} & \end{array}$$

The H_2SO_4 and the $Ba(OH)_2$ react completely producing insoluble $BaSO_4$ and H_2O. No ions are present in solution.

(f) $$0.500 \text{ L NaCl} \times \frac{2.0 \text{ mol NaCl}}{\text{L NaCl}} = 1.0 \text{ mol NaCl}$$

$$1.00 \text{ L AgNO}_3 \times \frac{1.0 \text{ mol AgNO}_3}{\text{L AgNO}_3} = 1.0 \text{ mol AgNO}_3$$

$$\begin{array}{lllll} NaCl & + \; AgNO_3 & \rightarrow & AgCl(s) & + \; NaNO_3 \\ 1.0 \text{ mol} & 1.0 \text{ mol} & & 1.0 \text{ mol} & 1.0 \text{ mol} \end{array}$$

The AgCl is insoluble and produces no ions. The 1.0 mol of $NaNO_3$ will give 1.0 mol of Na^+ ions and 1.0 mol of NO_3^- ions. Final volume of solution is 1.5 L. Concentrations of ions are:

$$\frac{1.0 \text{ mol Na}^+}{1.5 \text{ L}} = 0.67 \text{ M Na}^+$$

$$\frac{1.0 \text{ mol NO}_3^-}{1.5 \text{ L}} = 0.67 \text{ M NO}_3^-$$

5. Dilution problems. $V_1M_1 = V_2M_2$; $100 \text{ ml} \times 12 \text{ M} = V_2 \times 0.40 \text{ M}$

$$V_2 = \frac{100 \text{ ml} \times 12 \text{ M}}{0.40 \text{ M}} = 3.0 \times 10^3 \text{ mL}$$

6. The reaction of HCl and NaOH occurs on a 1 to 1 mole ratio. $HCl + NaOH \rightarrow NaCl + H_2O$. At the end point in these titration reactions equal moles of HCl and NaOH will have reacted. Moles = M × L. At the end point moles HCl = moles NaOH. Therefore, at the end point, L acid × M acid = L base × M base, or using mL the expression becomes, mL acid × M acid = mL base × M base. This last expression is used in calculating the molarity of the HCl in parts (a), (b), and (c), and the molarity of the NaOH in parts (d), (e), and (f).

(a) $37.70\ mL\ NaOH \times 0.728\ M\ NaOH = 40.13\ mL\ HCl \times M\ HCl$

$$M\ HCl = \frac{37.70\ mL \times 0.728\ M}{40.13\ mL} = 0.684\ M\ HCl$$

(b) $$\frac{33.66\ mL\ NaOH \times 0.306\ M\ NaOH}{19.00\ mL\ HCl} = 0.542\ M\ HCl$$

(c) $$\frac{18.00\ mL\ HaOH \times 0.555\ M\ NaOH}{27.25\ mL\ HCl} = 0.367\ M\ HCl$$

(d) $$\frac{37.19\ mL\ HCl \times 0.126\ M\ HCl}{31.91\ mL\ NaOH} = 0.147\ M\ NaOH$$

(e) $$\frac{48.04\ mL\ HCl \times 0.482\ M\ HCl}{24.02\ mL\ NaOH} = 0.964\ M\ NaOH$$

(f) $$\frac{13.13\ mL\ HCl \times 1.425\ M\ HCl}{39.39\ mL\ NaOH} = 0.4750\ M\ NaOH$$

7. $Ba(OH)_2 + 2\ HCl \rightarrow BaCl_2 + 2\ H_2O$

$M\ HCl \rightarrow mol\ HCl \rightarrow mol\ Ba(OH)_2 \rightarrow M\ Ba(OH)_2$

$$\frac{0.430\ mol\ HCl}{L} \times 29.26\ mL \times \frac{1\ L}{1000\ mL} = 0.0126\ mol\ HCl$$

$$0.0126\ mol\ HCl \times \frac{1\ mol\ Ba(OH)_2}{2\ mol\ HCl} = 0.00630\ mol\ Ba(OH)_2$$

$$\frac{0.00630\ mol\ Ba(OH)_2}{0.02040\ L} = 0.309\ M\ Ba(OH)_2$$

8. (a) $2\ HCl + Ca(OH)_2 \rightarrow CaCl_2 + 2\ H_2O$

$M\ Ca(OH)_2 \rightarrow mol\ Ca(OH)_2 \rightarrow mol\ HCl \rightarrow mL\ HCl$

$$0.0500\ L\ Ca(OH)_2 \times \frac{0.100\ mol\ Ca(OH)_2}{L\ Ca(OH)_2} \times \frac{2\ mol\ HCl}{mol\ Ca(OH)_2}$$

$$\times \frac{1000\ mL\ HCl}{0.245\ mol\ HCl} = 40.8\ mL\ of\ 0.245\ M\ HCl$$

(b) $3\ HCl + Al(OH)_3 \rightarrow AlCl_3 + 3\ H_2O$

$g\ Al(OH)_3 \rightarrow mol\ Al(OH)_3 \rightarrow mol\ HCl \rightarrow mL\ HCl$

$$10.0\ g\ Al(OH)_3 \times \frac{1\ mol\ Al(OH)_3}{78.0\ g\ Al(OH)_3} \times \frac{3\ mol\ HCl}{mol\ Al(OH)_3}$$

$$\times \frac{1000\ mL}{0.245\ mol\ HCl} = 1.57 \times 10^3\ mL\ of\ 0.245\ M\ HCl$$

9. $Na_2CO_3 + 2\ HCl \rightarrow 2\ NaCl + CO_2 + H_2O$

$$\frac{g\ Na_2CO_3}{L\ HCl} \rightarrow \frac{mol\ Na_2CO_3}{L\ HCl} \rightarrow \frac{mol\ HCl}{L\ HCl}$$

$$\frac{0.452\ g\ Na_2CO_3}{0.0424\ L\ HCl} \times \frac{1\ mol\ Na_2CO_3}{106.0\ g\ Na_2CO_3} \times \frac{2\ mol\ HCl}{1\ mol\ Na_2CO_3} = \frac{0.201\ mol\ HCl}{L}$$

$= 0.201$ M HCl

10. $2\ HCl + Ca(OH)_2 \rightarrow CaCl_2 + 2\ H_2O$

$g\ Ca(OH)_2 \rightarrow mol\ Ca(OH)_2 \rightarrow mol\ HCl \rightarrow mL\ HCl$

$$2.00\ g\ Ca(OH)_2 \times \frac{1\ mol\ Ca(OH)_2}{74.1\ g} \times \frac{2\ mol\ HCl}{1\ mol\ Ca(OH)_2} \times \frac{1000\ mL\ HCl}{0.1234\ mol\ HCl}$$

$= 437$ mL of 0.1234 M HCl

11. $KOH + HNO_3 \rightarrow KNO_3 + H_2O$

$$0.05000\ L\ HNO_3 \times \frac{0.240\ mol}{L} = 0.0120\ mol\ HNO_3$$

1 mol KOH reacts with 1 mol HNO_3, so 0.0120 mol KOH is needed to neutralize 0.0120 mol HNO_3.

$$0.0120\ mol\ KOH \times \frac{56.1\ g\ KOH}{mol} = 0.673\ g\ KOH$$

12. $NaOH + HCl \rightarrow NaCl + H_2O$

$L\ HCl \rightarrow mol\ HCl \rightarrow mol\ NaOH \rightarrow g\ NaOH$

$$0.01825\ L\ HCl \times \frac{0.2406\ mol\ HCl}{L\ HCl} \times \frac{1\ mol\ NaOH}{1\ mol\ HCl} \times \frac{40.0\ g\ NaOH}{mol\ NaOH}$$

$= 0.176$ g NaOH that reacts with the HCl

The 0.200 g sample contains 0.176 g NaOH.

$$\frac{0.176\ g\ NaOH}{0.200\ g\ sample} \times 100\% = 88.0\%\ NaOH$$

13. $NaOH + HCl \rightarrow NaCl + H_2O$

$L\ HCl \rightarrow mol\ HCl \rightarrow mol\ NaOH \rightarrow g\ NaOH$

$$0.04990\ L\ HCl \times \frac{0.466\ mol\ HCl}{L\ HCl} \times \frac{1\ mol\ NaOH}{1\ mol\ HCl} \times \frac{40.0\ g\ NaOH}{mol\ NaOH}$$

$= 0.930$ g NaOH in the sample

1.00 g sample - 0.930 g NaOH = 0.070 g NaCl in the sample

$$\frac{0.070\ g\ NaCl}{1.00\ g\ sample} \times 100\% = 7.0\%\ NaCl\ in\ the\ sample$$

14. $Zn + 2\ HCl \rightarrow ZnCl_2 + H_2$

$$5.00\ g\ Zn \times \frac{1\ mol\ Zn}{65.4\ g\ Zn} = 0.0765\ mol\ Zn$$

(a) $$100\ mL \times \frac{1\ L}{1000\ mL} \times \frac{0.350\ mol\ HCl}{L} = 0.0350\ mol\ HCl$$

The HCl is the limiting reagent.

$$0.0350\ mol\ HCl \times \frac{1\ mol\ H_2}{2\ mol\ HCl} = 0.0175\ mol\ H_2$$

$$27°C = 300\ K;\quad P = \frac{700\ torr}{760\ torr} \times 1\ atm = 0.921\ atm$$

$$L\ H_2 = \frac{nRT}{P} = \frac{0.0175\ mol\ H_2 \times 0.0821\ L\text{-}atm \times 300\ K}{0.921\ atm \times mol\ K} = 0.468\ L\ H_2$$

(b) $$200\ mL\ HCl \times \frac{1\ L}{1000\ mL} \times \frac{0.350\ mol\ HCl}{L\ HCl} = 0.0700\ mol\ HCl$$

The HCl is the limiting reactant.

$$0.0700\ mol\ HCl \times \frac{1\ mol\ H_2}{2\ mol\ HCl} = 0.0350\ mol\ H_2$$

$27°C = 300\ K;\quad P = 0.921\ atm$ (part a)

$$L\ H_2 = \frac{nRT}{P} = \frac{0.0350\ mol\ H_2 \times 0.0821\ L\text{-}atm \times 300\ K}{0.921\ atm \times mol\ K} = 0.936\ L\ H_2$$

15. Calculation of the pH of solutions.

(a) $H^+ = 0.01\ M = 10^{-2}\ M;\quad pH = -\log 10^{-2} = 2$

(b) $H^+ = 1.0\ M;\quad pH = -\log 1.0 = 0$

(c) $H^+ = 6.5 \times 10^{-9}\ M;\quad pH = -\log 6.5 \times 10^{-9} = 9 - \log 6.5$
$= 9 - 0.81 = 8.19$

(d) $H^+ = 1 \times 10^{-7}\ M;\quad pH = -\log 1 \times 10^{-7} = 7$

(e) $H^+ = 0.50\ M;\ pH = -\log 5.0 \times 10^{-1} = 1 - \log 5.0 = 1 - 0.70 = 0.30$

(f) $H^+ = 0.00010\ M = 1.0 \times 10^{-4};\quad pH = -\log 1.0 \times 10^{-4} = 4$

16. (a) Orange juice, $3.7 \times 10^{-4}\ M\ H^+$

$pH = -\log 3.7 \times 10^{-4} = 4 - \log 3.7 = 4 - 0.57 = 3.4$

(b) Vinegar, $2.8 \times 10^{-3}\ M\ H^+$

$pH = -\log 2.8 \times 10^{-3} = 3 - \log 2.8 = 3 - 0.45 = 2.6$

(c) Black coffee, 5.0×10^{-5} M H^+

$$pH = -\log 5.0 \times 10^{-5} = 5 - \log 5.0 = 5 - 0.70 = 4.3$$

(d) Limewater, 3.4×10^{-11} M H^+

$$pH = -\log 3.4 \times 10^{-11} = 11 - \log 3.4 = 11 - 0.53 = 10.5$$

17. pH of 1.0 L solution containing 0.1 mL of 1.0 M HCl.
Calculate the molarity of the solution

$$0.1 \text{ mL} \times \frac{1 \text{ L}}{1000 \text{ mL}} \times \frac{1.0 \text{ mol HCl}}{\text{L}} = 1 \times 10^{-4} \text{ M HCl}$$

1.0×10^{-4} M HCl produces 1×10^{-4} M H^+

$$pH = -\log 1 \times 10^{-4} = 4$$

18. Dilution problem: $V_1M_1 = V_2M_2$; $V_1 \times 18.0 \text{ M} = 50.0 \text{ L} \times 5.00 \text{ M}$

$$V_1 = \frac{50.0 \text{ L} \times 5.00 \text{ M}}{18.0 \text{ M}} = 13.9 \text{ L of } 18.0 \text{ M } H_2SO_4$$

19. $NaOH + HCl \rightarrow NaCl + H_2O$

$$3.0 \text{ g NaOH} \times \frac{1 \text{ mol NaOH}}{40.0 \text{ g NaOH}} = 0.075 \text{ mol NaOH}$$

$$500 \text{ mL HCl} \times \frac{1 \text{ L}}{1000 \text{ mL}} \times \frac{0.10 \text{ mol HCl}}{\text{L}} = 0.050 \text{ mol HCl}$$

The solution is basic. The NaOH will neutralize the HCl with an excess of 0.025 moles of NaOH remaining.

20. $V_AN_A = V_BN_B$; $28.92 \text{ mL} \times 0.1240 \text{ N} = 10.00 \text{ mL} \times N_B$

$$N_B = \frac{28.92 \text{ mL} \times 0.1240 \text{ N}}{10.00 \text{ mL}} = 0.3586 \text{ N NaOH}$$

21. $V_AN_A = V_BN_B$; $V_A \times 0.325 \text{ N} = 32.8 \text{ mL} \times 0.225 \text{ N}$

$$V_A = \frac{32.8 \text{ mL} \times 0.225 \text{ N}}{0.325 \text{ N}} = 22.7 \text{ mL of } 0.325 \text{ N } HNO_3$$

22. $V_AN_A = V_BN_B$; $V_A \times 0.325 \text{ N} = 32.8 \text{ mL} \times 0.225 \text{ N}$

$$V_A = \frac{32.8 \text{ mL} \times 0.225 \text{ N}}{0.325 \text{ N}} = 22.7 \text{ mL of } 0.325 \text{ N } H_2SO_4$$

23. $V_AN_A = V_BN_B$; $25.00 \text{ mL} \times N_A = 22.68 \text{ mL} \times 0.5000 \text{ N}$

$$N_A = \frac{22.68 \text{ mL} \times 0.5000 \text{ N}}{25.00 \text{ mL}} = 0.4536 \text{ N } H_3PO_4$$

$$\frac{0.4536 \text{ eq } H_3PO_4}{L} \times \frac{1 \text{ mol}}{3 \text{ eq}} = 0.1512 \text{ M } H_3PO_4$$

24. (a) Wt. percent: $125 \text{ mL} \times \frac{1.78 \text{ g}}{1.00 \text{ mL}} = 147 \text{ g solution}$

$$\frac{29.4 \text{ g } CaCl_2}{147 \text{ g solution}} \times 100\% = 20.0\% \text{ } CaCl_2$$

(b) $$\frac{29.4 \text{ g } CaCl_2}{125 \text{ mL}} \times \frac{1000 \text{ mL}}{L} \times \frac{1 \text{ mol } CaCl_2}{111.1 \text{ g}} = 2.12 \text{ M } CaCl_2$$

(c) 147 g solution - 29.4 g $CaCl_2$ = 118 g H_2O in the solution

$$\frac{29.4 \text{ g } CaCl_2}{118 \text{ g } H_2O} \times \frac{1 \text{ mol } CaCl_2}{111.1 \text{ g } CaCl_2} \times \frac{1000 \text{ g } H_2O}{1 \text{ kg } H_2O} = 2.24 \text{ m } CaCl_2$$

B. 1. Both tubes would appear the same and contain more molecules in the gaseous state than the tube at 0°C and less molecules in the gaseous state than the tube at 80°C.

2. The reaction is endothermic because the increased temperature increases the concentration of product present at equilibrium.

3. At equilibrium the rate of the forward reaction equals the rate of the reverse reaction.

4. The yield of NH_3 would be greater if the reaction were carried out in a one liter vessel. Increased pressure favors more product because the reaction of four moles of reactants (3 moles H_2 and 1 mole N_2) produces 2 moles of product, NH_3.

5. Acids stronger than acetic acid are: benzoic, cyanic, formic, hydrofluoric, and nitrous acids (all have equilibrium constants greater than the equilibrium constant for acetic acid). Acids weaker than acetic acid are: carbolic, hydrocyanic and hypochlorous acids (all have equilibrium constants smaller than the equilibrium constant for acetic acid). All these acids have one ionizable hydrogen atom.

6. The sum of the pH and the pOH is 14. A solution whose pH is -1 would have a pOH of 15.

7. The order of solubility will correspond to the order of the values of the solubility product constants of the salts being compared. This is so because each salt in the comparison produces the same number of ions (two in this case), for each formula unit of salt that dissolves. This kind of comparison would not necessarily be valid if the salts being compared gave different numbers of ions per formula unit of salt dissolving. The order is: $AgC_2H_3O_2$, $PbSO_4$, AgCl, $BaCrO_4$, AgBr, AgI, PbS.

8. Molar solubilities.

(a) $K_{sp}Mn(OH)_2 = 2.0 \times 10^{-13}$; $K_{sp}Ag_2CrO_4 = 1.9 \times 10^{-12}$

Each salt gives 3 ions per formula unit of salt dissolving; therefore, the salt with the largest K_{sp} is more soluble. This is Ag_2CrO_4.

(b) $K_{sp}\ BaCrO_4 = 8.5 \times 10^{-11}$; $K_{sp}Ag_2CrO_4 = 1.9 \times 10^{-12}$

$K_{sp}BaCrO_4$ has a greater molar solubility than $BaCrO_4$ even though its K_{sp} is smaller. This is because the Ag_2CrO_4 gives more ions per formula unit of salt dissolving than the $BaCrO_4$.

$BaCrO_4(s) \rightleftharpoons Ba^{2+} + CrO_4^{2-}$

Molar solubility = $\sqrt{8.5 \times 10^{-12}}$ = 9.2×10^{-5} mol $BaCrO_4$/L

$Ag_2CrO_4(s) \rightleftharpoons 2\ Ag^+ + CrO_4^{2-}$

Molar solubility = $\sqrt[3]{\frac{1.9 \times 10^{-12}}{4}}$ = 7.8×10^{-5} mol Ag_2CrO_4/L

9. $HC_2H_3O_2 \rightleftharpoons H^+ + C_2H_3O_2^-$

Initial Concentrations		Added	Concentration After Equilibrium Shifts
$HC_2H_3O_2$	1.00 M	---	1.01 M
H^+	1.8×10^{-5} M	0.01 M	1.9×10^{-5}
$C_2H_3O_2^-$	1.00 M	---	0.99 M

The initial concentration of H^+ in the buffer solution is very low (1.8×10^{-5} M) because of the large excess of acetate ions. When 0.010 mole of HCl is added to one liter of the buffer solution, this supplies 0.010 M H^+. The added H^+ is a stress to the right side of the equation. The equilibrium shifts to the left, using up almost all the added H^+, reducing the acetate ion by approximately 0.010 M, and increasing the acetic acid by approximately 0.010 M. The concentration of H^+ which will be in equilibrium with these concentrations calculates to be 1.9×10^{-5} M (which can be verified by experiment).

10. If the reaction shown in Figure 16.4 were exothermic, the figure should be modified to show the energy level of the product as lower than the energy level of reactants.

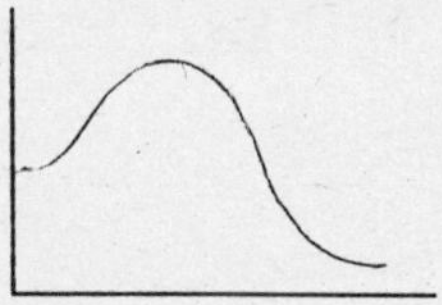

rather than

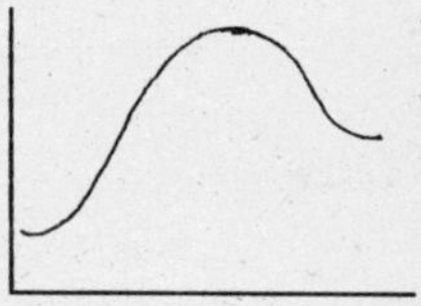

C. Review questions.

1. Reversible systems

(a) $H_2O(s) \xrightleftharpoons{0°C} H_2O(l)$

(b) $H_2O(l) \xrightleftharpoons{100°C} H_2O(g)$

(c) $Na_2SO_4(s) \rightleftharpoons 2\ Na^+(aq) + SO_4^{2-}(aq)$

(d) $SO_2(l) \rightleftharpoons SO_2(g)$

2. If a sodium chloride solution is saturated, the equilibrium is

   $Na^+(aq) + Cl^-(aq) \rightleftharpoons NaCl(s)$

   Bubbling in HCl gas increases the concentration of Cl^-, creating a stress which will cause the equilibrium to shift to the right, precipitating solid NaCl.

3. The rate of a reaction increases when the concentration of one of the reactants increases because the increase in concentration causes the number of collisions between reactants to increase.

4. Equilibrium system.

   $4\ NH_3(g) + 3\ O_2(g) \rightleftharpoons 2\ N_2(g) + 6\ H_2O(g) + 1531\ kJ$

   (a) The reaction is exothermic, heat is evolved.

   (b) Addition of O_2 will shift the reaction to the right in order to reestablish equilibrium. The concentrations of N_2, H_2O, and O_2 will be increased. The concentration of the NH_3 will be decreased.

   (c) Addition of heat will shift the reaction to the left. This uses up some of the heat added.

5. $N_2 + 3\ H_2 \rightleftharpoons 2\ NH_3 + 92.5\ kJ$

Change or stress imposed on the system at equilibrium	Direction of reaction, left or right, to reestablish equilibrium	Change in number of moles		
		N_2	H_2	NH_3
(a) Add N_2	right	I	D	I
(b) Remove H_2	left	I	D	D
(c) Decrease volume of reaction vessel	right	D	D	I
(d) Increase volume of reaction vessel	left	I	I	D
(e) Increase temperature	left	I	I	D
(f) Add a catalyst	no change	N	N	N

(g) Add H_2 and NH_3	?	?	I	I

I = increase; D = Decrease; N = No change;
? = Not sufficient information to determine.

6. If pure HI is placed in a vessel at 700 K, some of it will decompose. The mechanism would be for two molecules to collide to form the activated complex, which would then decompose to form H_2 and I_2.

7. Direction of shift in equilibrium, left or right, when changes are imposed on a system at equilibrium.

	Kind of Change Made		
Reaction	Increased Temperature	Increase Pressure (Volume Decrease)	Add Catalyst
(a)	right	right	no change
(b)	left	no change	no change
(c)	left	right	no change
(d)	right	left	no change
(e)	left	left	no change

8. Increase in temperature causes the rate of reaction to increase because it speeds up the molecules. With the molecules moving faster, more of the collisions will be energetic enough to react.

9. $A + B \rightleftarrows C + D$ When A and B are first mixed the rate of the forward reaction to produce C and D is at its maximum. As the reaction proceeds, the rate of production of C and D will decrease because the concentrations of A and B have decreased. And as C and D are produced, some of the collisions between C and D will result in the reverse reaction, forming A and B. Finally an equilibrium is achieved in which the forward rate exactly equals the reverse rate.

10. $HC_2H_3O_2 + H_2O \rightleftarrows H_3O^+ + C_2H_3O_2^-$ As water is added (diluting the solution from 1.0 M to 0.10 M) the equilibrium shifts to the right, giving a higher percent ionization.

11. The statement does not contradict Le Chatelier's principle. The previous question deals with the case of dilution. If you go the other way, adding pure acetic acid to a dilute solution, the reaction will shift to the right, giving more ions in accordance with Le Chatelier's principle. But the concentration of the unionized acetic acid will increase faster than the concentration of the ions thus giving a smaller percent ionization.

12. (a) $K = \dfrac{[Cl_2]^2[H_2O]}{[HCl]^4[O_2]}$ (b) $K = \dfrac{[NH_3]^2}{[N_2][H_2]^3}$ (c) $K = \dfrac{[PCl_3][Cl_2]}{[PCl_5]}$

(d) $K = \dfrac{[H^+][ClO_2^-]}{[HClO_2]}$ (e) $K = \dfrac{[NH_4^+][OH^-]}{[NH_4OH]}$ (f) $K = \dfrac{[NO]^4[H_2O]^6}{[NH_3]^4[O_2]^5}$

13. At different temperatures the degree of ionization of water varies, being higher at higher temperatures. Consequently, the pH of water can be different at different temperatures.

14. In pure water, H^+ and OH^-, are produced in equal quantities by the ionization of the water molecules, $H_2O \rightleftharpoons H^+ + OH^-$. Since $pH = -\log H^+$, and $pOH = -\log OH^-$, they will always be identical for pure water. At 25°C, they each have the value of 7, but at higher temperature, the degree of ionization is greater, so the pH and pOH would both be less than 7, but still equal.

15. If H^+ ion concentration is increased:

(a) pH is decreased (pH of 1 is more acidic than pH of 4)

(b) pOH is increased

(c) OH^- is decreased

(d) K_w remains unchanged. It is a constant at a given temperature.

16. (a) $K_{sp} = [Cu^{2+}][S^{2-}]$ (b) $K_{sp} = [Ba^{2+}][SO_4^{2-}]$

(c) $K_{sp} = [Pb^{2+}][Br^-]^2$ (d) $K_{sp} = [Ag^+]^3[AsO_4^{3-}]$

(e) $K_{sp} = [Fe^{3+}][OH^-]^3$ (f) $K_{sp} = [Sb^{5+}]^2[S^{2-}]^5$

(g) $K_{sp} = [Ca^{2+}][F^-]^2$ (h) $K_{sp} = [Ba^{2+}]^3[PO_4^{3-}]^2$

17. In water, the silver acetate dissociates until the equilibrium concentration of ions reached. In nitric acid solution, the acetate ions will react with hydrogen ions to form acetic acid molecules. Thus, since the acetate ion concentration is kept low, more silver acetate can dissolve. If HCl is used a precipitate of silver chloride would be formed since silver chloride is less soluble than

silver acetate. Thus more silver acetate would dissolve in HCl than in pure water. ($AgC_2H_3O_2(s) \rightleftharpoons Ag^+ + C_2H_3O_2^-$)

18. The basis for deciding whether or not a salt dissolved in water gives an acidic, a basic, or a neutral solution is whether or not the salt reacts with water (hydrolysis). Salts that contain an ion derived from a weak acid or base will hydrolyze to give an acidic or a basic solution.

(a) KCl, neutral solution (b) Na_2CO_3, basic solution

(c) K_2SO_4, neutral solution (d) $(NH_4)_2SO_4$, acidic solution

(e) $Ca(CN)_2$, basic solution (f) $BaBr_2$, neutral solution

(g) $NaNO_2$, basic solution (h) NaF, basic solution

See Questions 19, 20, and 21 for more details.

19. When the salt, sodium acetate, is dissolved in water, the solution becomes basic. First we have the dissolving reaction:

$$NaC_2H_3O_2(s) \xrightarrow{H_2O} Na^+ + C_2H_3O_2^-$$

The acetate ion reacts with water (hydrolysis). The reaction does not go to completion, but some OH^- ions are produced and the solution becomes basic.

$$C_2H_3O_2^- + H_2O \rightleftharpoons OH^- + HC_2H_3O_2$$

20. (a) $KNO_2 \longrightarrow K^+ + NO_2^-$; $NO_2^- + H_2O \rightleftharpoons OH^- + HNO_2$

(b) $Mg(C_2H_3O_2)_2 \longrightarrow Mg^{2+} + 2\ C_2H_3O_2^-$

$C_2H_3O_2^- + H_2O \rightleftharpoons OH^- + HC_2H_3O_2$

(c) $NH_4NO_3 \longrightarrow NH_4^+ + NO_3^-$; $NH_4^+ + H_2O \rightleftharpoons H^+ + NH_4OH$

(d) $Na_2SO_3 \longrightarrow 2\ Na^+ + SO_3^{2-}$

$SO_3^{2-} + 2\ H_2O \rightleftharpoons 2\ OH^- + H_2SO_3$

21. (a) $HCO_3^- + H_2O \rightleftharpoons OH^- + H_2CO_3$

(b) $NH_4^+ + H_2O \rightleftharpoons OH^- + NH_4OH$

(c) $OCl^- + H_2O \longrightarrow OH^- + HOCl$

(d) $ClO_2^- + H_2O \rightleftharpoons OH^- + HClO_2$

22. A buffer solution contains a weak acid or base plus a salt of that weak acid or base. For example, dilute acetic acid and sodium acetate.

$HC_2H_3O_2(aq) \rightleftharpoons H^+ + C_2H_3O_2^-$

$NaC_2H_3O_2(aq) \longrightarrow Na^+ + C_2H_3O_2^-$

When a small amount of a strong acid (H^+) is added to this buffer solution, the H^+ reacts with the acetate ions to form un-ionized acetic acid, thus neutralizing the added acid. When a strong base OH^- is added, the OH^- reacts with un-ionized acetic acid to neutralize the added base. As a result, in both cases, the approximate pH of the solution is maintained.

23. (a) When excess acid gets into the blood stream, bicarbonate ions react with it to form H_2CO_3, with almost no net increase in H^+, thus maintaining the pH.

(b) When excess base gets into the blood stream it reacts with H^+ to form water, but by the first reaction, more H_2CO_3 will ionize to replace the H^+ ions, maintaining the pH at a constant value.

24. The following statements are correct: c,d,g,i,j,k,l,n,q,s,t,u,v, and z.

D. Review problems.

1. $H_2 + I_2 \rightarrow 2\ HI$

The reaction is on a one to one mole ratio of H_2 to I_2. The H_2 is the limiting reactant.

$$2.10\ \text{mol}\ H_2 \times \frac{2\ \text{mol HI}}{1\ \text{mol}\ H_2} = 4.20\ \text{mol HI}$$

2. $H_2 + I_2 \rightarrow 2\ HI$

(a) $$2.00\ \text{mol}\ H_2 \times \frac{2\ \text{mol HI}}{1\ \text{mol}\ H_2} \times \frac{0.79\ \text{mol HI}}{1.00\ \text{mol HI}} = 3.16\ \text{mol HI}$$

(b) The addition of 0.27 mol of I_2 makes the I_2 present in excess and the 2.00 mole of H_2 the limiting reactant. The yield increases to 85%.

$$2.00\ \text{mol}\ H_2 \times \frac{2\ \text{mol HI}}{1\ \text{mol}\ H_2} \times \frac{0.85\ \text{mol HI}}{1.00\ \text{mol HI}} = 3.40\ \text{mol HI}$$

There will be 15% unreacted H_2 and I_2.

$0.15 \times 2.0 = 0.30$ mol H_2 present.

In addition to the 0.30 mol of unreacted I_2 will be the 0.27 mol of I_2 added.

$0.27 + 0.30 = 0.57$ mol I_2 present.

(c) $K = \frac{[HI]^2}{[H_2][I_2]}$

The formation of 3.16 mol of HI required the reaction of 1.58 mol of I_2. At equilibrium the concentrations are: 3.16 mol HI; $2.00 - 1.58 = 0.42$ mol $H_2 = 0.42$ mol I_2.

$$K_{eq} = \frac{(3.16)^2}{(0.42)(0.42)} = 57$$

In the calculation of the equilibrium constant the actual number of moles of reactants and products present at equilibrium can be used in the calculation in place of molar concentrations. This is so because the reaction is gaseous and the liters of HI produced equals the sum of the liters of H_2 and I_2 reacting. In the equilibrium expression the volumes will cancel.

3. $H_2 + I_2 \rightleftharpoons 2\,HI$

$$64.0 \text{ g HI} \times \frac{1 \text{ mol HI}}{127.9 \text{ g HI}} = 0.500 \text{ mol HI present}$$

$$0.500 \text{ mol HI} \times \frac{1 \text{ mol } I_2}{2 \text{ mol HI}} = 0.250 \text{ mol } I_2 \text{ reacted}$$

$$0.500 \text{ mol HI} \times \frac{1 \text{ mol } H_2}{2 \text{ mol HI}} = 0.250 \text{ mol } H_2 \text{ reacted}$$

$$6.00 \text{ g } H_2 \times \frac{1 \text{ mol}}{2.0 \text{ g}} = 3.00 \text{ mol } H_2 \text{ initially present}$$

$$200 \text{ g } I_2 \times \frac{1 \text{ mol}}{253.8 \text{ g}} = 0.788 \text{ mol } I_2 \text{ initially present}$$

At equilibrium, moles present are: 0.500 mol HI;
$3.00 - 0.250 = 2.75$ mol H_2; $0.788 - 0.250 = 0.538$ mol I_2

4. $PCl_3(g) + Cl_2(g) \rightleftharpoons PCl_5(g)$

$$K_{eq} = \frac{[PCl_5]}{[PCl_3[Cl_2]}$$

The concentrations are:

$$PCl_5 = \frac{0.22 \text{ mol}}{20 \text{ L}} = 0.011 \text{ mol/L}$$

$$PCl_3 = \frac{0.10 \text{ mol}}{20 \text{ L}} = 0.0050 \text{ mol/L}$$

$$Cl_2 = \frac{1.50 \text{ mol}}{20 \text{ L}} = 0.0750 \text{ mol/L}$$

$$K_{eq} = \frac{[0.011]}{[0.0050][0.0750]} = 29$$

5. 100°C - 30°C = 70°C temperature increase. The increase is equal to seven 10°C increments. The reaction rate will be increased $2 \times 2 \times 2 \times 2 \times 2 \times 2 \times 2 = 128$ times.

6. Calculation of ionization constants.

<u>Hypochlorous</u> <u>acid</u>, $HOCl \rightleftharpoons H^+ + OCl^-$

Equilibrium concentrations: $[H^+] = [OCl^-] = 5.95 \times 10^{-5}$ mol/L

$[HOCl] = 0.10 - 5.9 \times 10^{-5} = 0.10$ mol/L

$$K_a = \frac{[H^+][OCl^-]}{[HOCl]} = \frac{[5.9 \times 10^{-5}][5.9 \times 10^{-5}]}{0.10} = 3.5 \times 10^{-8}$$

<u>Propanoic</u> <u>acid</u>, $HC_3H_5O_2 \rightleftharpoons H^+ + C_3H_5O_2^-$

Equilibrium concentrations: $[H^+] = [C_3H_5O_2^-] = 1.4 \times 10^{-3}$ mol/L

$[HC_3H_5O_2] = 0.15 - 1.4 \times 10^{-3} = 0.15$ mol/L

$$K_a = \frac{[H^+][C_3H_5O_2^-]}{[HC_3H_5O_2]} = \frac{[1.4 \times 10^{-3}][1.4 \times 10^{-3}]}{[0.15]} = 1.3 \times 10^{-5}$$

<u>Hydrocyanic</u> <u>acid</u>, $HCN \rightleftharpoons H^+ + CN^-$

Equilibrium concentrations: $[H^+] = [CN^-] = 8.9 \times 10^{-6}$ M

$[HCN] = 0.20 - 8.9 \times 10^{-6} = 0.20$ M

$$K_a = \frac{[H^+][CN^-]}{[HCN]} = \frac{[8.9 \times 10^{-6}][8.9 \times 10^{-6}]}{[0.20]} = 4.0 \times 10^{-10}$$

7. (a) $HC_2H_3O_2 \rightleftharpoons H^+ + C_2H_3O_2^-$. Let X = mol/L of $HC_2H_3O_2$ ionizing to establish equilibrium. Equilibrium concentrations are:

$[H^+] = [C_2H_3O_2^-] = X$; $[HC_2H_3O_2] = 0.25 - X = 0.25$

$$K_a = \frac{[H^+][C_2H_3O_2^-]}{[HC_2H_3O_2]} = \frac{[X][X]}{0.25} = 1.8 \times 10^{-5}$$

$X^2 = 0.25 \times 1.8 \times 10^{-5}$; $X = \sqrt{0.25 \times 1.8 \times 10^{-5}}$

$= 2.1 \times 10^{-3}\ M = [H^+]$

(b) $pH = -\log [H^+] = -\log 2.1 \times 10^{-3} = 3 - \log 2.1 = 2.7$

(c) $\%\ \text{ionization} = \dfrac{[H^+]}{[HC_2H_3O_2]} \times 100\% = \dfrac{2.1 \times 10^{-3}}{0.25} \times 100\% = 0.84\%$

8. $HA \rightleftarrows H^+ + A^-$; $0.52\% = 0.0052$; $[HA] = 1.0\ M$ (initial)

$[H^+] = [A^-] = (1.0)(0.0052) = 5.2 \times 10^{-3}\ M$;

$[HA] = 1.0 - 0.0052 = 0.9948\ M$

$$K_a = \frac{[H^+][A^-]}{[HA]} = \frac{[5.2 \times 10^{-3}][5.2 \times 10^{-3}]}{[0.9948]} = 2.7 \times 10^{-5}$$

9. $HA \rightleftarrows H^+ + A^-$; $pH = 5 = -\log H^+$; $[H^+] = 1 \times 10^{-5}$

$[H^+] = [A^-]$; $[HA] = 0.15\ M$

$$K_a = \frac{[H^+][A^-]}{[HA]} = \frac{[1 \times 10^{-5}][1 \times 10^{-5}]}{0.15} = 7 \times 10^{-10}$$

10. $HC_2H_3O_2 \rightleftarrows H^+ + C_2H_3O_2^-$

$$K_a = 1.8 \times 10^{-5} = \frac{[H^+][C_2H_3O_2^-]}{[HC_2H_3O_2]}$$

Let X = moles/L of $HC_2H_3O_2$ that has ionized to establish equilibrium. Equilibrium concentrations will be:

$[H^+] = [C_2H_3O_2^-] = X$; $[HC_2H_3O_2]$ = initial concentration - X. Since K_a is small the degree of ionization is very small. Hence, the approximation, initial concentration - X = initial concentration, is a valid approximation.

(a) $[H^+] = [C_2H_3O_2^-] = X$; $[HC_2H_3O_2] = 1.0\ M$

$\dfrac{(X)(X)}{1.0} = 1.8 \times 10^{-5}$; $X^2 = (1.0)(1.8 \times 10^{-5})$

$X = \sqrt{1.8 \times 10^{-5}} = 4.2 \times 10^{-3}\ M$

$\dfrac{4.2 \times 10^{-3}\ M}{1.0\ M} \times 100\% = 0.42\%$ ionized

$pH = -\log 4.2 \times 10^{-3} = 3 - \log 4.2 = 2.4$

(b) $HC_2H_3O_2 = 0.10\ M$

$$\frac{(X)(X)}{[0.10]} = 1.8 \times 10^{-5};\ X^2 = (0.10)(1.8 \times 10^{-5}) = 1.8 \times 10^{-6}$$

$$X = \sqrt{1.8 \times 10^{-6}} = 1.3 \times 10^{-3}\ M$$

$$\frac{1.3 \times 10^{-3}\ M}{0.10\ M} \times 100\% = 1.3\%\ \text{ionized}$$

$$pH = -\log 1.3 \times 10^{-3} = 3 - \log 1.3 = 2.9$$

(c) $HC_2H_3O_2 = 0.010\ M$

$$\frac{(X)(X)}{[0.010]} = 1.8 \times 10^{-5};\ X^2 = (0.010)(1.8 \times 10^{-5}) = 1.8 \times 10^{-7}$$

$$X = \sqrt{1.8 \times 10^{-7}} = 4.2 \times 10^{-4}\ M;$$

$$\frac{4.2 \times 10^{-4}\ M}{0.010\ M} \times 100\% = 4.2\%\ \text{ionized}$$

$$pH = -\log 4.2 \times 10^{-4} = 4 - \log 4.2 = 3.4$$

11. $HA \rightleftarrows H^+ + A^-;\ K_a = \frac{[H^+][A^-]}{[HA]}$

We first need to find the $[H^+]$. To do this we work backwards using the pH expression.

$pH = -\log[H^+] = 3.7$

The 3 in 3.7 plus 1 gives us the exponent to which 10 is raised. The number is 4, but since the pH is a -log the exponent is -4.

$pH = -\log W \times 10^{-4}$; log W is a number which when subtracted from 4 gives us the pH value of 3.7 (4 - 0.3 = 3.7).

$4 - \log W = 3.7;\ \log W = 4 - 3.7 = 0.3$

$\log W = 0.3.\ \ W = 2.0$ (obtained from the log table)

$[H^+] = [A^-] = 2.0 \times 10^{-4};\ [HA] = 0.37$

$$K_a = \frac{(2.0 \times 10^{-4})(2.0 \times 10^{-4})}{0.37} = 1.1 \times 10^{-7}$$

12. Discussion related to solving this problem is given in Problem 11.

$HA \rightleftarrows H^+ + A^-;\ K_a = \frac{[H^+][A^-]}{[HA]}$

$pH = -\log\ [H^+] = 2.89$; power of $10 = -(2 + 1) = -3$

$pH = -\log W \times 10^{-3} = 2.89;\ pH = 3 - \log W = 2.89$

$\log W = 3 - 2.89 = 0.11;\ W = 1.3$ (estimated from the log table)

$[H^+] = [A^-] = 1.3 \times 10^{-3}$; $[HA] = 0.23$

$$K_a = \frac{(1.3 \times 10^{-3})(1.3 \times 10^{-3})}{0.23} = 7.3 \times 10^{-6}$$

13. 6.0 M HCl gives $[H^+] = 6.0$ M (100% ionized)

$pH = -\log 6.0 = -0.78$; $pOH = 14 - pH = 14 - (-0.78) = 14.78$

$$[OH^-] = \frac{K_w}{[H^+]} = \frac{1 \times 10^{-14}}{6.0} = 1.7 \times 10^{-15}$$

14. $pH + pOH = 14.0$; $pOH = 14.0 - pH$

(a) 0.00010 M HCl; $[H^+] = 0.00010\ M = 1.0 \times 10^{-4}$

$pH = -\log 1.0 \times 10^{-4} = 4.0$; $pOH = 14.0 - 4.0 = 10.0$

(b) 0.010 M NaOH; $[OH^-] = 0.010\ M = 1.0 \times 10^{-2}$

$pOH = -\log 1.0 \times 10^{-2} = 2.0$; $pH = 14.0 - 2.0 = 12.0$

(c) 0.0025 M NaOH; $[OH^-] = 2.5 \times 10^{-3}$

$pOH = -\log 2.5 \times 10^{-3} = 3 - \log 2.5 = 3 - 0.40 = 2.6$

$pH = 14.0 - 2.6 = 11.4$

(d) $HClO \rightleftharpoons H^+ + ClO^-$

0.10 M X X

$$K_a = \frac{[H^+][ClO^-]}{[HClO]} = 3.5 \times 10^{-8}$$

$\frac{(X)(X)}{0.10} = 3.5 \times 10^{-8}$; $X^2 = (0.10)(3.5 \times 10^{-8})$

$pH = -\log 5.9 \times 10^{-5} = 5 - \log 5.9 = 5 - 0.8 = 4.2$

$pOH = 14.0 - 4.2 = 9.8$

(e) $Fe(OH)_2(s) \rightleftharpoons Fe^{2+} + 2\ OH^-$

X X 2X

$$K_{sp} = [Fe^{2+}][OH^-]^2 = (X)(2X)^2 = 8.0 \times 10^{-16}$$

$$4X^3 = 8.0 \times 10^{-16};\quad X = \sqrt[3]{\frac{8.0 \times 10^{-16}}{4}} = 0.58 \times 10^{-5}$$

$[OH^-] = 2X = 2 \times 0.58 \times 10^{-5} = 1.2 \times 10^{-5}$

$pOH = -\log 1.2 \times 10^{-5} = 4.9$; $pH = 14.0 - 4.9 = 9.1$

15. Calculate the $[OH^-]$; $[OH^-] = \dfrac{K_w}{[H^+]}$

(a) $[H^+] = 1.0 \times 10^{-4}$; $[OH^-] = \dfrac{1.0 \times 10^{-14}}{1.0 \times 10^{-4}} = 1.0 \times 10^{-10}$

(b) $[H^+] = 2.8 \times 10^{-6}$; $[OH^-] = \dfrac{1.0 \times 10^{-14}}{2.8 \times 10^{-6}} = 3.6 \times 10^{-9}$

(c) $[H^+] = 4.0 \times 10^{-9}$; $[OH^-] = \dfrac{1.0 \times 10^{-14}}{4.0 \times 10^{-9}} = 2.5 \times 10^{-6}$

16. Calculate $[H^+]$; $[H^+] = \dfrac{K_w}{[OH^-]}$

(a) $[OH^-] = 6.0 \times 10^{-7}$; $[H^+] = \dfrac{1.0 \times 10^{-14}}{6.0 \times 10^{-7}} = 1.7 \times 10^{-8}$

(b) $[OH^-] = 1 \times 10^{-8}$; $[H^+]\ \dfrac{1 \times 10^{-14}}{1 \times 10^{-8}} = 1 \times 10^{-6}$

(c) $[OH^-] = 4.5 \times 10^{-6}$; $[H^+] = \dfrac{1.0 \times 10^{-14}}{4.5 \times 10^{-6}} = 2.2 \times 10^{-9}$

17. Calculation of solubility product constants. The molar solubilities of the salts and their ions will be indicated below the formulas in the equilibrium equations.

(a) $BaSO_4(s) \rightleftharpoons Ba^{2+} + SO_4^{2-}$

$3.9 \times 10^{-5} \qquad 3.9 \times 10^{-5} \qquad 3.9 \times 10^{-5}$

$K_{sp} = [Ba^{2+}][SO_4^{2-}] = (3.9 \times 10^{-5})(3.9 \times 10^{-5}) = 1.5 \times 10^{-9}$

(b) $Ag_2CrO_4(s) \rightleftharpoons 2\ Ag^+ + CrO_4^{2-}$

$7.8 \times 10^{-5} \qquad 2 \times 7.8 \times 10^{-5} \qquad 7.8 \times 10^{-5}$

$K_{sp} = [Ag^+]^2[CrO_4^{2-}] = (2 \times 7.8 \times 10^{-5})^2(7.8 \times 10^{-5})$

$= 1.9 \times 10^{-12}$

(c) $ZnS(s) \rightleftharpoons Zn + S^{2-}$

$3.5 \times 10^{-12} \qquad 3.5 \times 10^{-12} \qquad 3.5 \times 10^{-12}$

$K_{sp} = [Zn^{2+}][S^{2-}] = (3.5 \times 10^{-12})(3.5 \times 10^{-12}) = 1.2 \times 10^{-23}$

(d) $Pb(IO_3)_2(s) \rightleftarrows Pb^{2+} + 2\ IO_3^-$

4.0×10^{-5} $\quad 4.0 \times 10^{-5}$ $\quad 2 \times 4.0 \times 10^{-5}$

$K_{sp} = [Pb^{2+}][IO_3^-]^2 = (4.0 \times 10^{-5})(2 \times 4.0 \times 10^{-5})^2$

$= 2.6 \times 10^{-13}$

(e) $Bi_2S_3(s) \rightleftarrows 2\ Bi^{3+} + 3\ S^{2-}$

4.9×10^{-15} $\quad 2 \times 4.9 \times 10^{-15}$ $\quad 3 \times 4.9 \times 10^{-15}$

$K_{sp} = [Bi^{3+}]^2[S^{2-}]^3 = (2 \times 4.9 \times 10^{-15})^2(3 \times 4.9 \times 10^{-15})^3$

$= 3.1 \times 10^{-70}$

(f) $\frac{0.0019\ g\ AgCl}{L} \times \frac{1\ mol\ AgCl}{143.4\ g} = 1.3 \times 10^{-5}\ mol/L$

$AgCl(s) \rightleftarrows Ag^+ + Cl^-$

1.3×10^{-5} $\quad 1.3 \times 10^{-5}$ $\quad 1.3 \times 10^{-5}$

$K_{sp} = [Ag^+][Cl^-] = (1.3 \times 10^{-5})(1.3 \times 10^{-5}) = 1.7 \times 10^{-10}$

(g) $\frac{0.67\ g\ CaSO_4}{L} \times \frac{1\ mol}{136.2\ g} = 4.9 \times 10^{-3}\ mol/L$

$CaSo_4(s) \rightleftarrows Ca^{2+} + SO_4^{2-}$

4.9×10^{-3} $\quad 4.9 \times 10^{-3}$ $\quad 4.9 \times 10^{-3}$

$K_{sp} = [Ca^{2+}][SO_4^{2-}] = (4.9 \times 10^{-3})(4.9 \times 10^{-3}) = 2.4 \times 10^{-5}$

(h) $\frac{2.33 \times 10^{-4}\ g\ Zn(OH)_2}{L} \times \frac{1\ mol}{99.4\ g} = 2.34 \times 10^{-6}\ mol/L$

$Zn(OH)_2(s) \rightleftarrows Zn^{2+} + 2\ OH^-$

2.34×10^{-6} $\quad 2.34 \times 10^{-6}$ $\quad 2 \times 2.34 \times 10^{-6}$

$K_{sp} = [Zn^{2+}][OH^-]^2 = (2.34 \times 10^{-6})(2 \times 2.34 \times 10^{-6})^2$

$= 5.13 \times 10^{-17}$

(i) $\frac{6.73 \times 10^{-3}\ g\ Ag_3PO_4}{L} \times \frac{1\ mol}{418.7\ g} = 1.61 \times 10^{-5}\ mol/L$

$Ag_3PO_4(s) \rightleftarrows 3\ Ag^+ + PO_4^{3-}$

1.61×10^{-5} $\quad 3 \times 1.61 \times 10^{-5}$ $\quad 1.61 \times 10^{-5}$

$K_{sp} = [Ag^+]^3[PO_4^{3-}] = (3 \times 1.61 \times 10^{-5})^3(1.61 \times 10^{-5})$

$= 1.81 \times 10^{-18}$

18. Molar solubility. The molar solubilities of the salts and their ions will be represented in terms of X below their formulas in the equilibrium equations.

(a) $BaCO_3$, $K_{sp} = 2.0 \times 10^{-9}$; $\underset{X}{BaCO_3(s)} \rightleftharpoons \underset{X}{Ba^{2+}} + \underset{X}{CO_3^{2-}}$

$K_{sp} = [Ba^{2+}][CO_3^{2-}] = (X)(X) = 2.0 \times 10^{-9}$

$X = \sqrt{2.0 \times 10^{-9}} = 4.5 \times 10^{-5}$ mol/L = molar solubility of $BaCO_3$

(b) $AlPO_4$, $K_{sp} = 5.8 \times 10^{-19}$, $\underset{X}{AlPO_4(s)} \rightleftharpoons \underset{X}{Al^{3+}} + \underset{X}{PO_4^{3-}}$

$K_{sp} = [Al^{3+}][PO_4^{3-}]\ (X)(X) = 5.8 \times 10^{-19}$

$X = \sqrt{5.8 \times 10^{-19}} = 7.6 \times 10^{-10}$ mol/L

Molar solubility of $AlPO_4 = 7.6 \times 10^{-10}$ mol/L

(c) Ag_2SO_4, $K_{sp} = 1.5 \times 10^{-5}$; $\underset{X}{Ag_2SO_4(s)} \rightleftharpoons \underset{2X}{2\ Ag^+} + \underset{X}{SO_4^{2-}}$

$K_{sp} = [Ag^+]^2[SO_4^{2-}] = (2X)^2(X) = 4X^3 = 1.5 \times 10^{-5}$

$X = \sqrt[3]{\frac{1.5 \times 10^{-5}}{4}} = 1.6 \times 10^{-2}$ mol/L

Molar solubility of $Ag_2SO_4 = 1.6 \times 10^{-2}$ mol/L

(d) $Mg(OH)_2$, $K_{sp} = 7.1 \times 10^{-12}$; $\underset{X}{Mg(OH)_2(s)} \rightleftharpoons \underset{X}{Mg^{2+}} + \underset{2X}{2\ OH^-}$

$K_{sp} = [Mg^{2+}][OH^-]^2 = (X)(2X)^2 = 4X^3 = 7.1 \times 10^{-12}$

$X = \sqrt[3]{\frac{7.1 \times 10^{-12}}{4}} = 1.2 \times 10^{-4}$ mol/L

Molar solubility of $Mg(OH)_2 = 1.2 \times 10^{-4}$ mol/L

19. Calculate solubility in g/100 mL; $\frac{mol}{L} \times 0.100\ L \times \frac{g}{mol} = g$

(a) $\dfrac{4.5 \times 10^{-5} \text{ mol } BaCO_3}{L} \times 0.100 \text{ L} \times \dfrac{197.3 \text{ g } BaCO_3}{\text{mol}}$

$= 8.9 \times 10^{-4} \text{ g } BaCO_3$

(b) $\dfrac{7.6 \times 10^{-10} \text{ mol } AlPO_4}{L} \times 0.100 \text{ L} \times \dfrac{122.0 \text{ g } AlPO_4}{\text{mol}}$

$= 9.3 \times 10^{-9} \text{ g } AlPO_4$

(c) $\dfrac{1.6 \times 10^{-2} \text{ mol } Ag_2SO_4}{L} \times 0.100 \text{ L} \times \dfrac{311.9 \text{ g } Ag_2SO_4}{\text{mol}}$

$= 0.50 \text{ g } Ag_2SO_4$

(d) $\dfrac{1.2 \times 10^{-4} \text{ mol } Mg(OH)_2}{L} \times 0.100 \text{ L} \times \dfrac{58.3 \text{ g } Mg(OH)_2}{\text{mol}}$

$= 7.0 \times 10^{-4} \text{ g } Mg(OH)_2$

20. CaF_2, $K_{sp} = 3.9 \times 10^{-11}$; X = mol/L of CaF_2 dissolving.

(a) $CaF_2(s) \rightleftharpoons Ca^{2+} + 2F^-$

$\quad X \qquad\qquad X \qquad 2X$

$K_{sp} = [Ca^{2+}][F^-]^2 = (X)(2X)^2 = 4X^3 = 3.9 \times 10^{-11}$

$X = \sqrt[3]{\dfrac{3.9 \times 10^{-11}}{4}} = 2.1 \times 10^{-4}$

Molar solubility $CaF_2 = 2.1 \times 10^{-4}$ mol/L

$\dfrac{2.1 \times 10^{-4} \text{ mol } CaF_2}{L} \times \dfrac{1 \text{ mol } Ca^{2+}}{1 \text{ mol } CaF_2} = 2.1 \times 10^{-4} \text{ mol } Ca^{2+}/L$

$\dfrac{2.1 \times 10^{-4} \text{ mol } CaF_2}{L} \times \dfrac{2 \text{ mol } F^-}{\text{mol } CaF_2} = 4.2 \times 10^{-4} \text{ mol } F^-/L$

(b) $\dfrac{2.1 \times 10^{-4} \text{ mol } CaF_2}{L} \times 0.500 \text{ L} \times \dfrac{78.1 \text{ g } CaF_2}{\text{mol } CaF_2}$

$= 8.2 \times 10^{-3} \text{ g } CaF_2$

21. The molar concentrations of ions after mixing are calculated and these concentrations are substituted into the equilibrium expression. The value obtained by these substitutions is compared to the K_{sp} for the salt. If the value is greater than the K_{sp}, precipitation occurs. If the value is less than the K_{sp} no precipitation occurs.

(a) 100 mL 0.010 M Na_2SO_4 gives 100 mL 0.010 M SO_4^{2-}

100 mL 0.001 M $Pb(NO_3)_2$ gives 100 mL 0.001 M Pb^{2+}

Volume after mixing = 200 mL; concentrations after mixing:

SO_4^{2-} = 0.0050 M, Pb^{2+} = 0.0005 M

Ion product: $[Pb^{2+}][SO_4^{2-}] = (5 \times 10^{-3})(5 \times 10^{-4}) = 2.5 \times 10^{-6}$

$K_{sp} = 1.3 \times 10^{-8}$; Ion product is greater than K_{sp}.

Precipitation occurs.

(b) 50.0 mL 1.0×10^{-4} M $AgNO_3$ gives 50.0 mL 1.0×10^{-4} M Ag^+

100 mL 1.0×10^{-4} M NaCl gives 100 mL of 1.0×10^{-4} M Cl^-

Final volume = 150 mL. Concentrations after mixing:

$$1.0 \times 10^{-4}\ M\ Ag^+ \times \frac{50.0\ mL}{150\ mL} = 3.3 \times 10^{-5}\ M\ Ag^+$$

$$1.0 \times 10^{-4}\ M\ Cl^- \times \frac{100\ mL}{150\ mL} = 6.7 \times 10^{-5}\ M\ Cl^-$$

Ion product: $[Ag^+][Cl^-] = 3.3 \times 10^{-5} \times 6.7 \times 10^{-5} = 2.2 \times 10^{-9}$

$K_{sp} = 1.7 \times 10^{-10}$; Ion product is greater than K_{sp}.

Precipitation occurs.

(c) $$\frac{1.0\ g\ Ca(NO_3)_2}{0.150\ L} \times \frac{1\ mol\ Ca(NO_3)_2}{164.1\ g} \times \frac{1\ mol\ Ca^{2+}}{mol\ Ca(NO_3)_2} = 0.041\ M\ Ca^{2+}$$

250 mL 0.01 M NaOH gives 250 mL 0.01 M OH^-

Final volume = 400 mL. Concentrations after mixing:

$$0.041\ M\ Ca^{2+} \times \frac{150\ mL}{400\ mL} = 0.015\ M\ Ca^{2+}$$

$$0.01\ M\ OH^- \times \frac{250\ mL}{400\ mL} = 0.0063\ M\ OH^-$$

Ion product = $[Ca^{2+}][OH^-]^2 = (0.015)(0.0063)^2 = 6.0 \times 10^{-7}$

$K_{sp} = 1.3 \times 10^{-6}$; Ion product is less than the K_{sp}. No precipitation occurs.

22. With a known Ba^{2+} concentration, the SO_4^{2-} concentration can be calculated using the K_{sp} value.

$K_{sp} = [Ba^{2+}][SO_4^{2-}] = 1.5 \times 10^{-9}$; $Ba^{2+} = 0.050$ M

(a) $[SO_4^{2-}] = \dfrac{K_{sp}}{[Ba^{2+}]} = \dfrac{1.5 \times 10^{-9}}{0.050} = 3.0 \times 10^{-8}$ M SO_4^{2-}

(b) The molarity of SO_4^{2-} is equal to the mol/L of $BaSO_4$ that dissolves.

$$\frac{3.0 \times 10^{-8} \text{ mol } BaSO_4}{L} \times 0.100 \text{ L} \times \frac{233.4 \text{ g } BaSO_4}{\text{mol}}$$

$= 7.0 \times 10^{-7}$ g $BaSO_4$ remain in solution

23. $[Ba^{2+}][SO_4^{2-}] = 1.5 \times 10^{-9}$; $[Sr^{2+}][SO_4^{2-}] = 3.5 \times 10^{-7}$

Both cations are present in equal concentrations (0.10 M). Therefore, as SO_4^{2-} is added, the K_{sp} of $BaSO_4$ will be exceeded before that of $SrSO_4$. $BaSO_4$ precipitates first.

24. The concentration of Br^- = 0.10 M in 1.0 L of 0.10 M NaBr

$K_{sp} = [Ag^+][Br^-] = 5.0 \times 10^{-13}$

(a) $[Ag^+] = \dfrac{5.0 \times 10^{-13}}{[Br^-]} = \dfrac{5.0 \times 10^{-13}}{0.10} = 5.0 \times 10^{-12}$ M

$$\frac{5.0 \times 10^{-12} \text{ mol } Ag^+}{L} \times \frac{1 \text{ mol AgBr}}{1 \text{ mol } Ag^+} \times 1.0 \text{ L}$$

$= 5.0 \times 10^{-12}$ mol AgBr in solution

(b) $\dfrac{0.10 \text{ mol } MgBr_2}{L} \times \dfrac{2 \text{ mol } Br^-}{1 \text{ mol } MgBr_2} = \dfrac{0.20 \text{ mol } Br^-}{L} = 0.20$ M Br^-

$[Ag^+] = \dfrac{5.0 \times 10^{-13}}{[Br^-]} = \dfrac{5.0 \times 10^{-13}}{0.20} = 2.5 \times 10^{-12}$ M

$$\frac{2.5 \times 10^{-12} \text{ mol } Ag^+}{L} \times \frac{1 \text{ mol AgBr}}{1 \text{ mol } Ag^+} \times 1.0 \text{ L}$$

$= 2.5 \times 10^{-12}$ mol AgBr in solution

25. If $[Pb^{2+}][Cl^-]^2$ exceeds the K_{sp}, precipitation will occur

$K_{sp} = [Pb^{2+}][Cl^-]^2 = 2.0 \times 10^{-5}$

0.050 mol $Pb(NO_3)_2$/L gives 0.050 mol Pb^{2+}/L

0.010 mol NaCl/L gives 0.010 mol Cl^-/L

$[0.050][0.010]^2 = 5.0 \times 10^{-6}$

$[Pb^{2+}][Cl^-]^2$ is smaller than the K_{sp} value. Therefore, no precipitate of $PbCl_2$ will form.

26. $HC_2H_3O_2 \rightleftharpoons H^+ + C_2H_3O_2^-$

$$K_a = \frac{[H^+][C_2H_3O_2^-]}{[HC_2H_3O_2]} = 1.8 \times 10^{-5}; \; [H^+] = K_a \frac{[HC_2H_3O_2]}{[C_2H_3O_2^-]},$$

$[HC_2H_3O_2] = 0.20$ M

(a) $[C_2H_3O_2^-] = 0.10$ M; $[H^+] = 1.8 \times 10^{-5} \times \frac{0.20}{0.10} = 3.6 \times 10^{-5}$ M

$pH = -\log 3.6 \times 10^{-5} = 5 - \log 3.6 = 4.4$

(b) $[C_2H_3O_2^-] = 0.20$ M; $[H^+] = 1.8 \times 10^{-5} \times \frac{0.20}{0.20} = 1.8 \times 10^{-5}$ M

$pH = -\log 1.8 \times 10^{-5} = 5 - \log 1.8 = 4.7$

27. (a) Initially the solution of NaCl is neutral and $[H^+] = 1.0 \times 10^{-7}$;

$pH = -\log 1 \times 10^{-7} = 7.0$

Dilution: 1.0 M HCl × 1.0 mL = M_2 × 51 mL

$$M_2 = \frac{1.0 \text{ M} \times 1.0 \text{ mL}}{51 \text{ mL}} = 0.020 \text{ M HCl}$$

After addition of 1.0 mL 1.0 M HCl the $[H^+] = 2.0 \times 10^{-2}$

$pH = -\log 2.0 \times 10^{-2} = 2 - \log 2.0 = 1.7$

Change in pH = 7.0 - 1.7 = 5.3 units.

(b) Initially $[H^+] = 1.8 \times 10^{-5}$

$pH = -\log 1.8 \times 10^{-5} = 5 - \log 1.8 = 4.74$

Final $[H^+] = 1.9 \times 10^{-5}$; $pH = -\log 1.9 \times 10^{-5}$

$= 5 - \log 1.9 = 4.72$

change in pH = 4.74 - 4.72 = 0.02 units

B. 1. (a) Iodine is oxidized. Its oxidation number increases from 0 to +5.

(b) Chlorine is reduced. Its oxidation number decreases from 0 to -1.

2. The higher metal on the list is more reactive

(a) Al (b) Ba (c) Ni

3. If the free element is higher on the list than the ion with which it is paired then reaction occurs.

(a) Yes, $Zn + Cu^{2+} \longrightarrow Zn^{2+} + Cu$

(b) No

(c) Yes, $Sn + 2\ Ag^{+} \longrightarrow Sn^{2+} + 2\ Ag$

(d) No

(e) Yes, $Ba + FeCl_2 \longrightarrow BaCl_2 + Fe$

(f) No

(g) Yes, $Ni + Hg(NO_3)_2 \longrightarrow Ni(NO_3)_2 + Hg$

(h) Yes, $2\ Al + 3\ CuSO_4 \longrightarrow Al_2(SO_4)_3 + 3\ Cu$

4. (a) $2\ Al + Fe_2O_3 \longrightarrow Al_2O_3 + 2\ Fe$

(b) Al is above Fe in the activity series which indicates Al is more active than Fe.

(c) No. Iron is less active than aluminum and will not displace the aluminum ion from its compounds.

(d) Yes. Aluminum is above chromium in the activity series and will displace Cr^{3+} from its compounds.

5. The reactions are with dilute HCl and dilute H_2SO_4.

(a) $2\ Al + 6\ HCl \longrightarrow 2\ AlCl_3 + 3\ H_2$

$2\ Al + 3\ H_2SO_4 \longrightarrow Al_2(SO_4)_3 + 3\ H_2$

(b) $2\ Cr + 6\ HCl \longrightarrow 2\ CrCl_3 + 3\ H_2$

$2\ Cr + 3\ H_2SO_4 \longrightarrow Cr_2(SO_4)_3 + 3\ H_2$

(c) $Au + HCl \longrightarrow$ no reaction

$Au + H_2SO_4 \longrightarrow$ no reaction

(d) $Fe + 2\ HCl \longrightarrow FeCl_2 + H_2$

$Fe + H_2SO_4 \longrightarrow FeSO_4 + H_2$

(e) $Cu + HCl \longrightarrow$ no reaction

$Cu + H_2SO_4 \longrightarrow$ no reaction

(f) $Mg + 2\ HCl \longrightarrow MgCl_2 + H_2$

$Mg + H_2SO_4 \longrightarrow MgSO_4 + H_2$

(g) $Hg + HCl \longrightarrow$ no reaction

$Hg + H_2SO_4 \longrightarrow$ no reaction

(h) $Zn + 2\ HCl \longrightarrow ZnCl_2 + H_2$

$Zn + H_2SO_4 \longrightarrow ZnSO_4 + H_2$

6. (a) Oxidation occurs at the anode. The reaction is

$2\ Cl^- \longrightarrow Cl_2 + 2\ e^-$

(b) Reduction occurs at the cathode. The reaction is

$Ni^{2+} + 2\ e^- \longrightarrow Ni$

(c) The net electrochemical reaction is

$$Ni^{2+} + 2\ Cl^- \xrightarrow[\text{energy}]{\text{electrical}} Ni + Cl_2$$

7. In Figure 17.2 electrical energy is causing chemical reactions to occur. In Figure 17.3 chemical reactions are used to produce electrical energy.

8. (a) It would not be possible to monitor the voltage produced but the reactions in the cell would occur just the same.

(b) If the salt bridge were removed reaction would stop. Ions must be free to move to maintain electrical neutrality of ions in the solution. The two solutions would be isolated with no complete electrical circuit.

C. Review questions.

1. The oxidation number of the underlined element is indicated by the number following the formula.

(a) $\underline{Na}Cl$ +1 (b) $Fe\underline{Cl}_3$ -1 (c) $\underline{Pb}O_2$ +4 (d) $Na\underline{N}O_3$ +5

(e) $H_2\underline{S}O_3$ +4 (f) $\underline{N}H_4Cl$ -3 (g) $K\underline{Mn}O_4$ +7 (h) $\underline{I}_2$ 0

(i) $\underline{N}H_3$ -3 (j) $K\underline{Cl}O_3$ +5 (k) $K_2\underline{Cr}O_4$ +6 (l) $K_2\underline{Cr}_2O_7$ +6

2. The oxidation number of the underlined element in each formula is indicated by the number following the formula.

(a) $\underline{S}^{2-}$ -2 (b) $\underline{N}O_2^-$ +3 (c) $Na_2\underline{O}_2$ -1

(d) $\underline{Mn}O_4^-$ +7 (e) $\underline{Bi}^{3+}$ +3 (f) $\underline{O}_2$ 0

(g) $\underline{As}O_4^{3-}$ +5 (h) $Fe(\underline{O}H)_3$ -2 (i) $\underline{I}O_3^-$ +5

3. Half-reactions.

	Balanced half-reaction	Element changing oxidation number	Type of half-reaction
(a)	$Zn^{2+} + 2\ e^- \longrightarrow Zn$	Zn	reduction
(b)	$2\ Br^- \longrightarrow Br_2 + 2\ e^-$	Br	oxidation
(c)	$MnO_4^- + 8\ H^+ + 5\ e^- \longrightarrow Mn^{2+} + 4\ H_2O$	Mn	reduction
(d)	$Ni \longrightarrow Ni^{2+} + 2\ e^-$	Ni	oxidation
(e)	$SO_3^{2-} + H_2O \longrightarrow SO_4^{2-} + 2\ H^+ + 2\ e^-$	S	oxidation
(f)	$NO_3^- + 4\ H^+ + 3\ e^- \longrightarrow NO + 2\ H_2O$	N	reduction
(g)	$S_2O_4^{2-} + 2\ H_2O \longrightarrow 2\ SO_3^{2-} + 4\ H^+ + 2\ e^-$	S	oxidation
(h)	$Fe^{2+} \longrightarrow Fe^{3+} + 1\ e^-$	Fe	oxidation

4. (1) $Cr + HCl \longrightarrow CrCl_3 + H_2$

(a) element oxidized, Cr; element reduced, H

(b) oxidizing agent, HCl; reducing agent, Cr

(2) $SO_4^{2-} + I^- + H^+ \longrightarrow H_2S + I_2 + H_2O$

(a) element oxidized, I; element reduced, S

(b) oxidizing agent, SO_4^{2-}; reducing agent, I^-

(3) $AsH_3 + Ag^+ + H_2O \longrightarrow H_3AsO_4 + Ag + H^+$

(a) element oxidized, As; element reduced, Ag

(b) oxidizing agent, Ag^+; reducing agent, AsH_3

(4) $Cl_2 + NaBr \longrightarrow NaCl + Br_2$

(a) element oxidized, Br; element reduced, Cl

(b) oxidizing agent, Cl_2; reducing agent, NaBr

5. Balancing oxidation-reduction equations by change in oxidation number method

(a) $Zn + S \longrightarrow ZnS$

0 0 +2 -2

oxid. $Zn \longrightarrow Zn^{2+} + 2\ e^-$ Add half-reactions; the $2e^-$ cancel

red. $S + 2\ e^- \longrightarrow S^{2-}$

$Zn + S \longrightarrow ZnS$

(b) $AgNO_3 + Pb \longrightarrow Pb(NO_3)_2 + Ag$

+1 0 +2 0

oxid. $Pb \longrightarrow Pb^{2+} + 2\ e^-$

red. $Ag^+ + 1\ e^- \longrightarrow Ag$ multiply by 2; add half-reactions; the 2 e^- cancel

$Pb + 2\ Ag^+ \longrightarrow Pb^{2+} + 2\ Ag$

Transfer coefficients to original equation; complete balancing by inspection.

$2\ AgNO_3 + Pb \longrightarrow Pb(NO_3)_2 + 2\ Ag$

(c) $Fe_2O_3 + CO \longrightarrow Fe + CO_2$

+3 +2 0 +4

$C^{2+} \longrightarrow C^{4+} + 2\ e^-$ multiply by 3

$Fe^{3+} + 3\ e^- \longrightarrow Fe$ multiply by 2; add; the 6 e^- cancel

$3\ C^{2+} + 2\ Fe^{3+} \longrightarrow 3\ C^{4+} + 2\ Fe$

Transfer coefficients to original equation (the coefficient 2 in front of the Fe^{3+} becomes the subscript 2 in Fe_2O_3); complete balancing by inspection.

$Fe_2O_3 + 3\ CO \longrightarrow 2\ Fe + 3\ CO_2$

(d) $H_2S + HNO_3 \longrightarrow S + NO + H_2O$

-2 +5 0 +2

$S^{2-} \longrightarrow S + 2\ e^-$ multiply by 3

$N^{5+} + 3\ e^- \longrightarrow N^{2+}$ multiply by 2; add; the 6 e^- cancel

$3\ S^{2-} + 2\ N^{5+} \longrightarrow 3\ S + 2\ N^{2+}$

Transfer coefficients to original equation; complete balancing by inspection. $3\ H_2S + 2\ HNO_3 \longrightarrow 3\ S + 2\ NO + 4\ H_2O$

(e) $MnO_2 + HBr \longrightarrow MnBr_2 + Br_2 + H_2O$

+4 -1 +2 0

$Br^{1-} \longrightarrow Br + 1\ e^-$ multiply by 2

$Mn^{4+} + 2\ e^- \longrightarrow Mn^{2+}$ add equations; the 2 e^- cancel

$Mn^{4+} + 2\ Br^- \longrightarrow Mn^{2+} + 2\ Br$

Transfer coefficients to original equation. The coefficient 2 in front of the Br^- becomes the subscript 2 in Br_2. Also, 2 more Br^- ions are required (making the coefficient 4) to account for the 2 Br^- ions that do not change oxidation numbers. These 2 are part of the compound $MnBr_2$.

$MnO_2 + 4\ HBr \longrightarrow MnBr_2 + Br_2 + 2\ H_2O$

(f) $Cl_2 + KOH \longrightarrow KCl + KClO_3$

0 -1 +5

$Cl^0 \longrightarrow Cl^{5+} + 5\ e^-$

$Cl^0 + e^- \longrightarrow Cl^-$ multiply by 5; add; the 5 e^- cancel

$3\ Cl_2 \longrightarrow Cl^{5+} + 5\ Cl^-$

Transfer the coefficients to the original equation; complete balancing by inspection.

$3\ Cl_2 + 6\ KOH \longrightarrow KClO_3 + 5\ KCl + H_2O$

(g) $Ag + HNO_3 \longrightarrow AgNO_3 + NO + H_2O$

0 +5 +1 +2

$Ag \longrightarrow Ag^+ + e^-$ multiply by 3; add; the 3 e^- cancel

$N^{5+} + 3\ e^- \longrightarrow N^{2+}$

$3\ Ag + N^{5+} \longrightarrow 3\ Ag^+ + N^{2+}$

Transfer the coefficients to the original equation; complete balancing by inspection.

$3\ Ag + 4\ HNO_3 \longrightarrow 3\ AgNO_3 + NO + 2\ H_2O$

(h) $CuO + NH_3 \longrightarrow N_2 + Cu + H_2O$

+2 -3 0 0

$N^{3-} \longrightarrow N^0 + 3\ e^-$ multiply by 2

$Cu^{2+} + 2\ e^- \longrightarrow Cu$ multiply by 3; add; the 6 e^- cancel

$2\ N^{3-} + 3\ Cu^{2+} \longrightarrow N_2 + 3\ Cu$

Transfer the coefficients to the original equation; complete balancing by inspection.

$3\ CuO + 2\ NH_3 \longrightarrow N_2 + 3\ Cu + 3\ H_2O$

(i) $PbO_2 + Sb + NaOH \longrightarrow PbO + NaSbO_2 + H_2O$
+4 0 +2 +3

$Sb \longrightarrow Sb^{3+} + 3\ e^-$ multiply by 2

$Pb^{4+} + 2\ e^- \longrightarrow Pb^{2+}$ multiply by 3; add; the 6 e^- cancel

$2\ Sb + 3\ Pb^{4+} \longrightarrow 2\ Sb^{3+} + 3\ Pb^{2+}$

Transfer the coefficients to the original equation; complete balancing by inspection.

$3\ PbO_2 + 2\ Sb + 2\ NaOH \longrightarrow 3\ PbO + 2\ NaSbO_2 + H_2O$

(j) $H_2O_2 + KMnO_4 + H_2SO_4 \longrightarrow O_2 + MnSO_4 + K_2SO_4 + H_2O$
-1 +7 0 +2

$O_2^{2-} \longrightarrow O_2^0 + 2\ e^-$ multiply by 5

$Mn^{7+} + 5\ e^- \longrightarrow Mn^{2+}$ multiply by 2; add; the 10 e^- cancel

$5\ O_2^{2-} + 2\ Mn^{7+} \longrightarrow 5\ O_2 + 2\ Mn^{2+}$

Transfer the coefficients to the original equation; complete balancing by inspection.

$5\ H_2O_2 + 2\ KMnO_4 + 3\ H_2SO_4 \longrightarrow 5\ O_2 + 2\ MnSO_4 + K_2SO_4 + 8\ H_2O$

6. The following steps will be used in balancing ionic redox equations in acid solution.

Step 1. Write two half-reaction equation; balance all elements except H and O.

Step 2. Balance H and O using H_2O and H^+

Step 3. Balance electrically with electrons.

Step 4. Multiply each half reaction by the appropriate numbers to equalize the loss and gain of electrons.

Step 5. Add the two half-reactions together, cancelling electrons and any other identical substances on opposite sides of the equations.

(a) $Zn + NO_3^- \longrightarrow Zn^{2+} + NH_4^+$

Step 1. Write half-reaction equations; balance except H and O

$Zn \longrightarrow Zn^{2+}$

$NO_3^- \longrightarrow NH_4^+$

Step 2. Balance H and O

$Zn \longrightarrow Zn^{2+}$

$$10\ H^+ + NO_3^- \longrightarrow NH_4^+ + 3\ H_2O$$

Step 3. Balance electrically

$$Zn \longrightarrow Zn^{2+} + 2\ e^-$$

$$10\ H^+ + NO_3^- + 8\ e^- \longrightarrow NH_4^+ + 3\ H_2O$$

Step 4. Equalize the loss and gain of electrons.

$$4\ [Zn \longrightarrow Zn^{2+} + 2\ e^-]$$

$$10\ H^+ + NO_3^- + 8\ e^- \longrightarrow NH_4^+ + 3\ H_2O$$

Step 5. Add half-reactions

$$10\ H^+ + 4\ Zn + NO_3^- \longrightarrow 4\ Zn^{2+} + NH_4^+ + 3\ H_2O$$

(b) $NO_3^- + S \longrightarrow NO_2 + SO_4^{2-}$

Step 1. Write half-reaction equations; balance except H and O

$$S \longrightarrow SO_4^{2-}$$

$$NO_3^- \longrightarrow NO_2$$

Step 2. Balance H and O

$$4\ H_2O + S \longrightarrow SO_4^{2-} + 8\ H^+$$

$$2\ H^+ + NO_3^- \longrightarrow NO_2 + H_2O$$

Step 3. Balance electrically

$$4\ H_2O + S \longrightarrow SO_4^{2-} + 8\ H^+ + 6\ e^-$$

$$2\ H^+ + NO_3^- + e^- \longrightarrow NO_2 + H_2O$$

Step 4 and 5. Equalize the loss and gain of electrons; add half-reactions

$$4\ H_2O + S \longrightarrow SO_4^{2-} + 8\ H^+ + 6\ e^-$$

$$6\ [2\ H^+ + NO_3^- + e^- \longrightarrow NO_2 + H_2O]$$

$$4\ H^+ + S + 6\ NO_3^- \longrightarrow 6\ NO_2 + SO_4^{2-} + 2\ H_2O$$

(c) $PH_3 + I_2 \longrightarrow H_3PO_2 + I^-$

Step 1. Write half-reaction equations; balance except H and O

$$PH_3 \longrightarrow H_3PO_2$$

$I_2 \longrightarrow 2\ I^-$

Step 2. Balance H and O

$2\ H_2O + PH_3 \longrightarrow H_3PO_2 + 4\ H^+$

$I_2 \longrightarrow 2\ I^-$

Step 3. Balance electrically

$2\ H_2O + PH_3 \longrightarrow H_3PO_2 + 4\ H^+ + 4\ e^-$

$I_2 + 2\ e^- \longrightarrow 2\ I^-$

Step 4 and 5. Equalize the loss and gain of electrons; add half-reactions

$2\ H_2O + PH_3 \longrightarrow H_3PO_2 + 4\ H^+ + 4\ e^-$

$2\ [I_2 + 2\ e^- \longrightarrow 2\ I^-]$

$PH_3 + 2\ H_2O + 2\ I_2 \longrightarrow H_3PO_2 + 4\ I^- + 4\ H^+$

(d) $Cu + NO_3^- \longrightarrow Cu^{2+} + NO$

Step 1. Write half-reaction equations; balance except H and O

$Cu \longrightarrow Cu^{2+}$

$NO_3^- \longrightarrow NO$

Step 2. Balance H and O

$Cu \longrightarrow Cu^{2+}$

$4\ H^+ + NO_3^- \longrightarrow NO + 2\ H_2O$

Step 3. Balance electrically

$Cu \longrightarrow Cu^{2+} + 2\ e^-$

$4\ H^+ + NO_3^- + 3\ e^- \longrightarrow NO + 2\ H_2O$

Step 4 and 5. Equalize the loss and gain of electrons; add half-reactions

$3\ [Cu \longrightarrow Cu^{2+} + 2\ e^-]$

$2\ [4\ H^+ + NO_3^- + 3\ e^- \longrightarrow NO + 2\ H_2O\ ]$

$3\ Cu + 8\ H^+ + 2\ NO_3^- \longrightarrow 3\ Cu^{2+} + 2\ NO + 4\ H_2O$

(e) $ClO_3^- + I^- \longrightarrow I_2 + Cl^-$

Step 1. Write half-reaction equations; balance except H and O

$$2\ I^- \longrightarrow I_2$$

$$ClO_3^- \longrightarrow Cl^-$$

Step 2. Balance H and O

$$2\ I^- \longrightarrow I_2$$

$$6\ H^+ + ClO_3^- \longrightarrow Cl^- + 3\ H_2O$$

Step 3. Balance electrically

$$2\ I^- \longrightarrow I_2 + 2\ e^-$$

$$6\ H^+ + ClO_3^- + 6\ e^- \longrightarrow Cl^- + 3\ H_2O$$

Step 4 and 5. Equalize the loss and gain of electrons; add half-reactions.

$$3\ [2\ I^- \longrightarrow I_2 + 2\ e^-]$$

$$6\ H^+ + ClO_3^- + 6\ e^- \longrightarrow Cl^- + 3\ H_2O$$

$$6\ H^+ + ClO_3^- + 6\ I^- \longrightarrow 3\ I_2 + Cl^- + 3\ H_2O$$

(f) $Cr_2O_7^{2-} + Fe^{2+} \longrightarrow Cr^{3+} + Fe^{3+}$

Step 1. Write half-reaction equations; balance except H and O.

$$Fe^{2+} \longrightarrow Fe^{3+}$$

$$Cr_2O_7^{2-} \longrightarrow 2\ Cr^{3+}$$

Step 2. Balance H and O

$$Fe^{2+} \longrightarrow Fe^{3+}$$

$$14\ H^+ + Cr_2O_7^{2-} \longrightarrow 2\ Cr^{3+} + 7\ H_2O$$

Step 3. Balance electrically

$$Fe^{2+} \longrightarrow Fe^{3+} + e^-$$

$$14\ H^+ + Cr_2O_7^{2-} + 6\ e^- \longrightarrow 2\ Cr^{3+} + 7\ H_2O$$

Step 4 and 5. Equalize the loss and gain of electrons; add half-reactions.

$$6\ [Fe^{2+} \longrightarrow Fe^{3+} + e^-]$$

$$14\ H^+ + Cr_2O_7^{2-} + 6\ e^- \longrightarrow 2\ Cr^{3+} + 7\ H_2O$$

$$14\ H^+ + Cr_2O_7^{2-} + 6\ Fe^{2+} \longrightarrow 2\ Cr^{3+} + 6\ Fe^{3+} + 7\ H_2O$$

(g) $MnO_4^- + SO_2 \longrightarrow Mn^{2+} + SO_4^{2-}$

Step 1. Write half-reaction equations; balance except H and O.

$SO_2 \longrightarrow SO_4^{2-}$

$MnO_4^- \longrightarrow Mn^{2+}$

Step 2. Balance H and O

$2\ H_2O + SO_2 \longrightarrow SO_4^{2-} + 4\ H^+$

$8\ H^+ + MnO_4^- \longrightarrow Mn^{2+} + 4\ H_2O$

Step 3. Balance electrically

$2\ H_2O + SO_2 \longrightarrow SO_4^{2-} + 4\ H^+ + 2\ e^-$

$8\ H^+ + MnO_4^- + 5\ e^- \longrightarrow Mn^{2+} + 4\ H_2O$

Steps 4 and 5. Equalize the loss and gain of electrons; add half-reactions.

$5\ [2\ H_2O + SO_2 \longrightarrow SO_4^{2-} + 4\ H^+ + 2\ e^-]$

$2\ [8\ H^+ + MnO_4^- + 5\ e^- \longrightarrow Mn^{2+} + 4\ H_2O]$

$2\ H_2O + 2\ MnO_4^- + 5\ SO_2 \longrightarrow 4\ H^+ + 2\ Mn^{2+} + 5\ SO_4^{2-}$

(h) $H_3AsO_3 + MnO_4^- \longrightarrow H_3AsO_4 + Mn^{2+}$

Step 1. Write half-reaction equations; balance except H and O.

$H_3AsO_3 \longrightarrow H_3AsO_4$

$MnO_4^- \longrightarrow Mn^{2+}$

Step 2. Balance H and O.

$H_2O + H_3AsO_3 \longrightarrow 2\ H^+ + H_3AsO_4$

$8\ H^+ + MnO_4^- \longrightarrow Mn^{2+} + 4\ H_2O$

Step 3. Balance electrically

$H_2O + H_3AsO_3 \longrightarrow 2\ H^+ + H_3AsO_4 + 2\ e^-$

$8\ H^+ + MnO_4^- + 5\ e^- \longrightarrow Mn^{2+} + 4\ H_2O$

Steps 4 and 5. Equalize the loss and gain of electrons; add half-reactions.

$5\ [H_2O + H_3AsO_3 \longrightarrow 2\ H^+ + H_3AsO_4 + 2\ e^-]$

$2\ [8\ H^+ + MnO_4^- + 5\ e^- \longrightarrow Mn^{2+} + 4\ H_2O]$

$6\ H^+ + 5\ H_3AsO_3 + 2\ MnO_4^- \longrightarrow 5\ H_3AsO_4 + 2\ Mn^{2+} + 3\ H_2O$

(i) $ClO_3^- + Cl^- \longrightarrow Cl_2$

Step 1. Write half-reaction equations; balance except H and O

$Cl^- \longrightarrow Cl^o$

$ClO_3^- \longrightarrow Cl^o$

Step 2. Balance H and O

$Cl^- \longrightarrow Cl^o$

$6\ H^+ + ClO_3^- \longrightarrow Cl^o + 3\ H_2O$

Step 3. Balance electrically

$Cl^- \longrightarrow Cl^o + e^-$

$6\ H^+ + ClO_3^- + 5\ e^- \longrightarrow Cl^o + 3\ H_2O$

Step 4 and 5. Equalize the loss and gain of electrons; add half-reactions.

$5\ [Cl^- \longrightarrow Cl^o + e^-]$

$6\ H^+ + ClO_3^- + 5\ e^- \longrightarrow Cl^o + 3\ H_2O$

$6\ H^+ + ClO_3^- + 5\ Cl^- \longrightarrow 3\ Cl_2 + 3\ H_2O$

(j) $Cr_2O_7^{2-} + H_3AsO_3 \longrightarrow Cr^{3+} + H_3AsO_4$

Step 1. Write half-reaction equations; balance except H and O

$H_3AsO_3 \longrightarrow H_3AsO_4$

$Cr_2O_7^{2-} \longrightarrow 2\ Cr^{3+}$

Step 2. Balance H and O

$H_2O + H_3AsO_3 \longrightarrow 2\ H^+ + H_3AsO_4$

$14\ H^+ + Cr_2O_7^{2-} \longrightarrow 2\ Cr^{3+} + 7\ H_2O$

Step 3. Balance electrically

$H_2O + H_3AsO_3 \longrightarrow 2\ H^+ + H_3AsO_4 + 2\ e^-$

$$14\ H^+ + Cr_2O_7^{2-} + 6\ e^- \longrightarrow 2\ Cr^{3+} + 7\ H_2O$$

Step 4 and 5. Equalize the loss and gain of electrons; add half-reactions.

$$3\ [H_2O + H_3AsO_3 \longrightarrow 2\ H^+ + H_3AsO_4 + 2\ e^-]$$

$$14\ H^+ + Cr_2O_7^{2-} + 6\ e^- \longrightarrow 2\ Cr^{3+} + 7\ H_2O$$

$$8\ H^+ + Cr_2O_7^{2-} + 3\ H_3AsO_3 \longrightarrow 2\ Cr^{3+} + 3\ H_3AsO_4 + 4\ H_2O$$

7. The following steps will be used in balancing ionic redox reactions in an alkaline solution.

Step 1. Write two half-reaction equations; balance all elements except H and O.

Step 2. Balance H and O using H_2O and H^+.

Step 3. Add as many OH^- ions to each side of the equations as there are H^+ ions.

Step 4. Combine H^+ and OH^- to form H_2O. Rewrite equations, cancelling equal numbers of H_2O that appear on opposite sides of the equation.

Step 5. Balance electrically with electrons.

Step 6. Multiply each half-reaction by the appropriate numbers to equalize the loss and gain of electrons.

Step 7. Add the two half-reactions together, cancelling electrons and any other identical substances on opposite sides of the equation.

(a) $Cl_2 + IO_3^- \longrightarrow Cl^- + IO_4^-$

Step 1. Write half-reaction equations; balance except H and O

$$IO_3^- \longrightarrow IO_4^-$$

$$Cl_2 \longrightarrow 2\ Cl^-$$

Step 2. Balance H and O

$$H_2O + IO_3^- \longrightarrow IO_4^- + 2\ H^+$$

$$Cl_2 \longrightarrow 2\ Cl^-$$

Step 3. Add OH^- ions

$$2\ OH^- + H_2O + IO_3^- \longrightarrow IO_4^- + 2\ H^+ + 2\ OH^-$$

$$Cl_2 \longrightarrow 2\ Cl^-$$

Step 4. Combine H^+ and OH^- to form H_2O

$2\ OH^- + H_2O + IO_3^- \longrightarrow IO_4^- + 2\ H_2O$

$Cl_2 \longrightarrow 2\ Cl^-$

$2\ OH^- + IO_3^- \longrightarrow IO_4^- + H_2O$

$Cl_2 \longrightarrow 2\ Cl^-$

Step 5. Balance electrically

$2\ OH^- + IO_3^- \longrightarrow IO_4^- + H_2O + 2\ e^-$

$Cl_2 + 2\ e^- \longrightarrow 2\ Cl^-$

Step 6. Electron loss and gain is balanced

Step 7. Add half-reactions

$2\ OH^- + IO_3^- + Cl_2 \longrightarrow IO_4^- + 2\ Cl^- + H_2O$

(b) $MnO_4^- + ClO_2^- \longrightarrow MnO_2 + ClO_4^-$

Step 1. Write half-reaction equations; balance except H and O

$ClO_2^- \longrightarrow ClO_4^-$

$MnO_4^- \longrightarrow MnO_2$

Step 2. Balance H and O

$2\ H_2O + ClO_2^- \longrightarrow ClO_2^- + 4\ H^+$

$MnO_4^- + 4\ H^+ \longrightarrow MnO_2 + 2\ H_2O$

Step 3. Add OH^- ions

$4\ OH^- + 2\ H_2O + ClO_2^- \longrightarrow ClO_4^- + 4\ H^+ + 4\ OH^-$

$4\ OH^- + MnO_4^- + 4\ H^+ \longrightarrow MnO_2 + 2\ H_2O + 4\ OH^-$

Step 4. Combine H^+ and OH^- to form H_2O

$4\ OH^- + 2\ H_2O + ClO_2^- \longrightarrow ClO_4^- + 4\ H_2O$

$4\ H_2O + MnO_4^- \longrightarrow MnO_2 + 2\ H_2O + 4\ OH^-$

$4\ OH^- + ClO_2^- \longrightarrow ClO_4^- + 2\ H_2O$

$2\ H_2O + MnO_4^- \longrightarrow MnO_2 + 4\ OH^-$

Step 5. Balance electrically

$4\ OH^- + ClO_2^- \longrightarrow ClO_4^- + 2\ H_2O + 4\ e^-$

$$2\ H_2O + MnO_4^- + 3\ e^- \longrightarrow MnO_2 + 4\ OH^-$$

Step 6 and 7. Equalize electron loss and gain; add half-reactions.

$$3\ [4\ OH^- + ClO_2^- \longrightarrow ClO_4^- + 2\ H_2O + 4\ e^-]$$

$$4\ [2\ H_2O + MnO_4^- + 3\ e^- \longrightarrow MnO_2 + 4\ OH^-]$$

$$2\ H_2O + 4\ MnO_4^- + 3\ ClO_2^- \longrightarrow 4\ MnO_2 + 3\ ClO_4^- + 4\ OH^-$$

(c) $Se \longrightarrow Se^{2-} + SeO_3^{2-}$

Step 1. Write half-reaction equations; balance except H and O

$$Se \longrightarrow SeO_3^{2-}$$

$$Se \longrightarrow Se^{2-}$$

Step 2. Balance H and O.

$$3\ H_2O + Se \longrightarrow SeO_3^{2-} + 6\ H^+$$

$$Se \longrightarrow Se^{2-}$$

Step 3. Add OH^- ions

$$6\ OH^- + 3\ H_2O + Se \longrightarrow SeO_3^{2-} + 6\ H^+ + 6\ OH^-$$

$$Se \longrightarrow Se^{2-}$$

Step 4. Combine H^+ and OH^- to form H_2O

$$6\ OH^- + 3\ H_2O + Se \longrightarrow SeO_3^{2-} + 6\ H_2O$$

$$6\ OH^- + Se \longrightarrow SeO_3^{2-} + 3\ H_2O$$

$$Se \longrightarrow Se^{2-}$$

Step 5. Balance electrically

$$6\ OH^- + Se \longrightarrow SeO_3^{2-} + 3\ H_2O + 4\ e^-$$

$$Se + 2\ e^- \longrightarrow Se^{2-}$$

Step 6 and 7. Equalize electron loss and gain; add half-reactions.

$$6\ OH^- + Se \longrightarrow SeO_3^{2-} + 3\ H_2O + 4\ e^-$$

$$2\ [Se + 2\ e^- \longrightarrow Se^{2-}]$$

$$6\ OH^- + 3\ Se \longrightarrow SeO_3^{2-} + 2\ Se^{2-} + 3\ H_2O$$

(d) $MnO_4^- + SO_3^{2-} \longrightarrow MnO_2 + SO_4^{2-}$

Step 1. Write half-reaction equations; balance except H and O.

$SO_3^{2-} \longrightarrow SO_4^{2-}$

$MnO_4^- \longrightarrow MnO_2$

Step 2. Balance H and O

$H_2O + SO_3^{2-} \longrightarrow SO_4^{2-} + 2\ H^+$

$MnO_4^- + 4\ H^+ \longrightarrow MnO_2 + 2\ H_2O$

Step 3. Add OH^- ions

$2\ OH^- + H_2O + SO_3^{2-} \longrightarrow SO_4^{2-} + 2\ H^+ + 2\ OH^-$

$4\ OH^- + MnO_4^- + 4\ H^+ \longrightarrow MnO_2 + 2\ H_2O + 4\ OH^-$

Step 4. Combine H^+ and OH^- to form H_2O

$2\ OH^- + H_2O + SO_3^{2-} \longrightarrow SO_4^{2-} + 2\ H_2O$

$MnO_4^- + 4\ H_2O \longrightarrow MnO_2 + 2\ H_2O + 4\ OH^-$

$2\ OH^- + SO_3^{2-} \longrightarrow SO_4^{2-} + H_2O$

$MnO_4^- + 2\ H_2O \longrightarrow MnO_2 + 4\ OH^-$

Step 5. Balance electrically

$2\ OH^- + SO_3^{2-} \longrightarrow SO_4^{2-} + H_2O + 2\ e^-$

$3\ e^- + MnO_4^- + 2\ H_2O \longrightarrow MnO_2 + 4\ OH^-$

Step 6 and 7. Equalize electron loss and gain; add half-reactions.

$3\ [2\ OH^- + SO_3^{2-} \longrightarrow SO_4^{2-} + H_2O + 2\ e^-]$

$2\ [MnO_4^- + 2\ H_2O + 3\ e^- \longrightarrow MnO_2 + 4\ OH^-]$

$H_2O + 2\ MnO_4^- + 3\ SO_3^{2-} \longrightarrow 2\ MnO_2 + 3\ SO_4^{2-} + 2\ OH^-$

(e) $ClO_2 + SbO_2^- \longrightarrow ClO_2^- + Sb(OH)_6^-$

Step 1. Write half-reaction equations; balance except H and O.

$SbO_2^- \longrightarrow Sb(OH)_6^-$

$ClO_2 \longrightarrow ClO_2^-$

Step 2. Balance H and O

$4\ H_2O + SbO_2^- \longrightarrow Sb(OH)_6^- + 2\ H^+$

$ClO_2 \longrightarrow ClO_2^-$

Step 3. Add OH^- ions

$2\ OH^- + 4\ H_2O + SbO_2^- \longrightarrow Sb(OH)_6^- + 2\ H^+ + 2\ OH^-$

$ClO_2 \longrightarrow ClO_2^-$

Step 4. Combine H^+ and OH^- to form H_2O

$2\ OH^- + 4\ H_2O + SbO_2^- \longrightarrow Sb(OH)_6^- + 2\ H_2O$

$2\ OH^- + 2\ H_2O + SbO_2^- \longrightarrow Sb(OH)_6^-$

$ClO_2 \longrightarrow ClO_2^-$

Step 5. Balance electrically

$2\ OH^- + 2\ H_2O + SbO_2^- \longrightarrow Sb(OH)_6^- + 2\ e^-$

$ClO_2 + e^- \longrightarrow ClO_2^-$

Steps 6 and 7. Equalize electron loss and gain; add half-reactions.

$2\ H_2O + 2\ OH^- + SbO_2^- \longrightarrow Sb(OH)_6^- + 2\ e^-$

$2\ [ClO_2 + e^- \longrightarrow ClO_2^-]$

$2\ H_2O + 2\ ClO_2 + 2\ OH^- + SbO_2^- \longrightarrow 2\ ClO_2^- + Sb(OH)_6^-$

(f) $Fe_3O_4 + MnO_4^- \longrightarrow Fe_2O_3 + MnO_2$

Step 1. Write half-reaction equations; balance except H and O

$2\ Fe_3O_4 \longrightarrow 3\ Fe_2O_3$

$MnO_4^- \longrightarrow MnO_2$

Step 2. Balance H and O

$H_2O + 2\ Fe_3O_4 \longrightarrow 3\ Fe_2O_3 + 2\ H^+$

$4\ H^+ + MnO_4^- \longrightarrow MnO_2 + 2\ H_2O$

Step 3. Add OH^- ions

$2\ OH^- + H_2O + 2\ Fe_3O_4 \longrightarrow 3\ Fe_2O_3 + 2\ H^+ + 2\ OH^-$

$4\ OH^- + 4\ H^+ + MnO_4^- \longrightarrow MnO_2 + 2\ H_2O + 4\ OH^-$

Step 4. Combine H^+ and OH^- to form H_2O

$2\ OH^- + H_2O + 2\ Fe_3O_4 \longrightarrow 3\ Fe_2O_3 + 2\ H_2O$

$4\ H_2O + MnO_4^- \longrightarrow MnO_2 + 2\ H_2O + 4\ OH^-$

$2\ OH^- + 2\ Fe_3O_4 \longrightarrow 3\ Fe_2O_3 + H_2O$

$2\ H_2O + MnO_4^- \longrightarrow MnO_2 + 4\ OH^-$

Step 5. Balance electrically

$2\ OH^- + 2\ Fe_3O_4 \longrightarrow 3\ Fe_2O_3 + H_2O + 2\ e^-$

$2\ H_2O + MnO_4^- + 3\ e^- \longrightarrow MnO_2 + 4\ OH^-$

Steps 6 and 7. Equalize loss and gain of electrons; add half-reactions.

$3\ [2\ OH^- + 2\ Fe_3O_4 \longrightarrow 3\ Fe_2O_3 + H_2O + 2\ e^-]$

$2\ [2\ H_2O + MnO_4^- + 3\ e^- \longrightarrow MnO_2 + 4\ OH^-]$

$H_2O + 6\ Fe_3O_4 + 2\ MnO_4^- \longrightarrow 9\ Fe_2O_3 + 2\ MnO_2 + 2\ OH^-$

(g) $BrO^- + Cr(OH)_4^- \longrightarrow Br^- + CrO_4^{2-}$

Step 1. Write half-reaction equations; balance except H and O

$Cr(OH)_4^- \longrightarrow CrO_4^{2-}$

$BrO^- \longrightarrow Br^-$

Step 2. Balance H and O

$Cr(OH)_4^- \longrightarrow CrO_4^{2-} + 4\ H^+$

$2\ H^+ + BrO^- \longrightarrow Br^- + H_2O$

Step 3. Add OH^- ions

$4\ OH^- + Cr(OH)_4^- \longrightarrow CrO_4^{2-} + 4\ H^+ + 4\ OH^-$

$2\ OH^- + 2\ H^+ + BrO^- \longrightarrow Br^- + H_2O + 2\ OH^-$

Step 4. Combine H^+ and OH^- to form H_2O

$4\ OH^- + Cr(OH)_4^- \longrightarrow CrO_4^{2-} + 4\ H_2O$

$2\ H_2O + BrO^- \longrightarrow Br^- + H_2O + 2\ OH^-$

$H_2O + BrO^- \longrightarrow Br^- + 2\ OH^-$

Step 5. Balance electrically

$4\ OH^- + Cr(OH)_4^- \longrightarrow CrO_4^{2-} + 4\ H_2O + 3\ e^-$

$H_2O + BrO^- + 2\ e^- \longrightarrow Br^- + 2\ OH^-$

Steps 6 and 7. Equalize loss and gain of electrons; add half-reactions.

$2\ [4\ OH^- + Cr(OH)_4^- \longrightarrow CrO_4^{2-} + 4\ H_2O + 3\ e^-]$

$3\ [H_2O + BrO^- + 2\ e^- \longrightarrow Br^- + 2\ OH^-]$

$2\ OH^- + 3\ BrO^- + 2\ Cr(OH)_4^- \longrightarrow 3\ Br^- + 2\ CrO_4^{2-} + 5\ H_2O$

(h) $P_4 \longrightarrow HPO_3^{2-} + PH_3$

Step 1. Write half-reaction equations; balance except H and O.

$P_4 \longrightarrow 4\ HPO_3^{2-}$

$P_4 \longrightarrow 4\ PH_3$

Step 2. Balance H and O

$12\ H_2O + P_4 \longrightarrow 4\ HPO_3^{2-} + 20\ H^+$

$12\ H^+ + P_4 \longrightarrow 4\ PH_3$

Step 3. Add OH^- ions

$20\ OH^- + 12\ H_2O + P_4 \longrightarrow 4\ HPO_3^{2-} + 20\ H^+ + 20\ OH^-$

$12\ OH^- + 12\ H^+ + P_4 \longrightarrow 4\ PH_3 + 12\ OH^-$

Step 4. Combine H^+ and OH^- to form H_2O

$20\ OH^- + 12\ H_2O + P_4 \longrightarrow 4\ HPO_3^{2-} + 20\ H_2O$

$12\ H_2O + P_4 \longrightarrow 4\ PH_3 + 12\ OH^-$

$20\ OH^- + P_4 \longrightarrow 4\ HPO_3^{2-} + 8\ H_2O$

Step 5. Balance electrically

$20\ OH^- + P_4 \longrightarrow 4\ HPO_3^{2-} + 8\ H_2O + 12\ e^-$

$12\ H_2O + P_4 + 12\ e^- \longrightarrow 4\ PH_3 + 12\ OH^-$

Steps 6 and 7. Equalize loss and gain of electrons; add half-reactions. Loss and gain of electrons are equal.

$8\ OH^- + 4\ H_2O + 2\ P_4 \longrightarrow 4\ HPO_3^{2-} + 4\ PH_3$ (divide by 2)

$$4\ OH^- + 2\ H_2O + P_4 \longrightarrow 2\ HPO_3^{2-} + 2\ PH_3$$

(i) $Al + OH^- \longrightarrow Al(OH)_4^- + H_2$

Step 1. Write half-reaction equations; balance except H and O

$$Al \longrightarrow Al(OH)_4^-$$

$$OH^- \longrightarrow H_2$$

Step 2. Balance H and O

$$4\ H_2O + Al \longrightarrow Al(OH)_4^- + 4\ H^+$$

$$3\ H^+ + OH^- \longrightarrow H_2 + H_2O$$

Step 3. Add OH^- ions

$$4\ OH^- + 4\ H_2O + Al \longrightarrow Al(OH)_4^- + 4\ H^+ + 4\ OH^-$$

$$3\ OH^- + 3\ H^+ + OH^- \longrightarrow H_2 + H_2O + 3\ OH^-$$

Step 4. Combine H^+ and OH^- to form H_2O

$$4\ OH^- + 4\ H_2O + Al \longrightarrow Al(OH)_4^- + 4\ H_2O$$

$$3\ H_2O + OH^- \longrightarrow H_2 + H_2O + 3\ OH^-$$

$$4\ OH^- + Al \longrightarrow Al(OH)_4^-$$

$$2\ H_2O + OH^- \longrightarrow H_2 + 3\ OH^-$$

Step 5. Balance electrically

$$4\ OH^- + Al \longrightarrow Al(OH)_4^- + 3\ e^-$$

$$2\ H_2O + OH^-\ 2\ e^- \longrightarrow H_2 + 3\ OH^-$$

Steps 6 and 7. Equalize loss and gain of electrons; add half-reactions.

$$2\ [4\ OH^- + Al \longrightarrow Al(OH)_4^- + 3\ e^-]$$

$$3\ [2\ H_2O + OH^- + 2\ e^- \longrightarrow H_2 + 3\ OH^-]$$

$$2\ Al + 6\ H_2O + 2\ OH^- \longrightarrow 2\ Al(OH)_4^- + 3\ H_2$$

(j) $Al + NO_3^- \longrightarrow NH_3 + Al(OH)_4^-$

Step 1. Write half-reaction equations; balance except H and O

$$Al \longrightarrow Al(OH)_4^-$$

$$NO_3^- \longrightarrow NH_3$$

Step 2. Balance H and O

$4\ H_2O + Al \longrightarrow Al(OH)_4^- + 4\ H^+$

$9\ H^+ + NO_3^- \longrightarrow NH_3 + 3\ H_2O$

Step 3. Add OH^- ions

$4\ OH^- + 4\ H_2O + Al \longrightarrow Al(OH)_4^- + 4\ H^+ + 4\ OH^-$

$9\ OH^- + 9\ H^+ + NO_3^- \longrightarrow NH_3 + 3\ H_2O + 9\ OH^-$

Step 4. Combine H^+ and OH^- to form H_2O

$4\ OH^- + 4\ H_2O + Al \longrightarrow Al(OH)_4^- + 4\ H_2O$

$9\ H_2O + NO_3^- \longrightarrow NH_3 + 3\ H_2O + 9\ OH^-$

$4\ OH^- + Al \longrightarrow Al(OH)_4^-$

$6\ H_2O + NO_3^- \longrightarrow NH_3 + 9\ OH^-$

Step 5. Balance electrically

$4\ OH^- + Al \longrightarrow Al(OH)_4^- + 3\ e^-$

$6\ H_2O + NO_3^- + 8\ e^- \longrightarrow NH_3 + 9\ OH^-$

Steps 6 and 7. Equalize loss and gain of electrons; add half-reactions.

$8\ [4\ OH^- + Al \longrightarrow Al(OH)_4^- + 3\ e^-]$

$3\ [6\ H_2O + NO_3^- + 8\ e^- \longrightarrow NH_3 + 9\ OH^-]$

$8\ Al + 3\ NO_3^- + 18\ H_2O + 5\ OH^- \longrightarrow 3\ NH_3 + 8\ Al(OH)_4^-$

8. Oxidation and reduction are complementary processes because one does not occur without the other.

9. $Ca^{2+} + 2\ e^- \longrightarrow Ca$ cathode reaction, reduction

   $2\ Br^- \longrightarrow Br + 2\ e^-$ anode reaction, oxidation

10. In the electroplating of metals, the metal is plated by reducing the positive ions of the metal in the solution, so the electrode where the plating will occur is the cathode, a source of electrons. With an alternating current, the polarity of the electrode would be constantly changing so that one instant the metal would be plating, and the next instant the metal would be dissolving (being oxidized).

11. (a) $Pb + SO_4^{2-} \longrightarrow PbSO_4 + 2\ e^-$

    $PbO_2 + SO_4^{2-} + 4\ H^+ + 2\ e^- \longrightarrow PbSO_4 + 2\ H_2O$

(b) The first reaction is oxidation (Pb is oxidized to Pb^{2+}).

The second reaction is reduction (Pb^{4+} is reduced to Pb^{2+}).

(c) The first reaction (oxidation) occurs at the anode of the battery.

12. The fact that lead dioxide and lead(II) sulfate are insoluble makes it unnecessary to have salt bridges in the cells of a lead storage battery.

13. The electrolyte in a lead storage battery is dilute sulfuric acid. In the discharge cycle, SO_4^{2-}, is removed from solution as it reacts with PbO_2 and H^+ to form $PbSO_4$ and H_2O. Therefore, the electrolyte solution contains less H_2SO_4 and becomes less dense.

14. If Hg^{2+} ions are reduced to metallic mercury, this would occur at the cathode, because reduction takes place at the cathode.

15. In both electrolysis and voltaic cells oxidation and reduction reactions occur. In an electrolysis cell an electric current is forced through the cell causing chemical change to occur. In voltaic cells spontaneous chemical change occurs generating an electric current.

16. In some voltaic cells the reactants at the electrodes are in solution. For the cell to function these reactants must be kept separated. A salt bridge permits movement of ions in the cell. This keeps the solution neutral with respect to charged particles (ions) in the solution.

17. The following statements are correct: a,c,e,g,j,k,m,p,q,r, and s.

D. Review problems.

1. $3\ Ag + 4\ HNO_3 \longrightarrow 3\ AgNO_3 + NO + 2\ H_2O$

$g\ Ag \longrightarrow mol\ Ag \longrightarrow mol\ NO$

$$25.0\ g\ Ag \times \frac{1\ mol\ Ag}{107.9\ g\ Ag} \times \frac{1\ mol\ NO}{3\ mol\ Ag} = 0.0772\ mol\ NO$$

2. $3\ Cl_2 + 6\ KOH \longrightarrow KClO_3 + 5\ KCl + 3\ H_2O$

$mol\ KClO_3 \longrightarrow mol\ Cl_2 \longrightarrow L\ Cl_2$

$$0.300\ mol\ KClO_3 \times \frac{3\ mol\ Cl_2}{1\ mol\ KClO_3} \times \frac{22.4\ L\ Cl_2}{mol\ Cl_2} = 20.2\ L\ Cl_2$$

3. $5\ H_2O_2 + 2\ KMnO_4 + 3\ H_2SO_4 \longrightarrow 5\ O_2 + 2\ MnSO_4 + K_2SO_4 + 8\ H_2O$

$mL\ H_2O_2 \longrightarrow g\ H_2O_2 \longrightarrow mol\ H_2O_2 \longrightarrow mol\ KMnO_4 \longrightarrow g\ KMnO_4$

$$100 \text{ mL } H_2O_2 \text{ sol'n} \times \frac{1.031 \text{ g } H_2O_2 \text{ sol'n}}{\text{mL sol'n}} \times \frac{9.0 \text{ g } H_2O_2}{100 \text{ g sol'n}} \times \frac{1 \text{ mol } H_2O_2}{34.0 \text{ g } H_2O_2}$$

$$\times \frac{2 \text{ mol } KMnO_4}{5 \text{ mol } H_2O_2} \times \frac{158.0 \text{ g } KMnO_4}{\text{mol } KMnO_4} = 17 \text{ g } KMnO_4$$

4. $Cr_2O_7^{2-} + 3\ H_3AsO_3 + 8\ H^+ \longrightarrow 2\ Cr^{3+} + 3\ H_3AsO_4 + 4\ H_2O$

$\text{g } H_3AsO_3 \longrightarrow \text{mol } H_3AsO_3 \longrightarrow \text{mol } Cr_2O_7^{2-} \longrightarrow \text{mL } Cr_2O_7^{2-}$

$$5.00 \text{ g } H_3AsO_3 \times \frac{1 \text{ mol } H_3AsO_3}{125.9 \text{ g } H_3AsO_3} \times \frac{1 \text{ mol } Cr_2O_7^{2-}}{3 \text{ mol } H_3AsO_3} \times \frac{1000 \text{ mL } Cr_2O_7^{2-}}{0.200 \text{ mol } Cr_2O_7^{2-}}$$

$= 66.2 \text{ mL } 0.200 \text{ M } K_2Cr_2O_7$ solution

5. $Cr_2O_7^{2-} + 6\ Fe^{2+} + 14\ H^+ \longrightarrow 2\ Cr^{3+} + 6\ Fe^{3+} + 7\ H_2O$

$\text{mL } FeSO_4 \longrightarrow \text{mol } FeSO_4 \longrightarrow \text{mol } Cr_2O_7^{2-} \longrightarrow \text{mL } Cr_2O_7^{2-}$

$$60.0 \text{ mL } FeSO_4 \times \frac{0.200 \text{ mol } FeSO_4}{1000 \text{ mL } FeSO_4} \times \frac{1 \text{ mol } Cr_2O_7^{2-}}{6 \text{ mol } FeSO_4} \times \frac{1000 \text{ mL } Cr_2O_7^{2-}}{0.200 \text{ mol } Cr_2O_7^{2-}}$$

$= 10.0 \text{ mL } 0.200 \text{ M } K_2Cr_2O_7$ solution

6. $8\ KI + 5\ H_2SO_4 \longrightarrow 4\ I_2 + H_2S + 4\ K_2SO_4 + 4\ H_2O$

$\text{g } I_2 \longrightarrow \text{mol } I_2 \longrightarrow \text{mol KI} \longrightarrow \text{g KI}$

$$2.79 \text{ g I} \times \frac{1 \text{ mol } I_2}{253.8 \text{ g } I_2} \times \frac{8 \text{ mol KI}}{4 \text{ mol } I_2} \times \frac{166.0 \text{ g KI}}{1 \text{ mol KI}} = 3.65 \text{ g KI in sample.}$$

$$\frac{3.65 \text{ g KI}}{4.00 \text{ g sample}} \times 100\% = 91.3\% \text{ KI}$$

7. $3\ CuO + 2\ NH_3 \longrightarrow N_2 + 3\ Cu + 3\ H_2O$

$\text{L } NH_3 \longrightarrow \text{mol } NH_3 \longrightarrow \text{mol Cu} \longrightarrow \text{g Cu}$

$$35.0 \text{ L } NH_3 \times \frac{1 \text{ mol } NH_3}{22.4 \text{ L } NH_3} \times \frac{3 \text{ mol Cu}}{2 \text{ mol } NH_3} \times \frac{63.5 \text{ g Cu}}{1 \text{ mol Cu}} = 149 \text{ g Cu}$$

8. $3\ Ag + 4\ HNO_3 \longrightarrow 3\ AgNO_3 + NO + 2\ H_2O$

mol Ag $\longrightarrow$ mol NO

$$0.500 \text{ mol Ag} \times \frac{1 \text{ mol NO}}{3 \text{ mol Ag}} = 0.167 \text{ mol NO}$$

$PV = nRT \qquad V = \frac{nRT}{P}$

$P = \frac{744 \text{ torr} \times 1 \text{ atm}}{760 \text{ torr}} = 0.979 \text{ atm}; \quad T = 301 \text{ K}$

$V = \frac{0.167 \text{ mol NO} \times 301 \text{ K} \times 0.081 \frac{\text{L-atm}}{\text{mol K}}}{0.979 \text{ atm}}$

$V = 4.22 \text{ L NO}$

9. $2 \text{ Al} + 2 \text{ OH}^- + 6 \text{ H}_2\text{O} \longrightarrow 2 \text{ Al(OH)}_4^- + 3 \text{ H}_2$

$\text{g Al} \longrightarrow \text{mol Al} \longrightarrow \text{mol H}_2$

$100 \text{ g Al} \times \frac{1 \text{ mol Al}}{27.0 \text{ g}} \times \frac{3 \text{ mol H}_2}{2 \text{ mol Al}} = 5.56 \text{ mol H}_2$

B. 1. (a) Gamma radiation requires the most shielding
(b) Alpha radiation requires the least shielding.

2. Alpha particles are deflected less than beta particles in passing through a magnetic field because they are much heavier (more than 7,000 times heavier) than beta particles.

3. The first five isotopes of Table 18.3 in order of increasing half-life are:

$^{131}_{53}I$, 8.1 days; $^{32}_{15}P$, 14.3 days; $^{35}_{16}S$, 87 days; $^{60}_{25}Co$, 5.3 years;

$^{90}_{38}Sr$, 28 years.

4. The half-life of strontium-90 is 28 years.

5. Pairs of nuclides that would be found in the fission reaction of U-235: Any two whose atomic numbers add up to 92 and whose mass numbers (in the range of 70-160) add up to 230-234. Examples:

$^{90}_{38}Sr$ and $^{141}_{56}Xe$; $^{139}_{56}Ba$ and $^{94}_{36}Kr$; $^{101}_{42}Mo$ and $^{131}_{50}Sn$

6. The transuranium elements:

(a) Named for people: Curium (Marie Curie), Einsteinium (Albert Einstein), Fermium (Enrico Fermi), Mendelevium (Dmitri Mendeleev), Nobelium (Alfred Nobel), Lawrencium (Ernest O. Lawrence).

(b) Named for geographic places: Americum (America), Berkelium (Berkeley), Californium (California).

(c) Named for planets: Neptunium (Neptune), Plutonium (Pluto).

C. Review Questions.

1. Contributions to the early history of radioactivity:

(a) Henri Becquerel: He discovered radioactivity

(b) Marie and Pierre Curie: They discovered the elements polonium and radium.

(c) Wilhelm Roentgen: He discovered X rays and developed the technique of producing them. While this was not a radioactive phenomenon, it triggered Becquerel's discovery of radioactivity.

(d) Ernest Rutherford: He discovered alpha and beta rays. He worked out the link between radioactivity and transmutation. He produced the first man-made transmutation.

(e) Paul Villard: He discovered gamma rays.

2. Chemical reactions are caused by atoms or ions coming together, so are greatly influenced by temperature and concentration which affect

the number of collisions. Radioactivity is a spontaneous reaction of an individual nucleus, and is not affected by these things.

3. We use the term isotope when we speak about atoms of the same element that have different masses. For example, carbon-12 and carbon-14, $^{12}_{6}C$ and $^{14}_{6}C$. The term nuclide is used in nuclear chemistry to mean any isotope of any atom.

4.

	Protons	Neutrons	Nucleons
(a) $^{35}_{17}Cl$	17	18	35
(b) $^{226}_{88}Ra$	88	138	226
(c) $^{235}_{92}U$	92	143	235
(d) $^{82}_{35}Br$	35	47	82

5. "Half-life" is the length of time required for one-half of a specific amount of a radionuclide to disintegrate. It is a convenient way of comparing stabilities of radioisotopes.

6. $5 \times 10^9 \text{ years} \times \frac{1 \text{ half-life}}{7.6 \times 10^7 \text{ years}} = 66 \text{ half-lives}$

Even if plutonium-224 had been present in large quantities five billion years ago, no measurable amount would survive after 66 half-lives.

7. Alpha particle: charge, +2; relative mass, 4 amu; nature of particle, a helium nucleus; penetrating power, very low, can be stopped by paper.

Beta particle: charge, -1; relative mass, 1/1837 amu; nature of particle, an electron; penetrating power, moderate, requires a thin sheet of aluminum to block it.

Gamma ray: charge, 0; relative mass, 0; nature of ray, high energy electromagnetic radiation; penetrating power, highest, requires 5 cm of lead or more to stop it.

8. (a) When a nucleus loses an alpha particle, its atomic number decreases by two, and its mass number decreases by four.

(b) When a nucleus loses a beta particle, its atomic number increases by one, and its mass number remains unchanged.

9. Natural radioactivity is the spontaneous disintegration of those radioactive isotopes found in nature. Artificial radioactivity is the spontaneous disintegration of radioactive isotopes produced by man (in a nuclear reactor or an accelerator).

10. A disintegration series starting with a particular radionuclide, progressing stepwise by alpha and beta emissions to other radionuclides, and ending at a stable nuclide.

For example: ${}^{238}_{92}U \longrightarrow {}^{206}_{82}Pb$ (stable).

11. Transmutation is the conversion of one element into another by natural or artificial means. The nucleus of the atom is bombarded by various particles (alpha, beta, protons, etc.). The fast moving particles are captured by the nucleus to generally form an unstable nucleus which subsequently decays to another kind of atom. For example, ${}^{9}_{4}Be + {}^{4}_{2}He \longrightarrow {}^{12}_{6}C + {}^{1}_{0}n$

12. Equations for alpha decays:

(a) ${}^{218}_{85}At \longrightarrow {}^{4}_{2}He + {}^{214}_{83}Bi$ (b) ${}^{221}_{87}Fr \longrightarrow {}^{4}_{2}He + {}^{217}_{85}At$

(c) ${}^{192}_{78}Pt \longrightarrow {}^{4}_{2}He + {}^{188}_{76}Os$ (d) ${}^{210}_{84}Po \longrightarrow {}^{4}_{2}He + {}^{206}_{82}Pb$

13. Equations for beta decays:

(a) ${}^{14}_{6}C \longrightarrow {}^{0}_{-1}e + {}^{14}_{7}N$ (b) ${}^{137}_{55}Cs \longrightarrow {}^{0}_{-1}e + {}^{137}_{56}Ba$

(c) ${}^{239}_{93}Np \longrightarrow {}^{0}_{-1}e + {}^{239}_{94}Pu$ (d) ${}^{90}_{38}Sr \longrightarrow {}^{0}_{-1}e + {}^{90}_{39}Y$

14. ${}^{232}_{90}Th \xrightarrow{-\alpha} {}^{228}_{88}Ra \xrightarrow{-\beta} {}^{228}_{89}Ac \xrightarrow{-\beta} {}^{228}_{90}Th \xrightarrow{-\alpha} {}^{224}_{88}Ra \xrightarrow{-\alpha}$

${}^{220}_{86}Rn \xrightarrow{-\alpha} {}^{216}_{84}Po \xrightarrow{-\alpha} {}^{212}_{82}Pb \xrightarrow{-\beta} {}^{212}_{83}Bi \xrightarrow{-\beta} {}^{212}_{84}Po \xrightarrow{-\alpha} {}^{208}_{82}Pb$

15. ${}^{237}_{93}Np$ loses seven alpha particles, four beta particles. For final product, nuclear charge = 93 - 7(2) + 4(1) = 83, and mass = 237 - 7(4) = 209. Therefore, the product is ${}^{209}_{83}Bi$.

16. Decay of bismuth-211

${}^{211}_{83}Bi \longrightarrow {}^{4}_{2}He + {}^{207}_{81}Tl$ ${}^{207}_{81}Tl \longrightarrow {}^{0}_{-1}e + {}^{207}_{82}Pb$

17. (a) ${}^{13}_{6}C + {}^{1}_{0}n \longrightarrow {}^{14}_{6}C$ (b) ${}^{30}_{15}P \longrightarrow {}^{30}_{14}Si + {}^{0}_{-1}e$

There are possibly other ways to make each of these reactions occur.

18. (a) ${}^{27}_{13}Al + {}^{4}_{2}He \longrightarrow {}^{30}_{15}P + {}^{1}_{0}n$

(b) ${}^{27}_{14}Si \longrightarrow {}^{0}_{-1}e + {}^{27}_{13}Al$

(c) $^{12}_{6}C + ^{2}_{1}H \longrightarrow ^{13}_{7}N + ^{1}_{0}n$

(d) $^{82}_{35}Br \longrightarrow ^{82}_{36}Kr + ^{0}_{-1}e$

(e) $^{66}_{29}Cu \longrightarrow ^{66}_{30}Zn + ^{0}_{-1}e$

(f) $^{0}_{-1}e + ^{7}_{4}Be \longrightarrow ^{7}_{3}Li$

(g) $^{27}_{13}Al + ^{4}_{2}He \longrightarrow ^{30}_{14}Si + ^{1}_{1}H$

(h) $^{85}_{37}Rb + ^{1}_{0}n \longrightarrow ^{82}_{35}Br + ^{4}_{2}He$

(i) $^{214}_{83}Bi \longrightarrow ^{4}_{2}He + ^{210}_{81}Tl$

19. Shorthand notations for nuclear equations

(a) $^{27}_{13}Al(\alpha,n)^{30}_{15}P$

(b) $^{14}_{7}N(\alpha,p)^{17}_{8}O$

(c) $^{59}_{27}Co(n,\gamma)^{60}_{27}Co$

(d) $^{249}_{98}Cf(^{12}_{6}C,4n)^{257}_{104}Unq$

20. Nuclear equations from shorthand notations.

(a) $^{23}_{11}Na + ^{1}_{1}H \longrightarrow ^{23}_{12}Mg + ^{1}_{0}n$

(b) $^{209}_{83}Bi + ^{2}_{1}H \longrightarrow ^{210}_{84}Po + ^{1}_{0}n$

(c) $^{238}_{92}U + ^{16}_{8}O \longrightarrow ^{249}_{100}Fm + 5\ ^{1}_{0}n$

(d) $^{9}_{4}Be + ^{4}_{2}He \longrightarrow ^{12}_{6}C + ^{1}_{0}n$

21. The Geiger counter is used to detect the presence of radioactive material. The rays which are emitted enter the Geiger tube and ionize the argon gas present in the tube. The ions cause an electrical discharge in the tube producing a current pulse, which is amplified and appears as a signal in the form of audible clicks, flashing light, and can be measured with a calibrated meter.

22. Units of radioactivity

(a) Curie (Ci). A unit of radioactivity indicating the rate of decay. 1 Ci = 2.7×10^{10} disintegrations per second.

(b) Roentgen. A unit of exposure to gamma or X ray radiation.

(c) Rad or Gray. A unit of absorbed dose radiation.

1 rad = 0.01 J of energy absorbed per 1 kg of matter.

Grays (Gy) represent energy absorbed by tissue. 1 Gy = 100 rad

(d) Rem. A unit of radiation dose equivalent. The rem takes into account that the same dose from different sources of radiation does not produce the same degree of biological affect.

23. The Germans, Otto Hahn and Fritz Strassmann, were the first scientists to report nuclear fission. The fission resulted from bombarding uranium nuclei with neutrons.

24. Natural U is 99+% U-238. Commercial nuclear reactors use U-235 enriched uranium as a fuel. Slow neutrons will cause the fission of U-235 but not U-238. Fast neutrons are capable of nuclear reaction with U-238 to produce fissionable Pu-239. A breeder reactor converts nonfissionable U-238 to fissionable Pu-239 and in the process manufactures more fuel than it consumes.

25. The fission reactions in a nuclear reactor and in an atomic bomb are essentially the same, the difference being that the fissioning is "wild" or uncontrolled in the bomb. In a nuclear reactor the fissioning rate is controlled by means of moderators such as graphite to slow the neutrons and control rods of cadmium or boron to absorb some of them.

26. A certain amount of fissionable material (a critical mass) must be present before a self-supporting chain reaction can occur. Without a critical mass, too many neutrons from fissions will escape and the reaction cannot reach a chain reaction status unless at least one neutron is captured for every fission that occurs.

27. The terms "atomic bomb" and "hydrogen bomb" are not synonymous. The energy released by the explosion of an atomic bomb results from the fission of heavy atoms. The major part of the energy released by the explosion of a hydrogen bomb results from the fusion of light atoms.

28. The mass defect is the difference between the mass of an atom and the mass of the number of protons, neutrons and electrons in that atom. The energy equivalent to this difference in mass is known as the nuclear binding energy.

29. When radioactive rays pass through normal matter, they cause that matter to become ionized (usually by knocking out electrons). Therefore, the radioactive rays are classified as ionizing radiation.

30. Some biological hazards associated with radioactivity are:

    (a) High levels of radiation can cause nausea, vomiting, diarrhea, and death. The radiation produces ionization in the cells, particularly in the nucleus of the cells.

    (b) Long-term exposure to low levels of radiation can weaken the body and cause malignant tumors.

    (c) Radiation can damage DNA molecules in the body causing mutations, by reproduction, to be passed on.

31. Strontium-90 has two characteristics which create much concern. Its half-life is 28 years so it remains active for a long period of time. The other characteristic is that chemically it is similar to calcium, so the human body uses it in bones. In the bone marrow red

corpuscles are produced. If this area is subject to beta radiation from strontium-90, red corpuscles are destroyed, and the incidence of leukemia and bone cancer is increased.

32. A radioactive "tracer" is radioactive material whose presence is traced by a Geiger counter or other detecting device. Tracers are often injected into the human body, animals, plants, or industrial processes to determine the path and speed of circulation. For example, use of a tracer could determine the length of time for material to travel from the root system to the leaves in a tree.

33. In living species, the ratio of carbon-14 to carbon-12 is constant due to the constant C-14/C-12 ratio in the atmosphere and food sources. When the species dies and life processes stop the C-14/C-12 ratio decreases with time because C-14 is radioactive and decays according to its half-life while the amount of C-12 in the species remains constant. Thus the age of an archaeological artifact containing carbon can be calculated by comparing the C-14/C-12 ratio in the artifact with the C-14/C-12 ratio in the living species.

34. Radioactivity could be used to locate a leak in an underground pipe by using a water soluble tracer element. Dissolve the tracer in water and pass the water through the pipe. Test the ground along the path of the pipe with a Geiger counter until radioactivity from the leak is detected. Dig.

35. Half-life carbon-14 is 5668 years

$$4 \times 10^6 \text{ years} \times \frac{1 \text{ half-life}}{5668 \text{ years}} = 7 \times 10^2 \text{ half-lives}$$

700 half-lives would pass in 4 million years. Too little C-14 would remain to detect with any reliability. Such dating would not prove useful.

36. The following statements are correct: c,e,f,g,j,k,n,o,q,r, and t.

D. Review problems.

1. $$112 \text{ years} \times \frac{\text{one half-life}}{28 \text{ years}} = 4 \text{ half-lives}$$

   In 4 half-lives (1/2 × 1/2 × 1/2 × 1/2) 1/16th of the starting amount would remain.

   $$\frac{1.00 \text{ mg Sr-90}}{16} = 0.0625 \text{ mg Sr-90 remains after 112 years}$$

2. $$0.0100 \text{ g } RaCl_2 \times \frac{226 \text{ g Ra}}{297 \text{ g } RaCl_2} \times \frac{\$50{,}000}{1 \text{ g Ra}} = \$380$$

3. $\frac{240}{2} = 120$; $\frac{120}{2} = 60$; $\frac{60}{2} = 30$. 3 half-lives required to reduce the count from 240 to 30 counts/min.

   1980 + (3 × 28) = 2064 A.D one-eighth of the starting amount remains.

4. To go from 100% to 25% requires 2 half-lives. C-14 half-life is 5668 years. The specimen will be the age of two half-lives,

$2 \times 5668 \text{ years} = 1.1 \times 10^4$ years old (11,000).

5. $16.0 \text{ g} \longrightarrow 8.0 \text{ g} \longrightarrow 4.0 \text{ g} \longrightarrow 2.0 \text{ g} \longrightarrow 1.0 \text{ g} \longrightarrow 0.50 \text{ g}$

Five half-lives required to go from 16.0 g to 0.50 g.

$$\frac{90 \text{ minutes}}{5 \text{ half-lives}} = 18 \text{ minutes/half-life}$$

6. (a) Calculated mass:

3 protons	3 × 1.0073 g	= 3.0219 g
4 neutrons	4 × 1.0087 g	= 4.0348 g
3 electrons	3 × 0.00055 g	= 0.0017 g
calculated mass		7.0584 g

Mass defect = calculated mass - actual mass

Mass defect = 7.0584 g - 7.0160 g = 0.0424 g/mol

(b) Binding energy

$$\frac{0.0424 \text{ g}}{\text{mol}} \times \frac{9.2 \times 10^{13} \text{ J}}{\text{g}} = 3.9 \times 10^{12} \text{ J/mol}$$

7. (a) $^{235}_{92}\text{U} + {}^{1}_{0}\text{n} \longrightarrow {}^{94}_{38}\text{Sr} + {}^{139}_{54}\text{Xe} + 3\,{}^{1}_{0}\text{n} + \text{energy}$

Mass loss = mass of reactants - mass of products

Mass of reactants = 235.0439 + 1.0087 = 236.0526 amu

Mass of products = 93.9154 + 138.9179 + 3 × 1.0087
= 235.8594 amu

Mass loss = 236.0526 amu - 235.8594 amu = 0.1932 amu.

$$0.1932 \text{ amu} \times \frac{1.000 \text{ g}}{6.022 \times 10^{23} \text{ amu}} \times \frac{9.2 \times 10^{13} \text{ J}}{1.00 \text{ g}}$$

$= 3.0 \times 10^{-11}$ J/atom U-235

(b) $$\frac{3.0 \times 10^{-11} \text{ J}}{\text{atom}} \times \frac{6.022 \times 10^{23} \text{ atoms}}{\text{mol}} = 1.8 \times 10^{13} \text{ J/mol}$$

(c) $$\frac{0.1932 \text{ amu}}{236.0526 \text{ amu}} \times 100\% = 0.08185\% \text{ mass loss}$$

8. (a) $^{1}_{1}\text{H} + {}^{2}_{1}\text{H} \longrightarrow {}^{3}_{2}\text{He} + \text{energy}$

Mass loss = mass reactants - mass products

Mass reactants = 1.00782 + 2.01410 = 3.02192 g/mol

Mass products = 3.01603 g/mol

Mass loss = 3.02192 - 3.01603 = 0.00589 g/mol

$$\frac{0.00589\ g}{mol} \times \frac{9.2 \times 10^{13}\ J}{g} = 5.4 \times 10^{11}\ J/mol$$

(b) $\frac{0.00589\ g}{3.02192\ g} \times 100\% = 0.195\%$ mass loss

9. $^{235}_{92}U \longrightarrow\ ^{207}_{82}Pb$

Mass loss: 235 - 207 = 28

Net proton loss (atomic number): 92 p - 82 p = 10 p

The mass loss is equivalent to 7 alpha particles (28/4)

A loss of 7 alpha particles gives a loss of 14 p

Decrease in atomic number due to loss of 7 alpha particles: 92 - 14 = 78. Therefore, a loss of 4 beta particles is required to increase the atomic number from 78 to 82.

Total loss = 7 alpha and 4 beta particles.

B. 1. To react both elements lose the outermost s electron. Because this electron is in the 4th principal energy level in K and in the second principal energy in Li, it is attracted less strongly by the K nucleus thus making K the more reactive element.

2. The ionization energy would be lower for Fr because the outermost electron is in the 7th principal energy level in Fr and in the 6th in Cs. The s electron in Fr is more highly shielded and farther from the nucleus.

3. Oxygen is oxidized ($-2 \longrightarrow 0$), potassium and hydrogen are reduced ($+1 \longrightarrow 0$).

4. Salt (NaCl) because it is cheap and readily available world wide.

5. Li_2O, NaOH, $KC_2H_3O_2$, $RbHCO_3$, Cs_2CO_3

6. K_2SO_4, because K is essential for plant growth. K_2SO_4 is a fertilizer.

7. The IIA element is less reactive because its two outermost s electrons are paired, and more energy is needed to remove them than is needed to remove the single outermost s electron from the IA element. Group IIA elements have higher ionization energies than Group IA elements.

8. Burning magnesium is hard to extinguish because Mg is a good reducing agent and will react exothermically with common fire extinguishers as water, carbon dioxide, or sand.

$$Mg + H_2O \longrightarrow MgO + H_2$$

$$2\ Mg + CO_2 \longrightarrow 2\ MgO + C$$

$$2\ Mg + SiO_2 \longrightarrow 2\ MgO + Si$$

9. Barium ions are toxic, but $BaSO_4$ is too insoluble to be absorbed. Barium chloride is soluble and therefore toxic.

10. (a) $Mg^{2+} + 2\ e^- \longrightarrow Mg$ Cathode reaction

(b) $2\ Cl^- \longrightarrow Cl_2 + 2\ e^-$ Anode reaction

(c) $MgCl_2 \xrightarrow{\text{electrolysis}} Mg + Cl_2$ Overall reaction

11. When an element is heated energy is absorbed and electrons are thereby promoted to higher or excited state energy levels. As the excited electrons fall back to lower levels, the absorbed energy is radiated in quanta corresponding to specific wavelengths or colors of light. for Groups IA and IIA, these wavelengths are in the visible spectrum; each element, due to its unique electronic structure, produces a flame of a different color.

12. Pig iron contains about 1% Si, 0.03% S, 0.27% P, 2.4% Mn, and 4.6% C. Steel contains controlled amounts of carbon up to 1.7% and possibly trace amounts of other material.

13. All three metals oxidize in the atmosphere. A thin, transparent and impervious layer of oxide forms on Al and Mg surfaces and protects these metals from further corrosion. But the oxide layer (rust) that forms on Fe is permeable to oxygen and water and permits the corrosion to continue.

14. When steel is given cathodic protection by magnesium, the anode reaction is:

$$Mg \longrightarrow Mg^{2+} + 2\ e^-$$

15. The anode is made of carbon. The oxygen set free immediately reacts with the C to form CO and CO_2.

16. The reaction of aluminum with nitrogen:

(a) $2\ Al + N_2 \longrightarrow 2\ AlN$

(b) $Al^{3+} :\ddot{\underset{..}{N}}:^{3-}$

17. Group VIIA elements react by gaining an electron in the outermost energy level. As we go downward in the group the attraction of the nucleus for the outermost electron decreases in each successive period. This decreased attraction and reactivity is because the outer electron is farther from the nucleus and shielded from its attraction by more electrons.

18. Covalent bonding $:\ddot{\underset{..}{X}}:\ddot{\underset{..}{X}}:$, $H:\ddot{\underset{..}{X}}:$; Ionic bonding $Na^{1+}\ :\ddot{\underset{..}{X}}:^{1-}$

19. The boiling point of HF is abnormally high compared to those of other hydrogen halides because of hydrogen bonding in HF.

20. Among the halogens, fluorine because of its small atomic size and limited electronic shielding has the greatest attraction for an additional electron. Oxidizing agents function by gaining electrons; fluorine, therefore, is the strongest oxidizing agent. The order of oxidizing agents is I_2, Br_2, Cl_2, and F_2.

21. (a) Lack of F^- would lead to increased dental caries.

(b) Lack of Cl^-, if complete, would eventually cause death as this ion is essential in metabolism.

(c) Lack of Br^- would have no known effect.

(d) Lack of I^- would cause the thyroid gland to enlarge, a condition known as goiter.

22. In the 160 - 187°C temperature range sulfur molecules, S_X, exist as long chains. High viscosity is caused by tangling and intertwining of these chains.

23. Sulfur can either gain electrons and act as an oxidizing agent or lose electrons and act as a reducing agent.

24. Sulfur is a Group VIA element with six electrons in the outermost principal energy level. Hence a sulfur atom can take two electrons from less electronegative elements or donate up to six electrons to more electronegative elements and assume oxidation states ranging from -2 to +6.

25. (a) $H_2SO_4 + Ca(OH)_2 \longrightarrow CaSO_4 + 2\ H_2O$

(b) $2\ H_2SO_4 + Cu \longrightarrow CuSO_4 + SO_2 + 2\ H_2O$

(c) $11\ H_2SO_4 + C_{12}H_{22}O_{11} \longrightarrow 12\ C + 11\ H_2SO_4 \cdot H_2O$

26. No, the acid would react with ammonia

$H_2SO_4 + 2\ NH_3 \longrightarrow (NH_4)_2SO_4$

27. When diluting concentrated sulfuric acid, <u>the acid should be added to the water slowly with stirring</u>. Wear safety goggles!

28. A Group VA atom has five electrons in the outermost principal energy level. It can therefore take up to three electrons from less electronegative elements or donate up to five electrons to more electronegative elements and thus assume oxidation numbers ranging from -3 to +5.

29. Yes, metallic character is directly related to the ease with which an element loses electrons. Group VA elements are more metallic from top to bottom because the outermost electrons are lost more easily with increasing atomic size.

30.
```
(a)  :N:::N:                (d)  ┌   H   ┐1+
                                 │   ..  │
      H                          │ H:N:H │
(b)  H:N:H                       │   ..  │
      ..                         └   H   ┘

      ..  3-                (e)  ┌      ..     ┐1-
(c)  :N:                         │     :O:     │
      ..                         │  ..  ..  .. │
                                 │ :O::N:O:    │
                                 │        ..   │
                                 └             ┘
```

31. The nitrogen cycle is the circulation of the element between the atmosphere and living organisms. Nitrogen is essential to all life. However, all animals and most plants are unable to utilize molecular nitrogen but require nitrogen compounds instead. The nitrogen compounds are formed by three "nitrogen fixing" routes: (a) soil bacteria, (b) lightning, and (c) industrial chemical processes. Higher plants take up nitrogen compounds from the soil; animals eat the plants (or one another) to obtain nitrogen compounds. As the plants and animals die and decompose, part of the nitrogen is recycled to the soil and part is returned to the atmosphere.

32. Nitrogen and phosphorus are the VA elements that are essential to life.

33. The correct statements are: c,d,e,i,j,m,n,o,r,s,t.

C. 1. $2\ NaCl \xrightarrow{Electrolysis} 2\ Na + Cl_2$

(a) $5.00\ kg\ Na \times \frac{1000\ g\ Na}{kg\ Na} \times \frac{1\ mol\ Na}{23.0\ g\ Na} \times \frac{1\ mol\ NaCl}{1\ mol\ Na} \times \frac{58.5\ g\ NaCl}{mol\ NaCl}$

$\times \frac{1\ kg\ NaCl}{1000\ g\ NaCl} = 12.7\ kg\ NaCl$

(b) $5.00\ kg\ Na \times \frac{1000\ g\ Na}{kg\ Na} \times \frac{1\ mol\ Na}{23.0\ g\ Na} \times \frac{1\ mol\ Cl_2}{2\ mol\ Na} = 109\ mol\ Cl_2$

2. Hematite = Fe_2O_3; magnetite = Fe_3O_4; siderite = $FeCO_3$

$\%\ Fe\ in\ Fe_3O_4 = \frac{3\ Fe}{Fe_3O_4} \times 100\% = \frac{167.4}{231.4} \times 100\% = 72.3\%$

The mineral, therefore, was magnetite. (Hematite and siderite contain 51.8% and 48.2% Fe, respectively).

3. $S + O_2 \longrightarrow SO_2$

$150\ g\ S \times \frac{1\ mol\ S}{32.1\ g\ S} \times \frac{1\ mol\ SO_2}{1\ mol\ S} \times \frac{22.4\ L\ SO_2}{mol\ SO_2} = 105\ L\ SO_2$

4. $\frac{300{,}000\ L\ H_2O}{1\ hr} \times \frac{1\ kg\ H_2O}{1\ L\ H_2O} \times \frac{0.20\ kg\ Cl}{1{,}000{,}000\ kg\ H_2O} \times \frac{1000\ g\ Cl}{kg\ Cl}$

$= 60\ g\ Cl/hr$

5. $Cl_2 + 2\ Br^- \longrightarrow 2\ Cl^- + Br_2$

$15.0\ L\ Cl_2 \times \frac{1\ mol\ Cl_2}{22.4\ L\ Cl_2} \times \frac{1\ mol\ Br_2}{1\ mol\ Cl_2} \times \frac{80}{100} = 0.536\ mol\ Br_2$

6. (a) $\frac{1000\ mL\ soln}{L\ soln} \times \frac{1.84\ g\ soln}{mL\ soln} \times \frac{96\ g\ H_2SO_4}{100\ g\ soln} \times \frac{1\ mol\ H_2SO_4}{98.1\ g\ H_2SO_4}$

$= 18\ M\ H_2SO_4$

(b) $\frac{1000\ mL\ soln}{L\ soln} \times \frac{1.42\ g\ soln}{mL\ soln} \times \frac{70\ g\ HNO_3}{100\ g\ soln} \times \frac{1\ mol\ HNO_3}{63.0\ g\ HNO_3}$

$= 16\ M\ HNO_3$

(c) $\frac{1000\ mL\ soln}{L\ soln} \times \frac{1.19\ g\ soln}{mL\ soln} \times \frac{37\ g\ HCl}{100\ g\ soln} \times \frac{1\ mol\ HCl}{36.5\ g\ HCl} = 12\ M\ HCl$

B. 1. The major reason for the larger number of organic compounds is that carbon has the ability to form short or very long chains of atoms covalently bonded together.

2. The carbon atom has only two unshared electrons, making two covalent bonds logical, but in CH_4, carbon forms four equivalent bonds.

   Promoting one 2s electron to the empty 2p orbital would make four bonds possible, but without hybridization, we could not explain the fact that all four bonds in CH are identical.

3. Electron dot structure:

   a. CCl_4 b. C_2Cl_6 c. $CH_3CH_2CH_3$

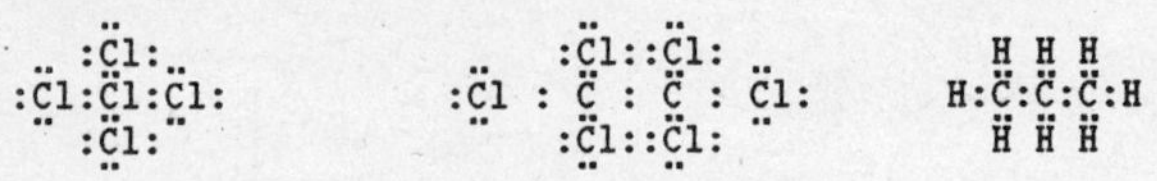

4. The first ten normal alkanes:

methane	CH_4	hexane	C_6H_{14}
ethane	C_2H_6	heptane	C_7H_{16}
propane	C_3H_8	octane	C_8H_{18}
butane	C_4H_{10}	nonane	C_9H_{10}
pentane	C_5H_{12}	decane	$C_{10}H_{22}$

5. Dodecane is $C_{12}H_{26}$

6. a. A molecule of ethane, C_2H_6, contains seven sigma bonds.

   b. A molecule of butane, C_4H_{10}, contains thirteen sigma bonds.

   c. A molecule of isobutane also contains thirteen sigma bonds.

7. Formula (a) and (i) are not isomers, they are the same compound. Formulas (b) and (c) are isomers of C_5H_{12}. Formulas (f) and (h) are isomers of C_5H_{10}. Formulas (d), (e), and (g) are isomers of C_6H_{14}.

8. a. The formulas in Question seven contain the following numbers of methyl groups: (a) 2; (b) 2; (c) 3; (d) 2; (e) 4; (f) 0; (g) 3; (h) 1; (i) 2.

   b. The formulas in Question seven contain the following numbers of ethyl groups: (a) 2; (b) 2; (c) 1; (d) 2; (e) 0; (f) 0; (g) 1; (h) 1; (i) 2.

9. Compounds (b), (e), and (f) are identical. The others are all different.

10. Isomers with these formulas:

a. CH_3Br, one

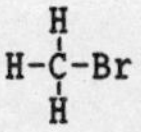

b. CH_2Cl_2, one

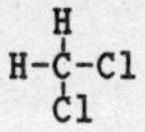

c. CH_3CH_2Cl, one

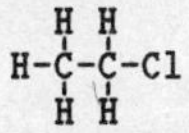

d. C_3H_7Br, two

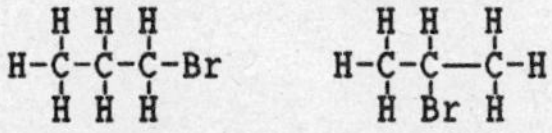

e. C_4H_9I, four

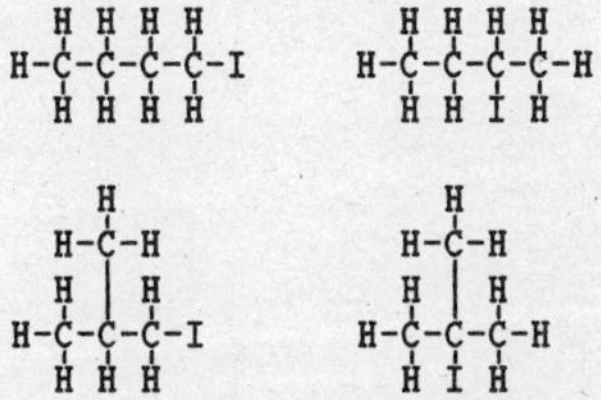

f. $C_3H_6Cl_2$, four

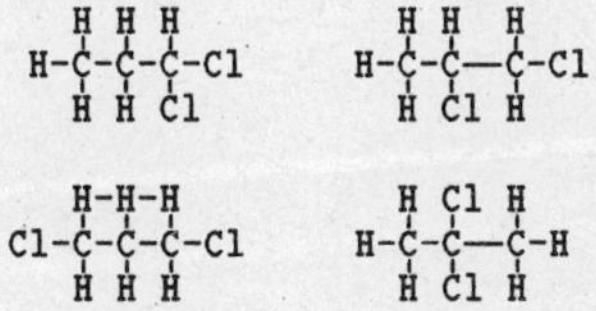

g. C_3H_6BrCl, five

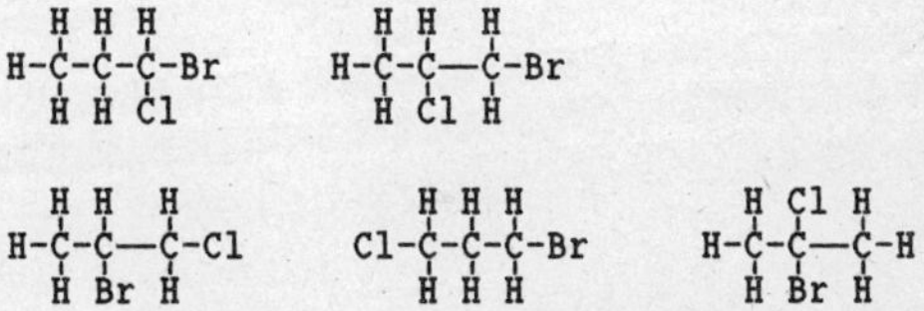

h. $C_4H_8Cl_2$, nine

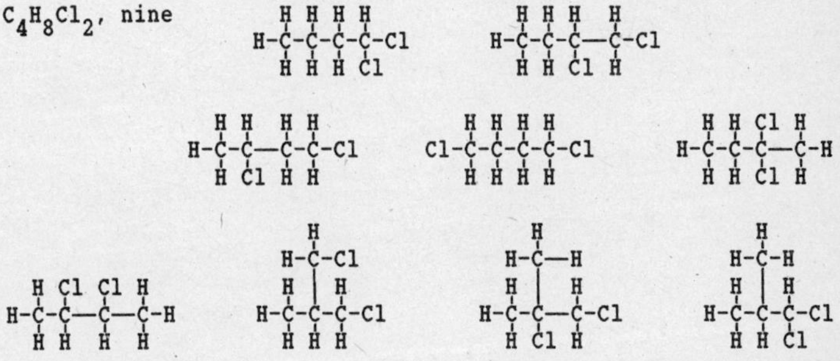

11. The formula of an alkane with 30 carbon atoms would be $C_{30}H_{62}$, so its molecular weight would be 30(12.0) + 62(1.0) = 422.

12. a. hexane $CH_3CH_2CH_2CH_2CH_2CH_3$ 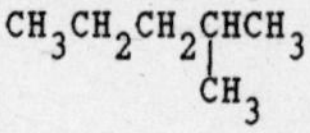

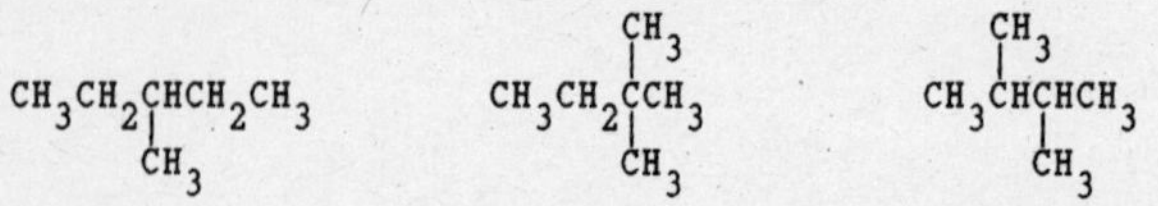

b. heptane $CH_3CH_2CH_2CH_2CH_2CH_2CH_3$ 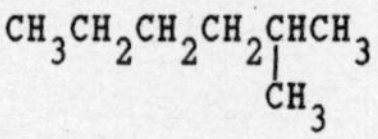

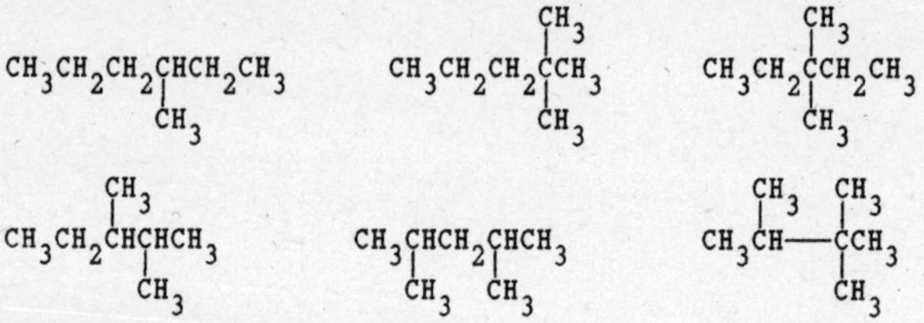

13. Common and IUPAC names:

a. CH_3CH_2Cl ethylchloride; chloroethane

b. $CH_3CHClCH_3$ isopropyl chloride; 2-chloropropane

c. $(CH_3)_2CHCH_2Cl$ isobutyl chloride; 1-chloro-2-methylpropane

d. $CH_3CH_2CH_2Cl$ n-propyl chloride; 1-chloropropane

e. $(CH_3)_3CCl$ t-butyl chloride; 2-chloro-2-methylpropane

f. $CH_3CHClCH_2CH_3$ sec-butyl chloride; 2-chlorobutane

14. a. 3-methylbutane $CH_3CH_2CH(CH_3)CH_3$

Numbering was done from the wrong end of the molecule. The correct name is 2-methylbutane

b. 2-ethylbutane 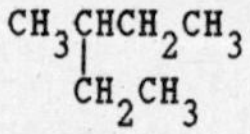

The name is not based on the longest carbon chain (5 carbons). The correct name is 3-methylpentane

c. 2-Dimethylpentane Each methyl group needs to be numbered. Depending on the structure, the correct name is 2,2-dimethylpentane, 2,3-dimethylpentane, or 2,4-dimethylpentane.

d. 3-methyl-5-ethyloctane 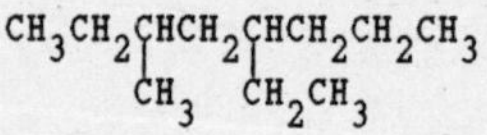

Ethyl should be named before methyl (alphabetical order). The numbering is correct. The correct name is 5-ethyl-3-methyloctane.

e. 3,5-triethylhexane

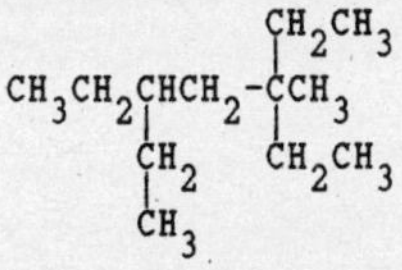

The name is not based on the longest carbon chain (7 carbons). The correct name is 3,5-diethyl-3-methylheptane.

15. Structural formulas:

a. 2,4-dimethylpentane $(CH_3)_2CHCH_2CH(CH_3)_2$

b. 2,2-dimethylpentane $(CH_3)_3CCH_2CH_2CH_3$

c. 3-isopropyloctane

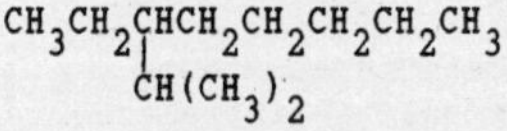

d. 4-ethyl-2-methylhexane

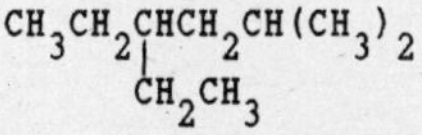

e. 4-t-butylheptane

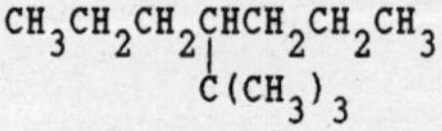

f. 4-ethyl-7-isopropyl-2,4,8-trimethyldecane

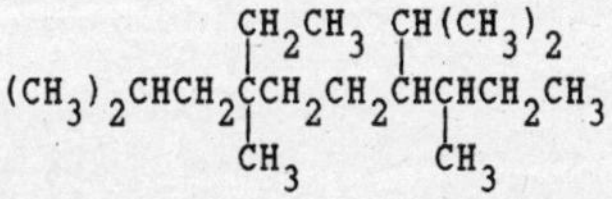

16. IUPAC names:

a. $CH_3CH_2CH(CH_3)_2$ 2-methylbutane

b. $(CH_3)_2CHCH_2CH(CH_3)_2$ 2,4-dimethylpentane

c. 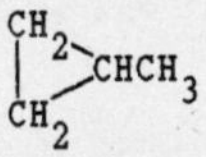 methylcyclopropane

d. 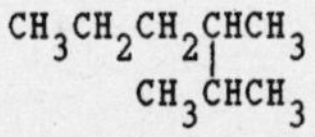 2,3-dimethylhexane

17. a. $CH_3CH_2CH_2CH_3 + Cl_2 \xrightarrow{h\nu} CH_3CH_2CHClCH_3 + CH_3CH_2CH_2CH_2Cl + HCl$

b. $2\ CH_3CH_2CH_2CH_3 + 13\ O_2 \xrightarrow{\Delta} 8\ CO_2 + 10\ H_2O$

18. $CH_3CH_2CH_2CH_2CH_2CH_2Cl$ $CH_3CH_2CH_2CH_2CHClCH_3$ $CH_3CH_2CH_2CHClCH_2CH_3$

19. Structures for the ten dichlorosubstituted compounds of 2-methylbutane:

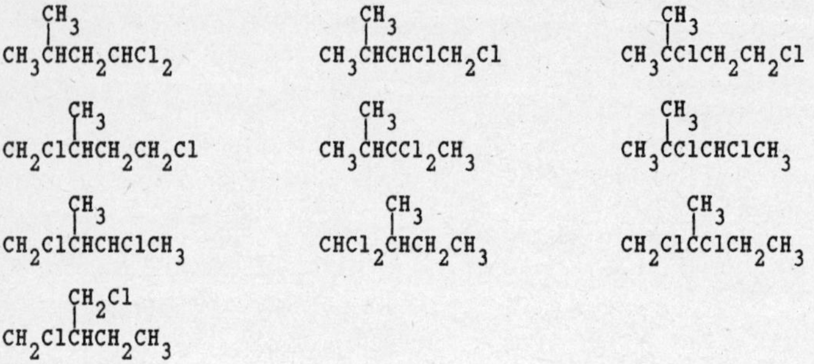

20. The sample had to be a mixture of the two butanes, n-butane and isobutane, to obtain four monobromo products. Each of the butanes yields only two monobromo compounds:

n-butane yields $CH_3CH_2CH_2CH_2Br$ and $CH_3CH_2CHBrCH_3$

isobutane yields

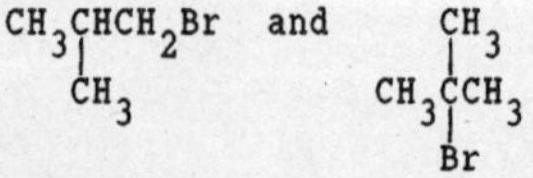

21. Two cycloalkanes having the formula C_4H_8:

cyclobutane methylcyclopropane

22. The following statements are correct: a, b, c, d, g, i, j, l, m, p, and s.

Chapter 21 Unsaturated Hydrocarbons--Alkenes, Alkynes, and Aromatic Hydrocarbons

B. 1. The sigma bond in the double bond of ethylene is formed by the overlap of two sp^2 electron orbitals and is symmetrical about a line drawn between the nuclei of the two carbon atoms. The pi bond is formed by the sidewise overlap of two p orbitals which are perpendicular to the carbon-carbon sigma bond. The pi bond consists of two electron clouds, one above and one below the plane of the carbon-carbon sigma bond.

2. a. ethane

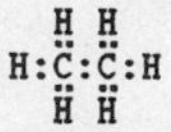

b. ethene

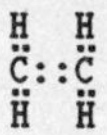

c. ethyne

H:C:::C:H

3. a. C_3H_5Cl

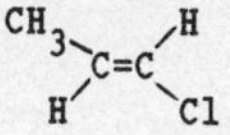

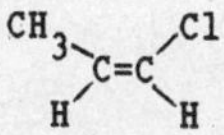

$CH_3CCl{=}CH_2$

$CH_2ClCH{=}CH_2$

b. chlorocyclopropane

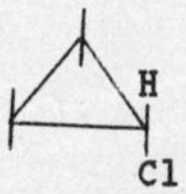

4. The seventeen hexenes (C_6H_{12}):

1-hexene $CH_2{=}CHCH_2CH_2CH_2CH_3$

cis-2-hexene

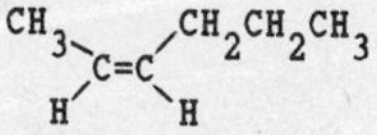

trans-2-hexene

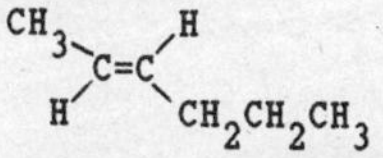

cis-3-hexene

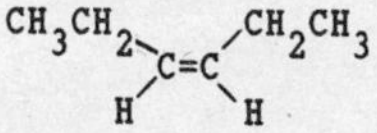

trans-3-hexene

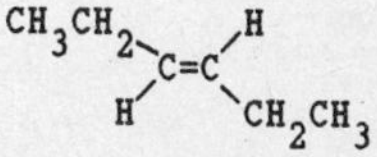

4-methyl-1-pentene

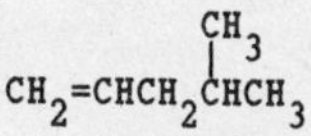

3-methyl-1-pentene

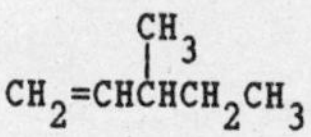

2-methyl-1-pentene

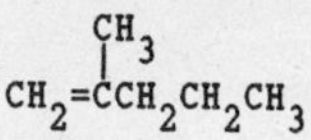

3,3-dimethyl-1-butene

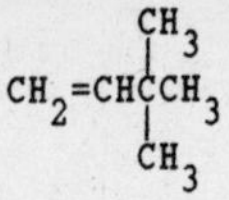

2,3-dimethyl-1-butene

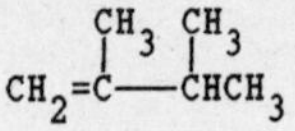

2-ethyl-1-butene

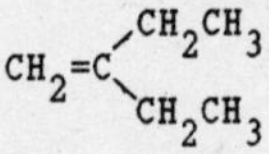

cis-3-methyl-2-pentene

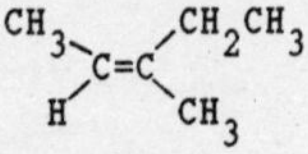

trans-3-methyl-2-pentene

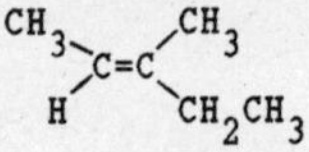

4-methyl-cis-2-pentene

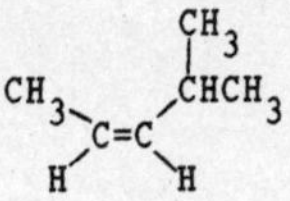

4-methyl-trans-2-pentene

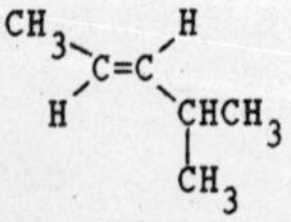

2-methyl-2-pentene

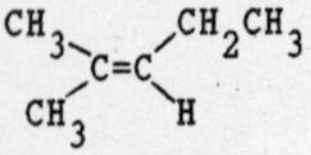

2,3-dimethyl-2-butene

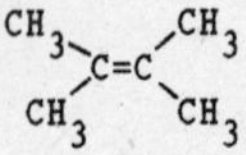

5. a. 2,5-dimethyl-3-hexene

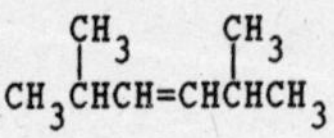

b. 2-ethyl-3-methyl-1-pentene

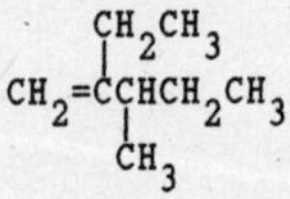

c. 4-methyl-cis-2-pentene

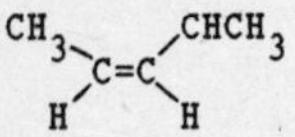

d. cis-1,2-diphenylethene

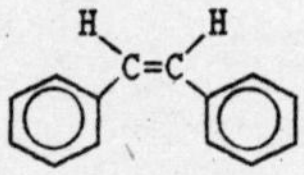

e. 3-penten-1-yne

$CH\equiv CCH{=}CHCH_3$

f. 3-phenyl-1-butyne

g. vinyl bromide

$CH_2{=}CHBr$

h. cyclopentene

i. trans-3-hexene

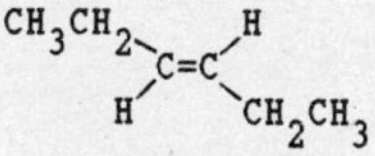

j. 1-methylcyclohexene

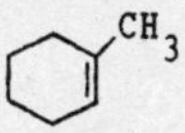

k. 3-methyl-1-pentyne

$CH\equiv CCHCH_2CH_3$
CH_3

l. 3-isopropylcyclopentene

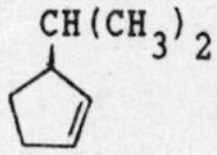

m. 3-methyl-2-phenylhexane

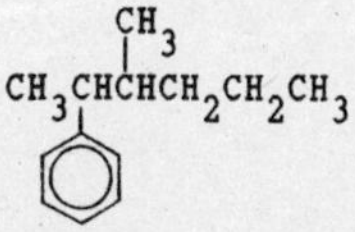

6. a. cis-4-methyl-2-hexene
b. 2,3-dimethyl-2-butene
c. 3-isopropyl-1-pentene
d. trans-3-methyl-3-hexene
e. 3-phenyl-1-propyne
f. 4,5-dibromo-2-hexyne

7. a. All the pentynes

1-pentyne

$CH_3CH_2CH_2C\equiv CH$

2-pentyne

$CH_3CH_2C\equiv CCH_3$

3-methyl-1-pentyne

CH_3
$CH_3CHC\equiv CH$

b. All the hexynes

1-hexyne

$CH_3CH_2CH_2CH_2C\equiv CH$

2-hexyne

$CH_3CH_2CH_2C\equiv CCH_3$

3-hexyne

$CH_3CH_2C\equiv CCH_2CH_3$

3-methyl-1-pentyne

CH_3
$CH_3CH_2CHC\equiv CH$

4-methyl-1-pentyne

CH_3
$CH_3CHCH_2C\equiv CH$

4-methyl-2-pentyne

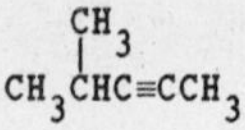

3,3-dimethyl-1-butyne

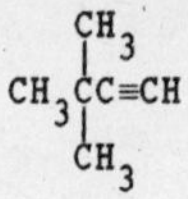

8. 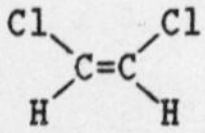

cis-1,2-dichloroethene

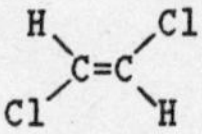

trans-1,2-dichloroethene

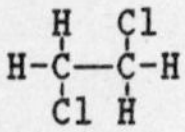

1,2-dichloroethane

We are able to get cis-trans isomers of ethene because there is no rotation around the carbon-carbon double bond, so the two structures shown are not the same. With 1,2-dichloroethane, there is free rotation of the carbon-carbon bond. In the structure shown, exchanging the chlorine on the right with either hydrogen atom on that carbon appears to make a different isomer, but rotation makes any of the three positions equivalent.

9. Only two of the given structures permit cis-trans isomers. They are:

(d) $CH_3CH{=}CHCl$ and (f) $CH_2ClCH{=}CHCH_2Cl$

10. a. 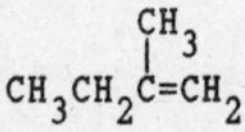

Numbering was started from the wrong end of the structure. 2-methyl-1-butene

b. $CH_3CH_2CH{=}CHCH_3$

Numbering was started from the wrong end of the structure. 2-pentene

c. 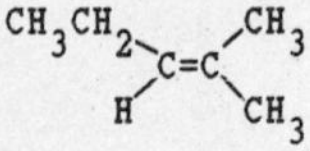

This compound does not have cis-trans isomers. 2-methyl-2-pentene

d. 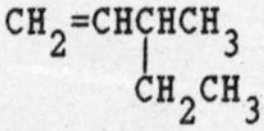

Longest chain contains five carbon atoms. 3-methyl-1-pentene

e. 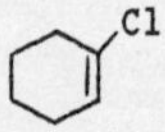

The C=C bond in cyclohexene is numbered so that substituted groups have the smallest numbers. 1-chlorocyclohexene

11. a. $CH_3CH_2CH_2CH{=}CH_2 + Br_2 \longrightarrow CH_3CH_2CH_2CHBrCH_2Br$

b. $CH_3CH_2CH_2CH{=}CH_2 + HCl \longrightarrow CH_3CH_2CH_2CHClCH_3$

c. $CH_3CH_2\underset{\large CH_3}{\underset{|}{C}}=CHCH_3 + HI \longrightarrow CH_3CH_2\underset{\large CH_3}{\underset{|}{C}}ICH_2CH_3$

d. $CH_3CH_2CH=CHCH_3 + HBr \longrightarrow CH_3CH_2CHBrCH_2CH_3 + CH_3CH_2CH_2CHBrCH_3$

e. $CH_3CH_2CH=CH_2 + H_2O \xrightarrow{H^+} CH_3CH_2CH(OH)CH_3$

f. 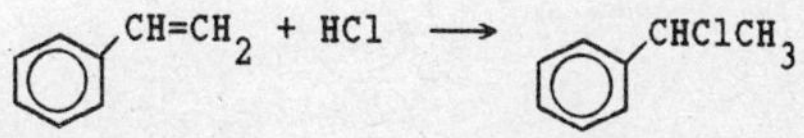

g. 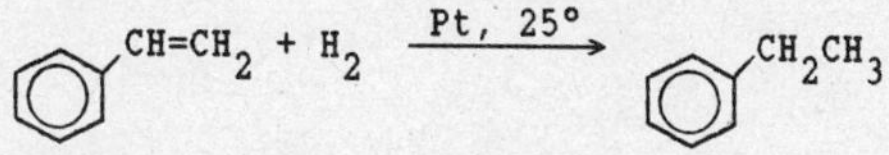

h. $CH_2=CHCl + HBr \longrightarrow CH_3CHClBr$

i. $CH_3CH{\equiv}CHCH_3 + KMnO_4 \xrightarrow[cold]{H_2O} CH_3CH(OH)CH(OH)CH_3 + MnO_2 + KOH$

12. a. $CH_3C{\equiv}CH + 2\ H_2 \xrightarrow[1\ atm]{Pt,\ 25°} CH_3CH_2CH_3$

b. $CH_3C{\equiv}CCH_3 + Br_2(1\ mole) \longrightarrow CH_3CBr=CBrCH_3$

c. $CH_3C{\equiv}CCH_3 + Br_2(2\ moles) \longrightarrow CH_3CBr_2CBr_2CH_3$

d. $CH{\equiv}CH + HCl \longrightarrow CH_2=CHCl + HCl \longrightarrow CH_3CHCl_2$

e. $CH_3C{\equiv}CH + HCl \longrightarrow CH_3CCl=CH_2 + HCl \longrightarrow CH_3CCl_2CH_3$

f. $CH_3CH_2C{\equiv}CH + Ag(NH_3)_2NO_3 \longrightarrow CH_3CH_2C{\equiv}CAg + NH_4NO_3 + NH_3$

13. When cyclohexene, 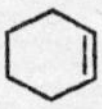, reacts with:

a. Br_2, the product is 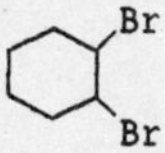

1,2-dibromocyclohexane

b. HI, the product is 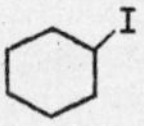

iodocyclohexane

c. H_2O, H^+, the product is 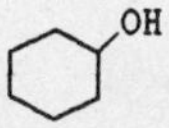

cyclohexanol

d. $KMnO_4(aq)$, the product is 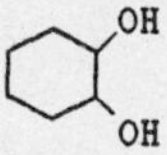

cyclohexene glycol or 1,2-dihydroxcyclohexane

14. Reactions to convert 2-butyne to:

a. 2,3-dibromobutane

$CH_3C{\equiv}CCH_3 + Br_2(\text{1 mole}) \longrightarrow CH_3CBr{=}CBrCH_3$

$CH_3CBr{=}CBrCH_3 + H_2 \xrightarrow[\text{1 atm}]{\text{Pt, 25°}} CH_3CHBrCHBrCH_3$

b. 2,2-dibromobutane

$CH_3C{\equiv}CCH_3 + HBr \longrightarrow CH_3CH{=}CBrCH_3 \xrightarrow{HBr} CH_3CH_2CBr_2CH_3$

c. 2,2,3,3-tetrabromobutane

$CH_3C{\equiv}CCH_3 + Br_2(\text{2 moles}) \longrightarrow CH_3CBr_2CBr_2CH_3$

15. When $CH_2{=}\overset{CH_3}{\overset{|}{C}}CH_2CH_2CH_3$ reacts with HBr, two products are possible $CH_3\overset{CH_3}{\overset{|}{C}}BrCH_2CH_2CH_3$ and $CH_2Br\overset{CH_3}{\overset{|}{C}}HCH_2CH_2CH_3$

The first will strongly predominate. This is according to Markovnikov's Rule and is because a tertiary carbocation is more stable than a primary one.

16. a. Add $Ag(NH_3)_2NO_3$ to a sample of each compound: Only 1-pentyne will react, giving a precipitate.

b. Add Br_2/CCl_4 to samples of the remaining two hydrocarbons: 1-pentene will react and decolorize the reddish Br_2 solution. The sample that does not decolorize the Br_2 is the pentane.

17. In the absence of sunlight, cyclohexane will not readily decolorize bromine; hexene will.

18. Rubber products deteriorate rapidly in smog-ridden areas because ozone causes oxidation of the carbon-carbon double bonds, which changes the properties of the rubber.

19. Acetylene presents two different explosion hazards:

a. It can form an explosive mixture with oxygen or air;
b. When highly compressed or liquefied, acetylene may decompose violently, either spontaneously or from a slight shock.

20. a. benzene

b. toluene

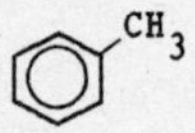

c. p-xylene

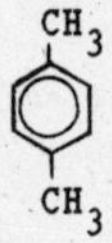

d. styrene

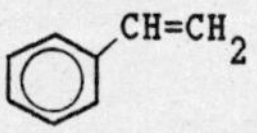

e. aniline

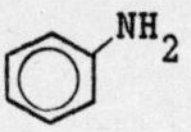

f. phenol

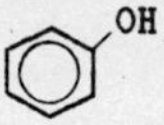

g. o-bromochlorobenzene

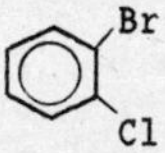

h. 1,3-dichloro-5-nitrobenzene

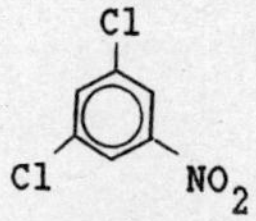

i. m-dinitrobenzene

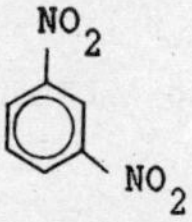

21. a. ethylbenzene

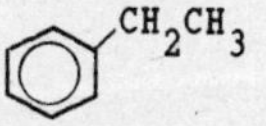

b. benzoic acid

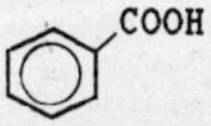

c. 1,3,5-tribromobenzene

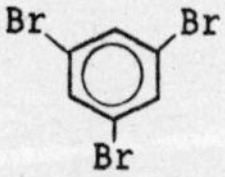

d. naphthalene

e. anthracene

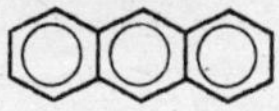

f. t-butylbenzene

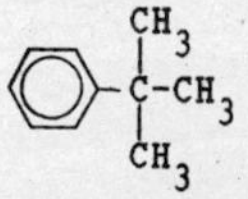

g. 1,1-diphenylethane

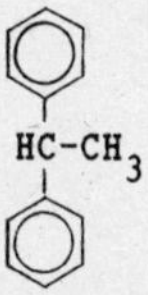

22. a. trichlorobenzene

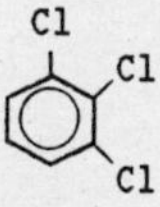

1,2,3-trichloro-benzene

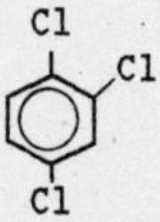

1,2,4-trichlor-benzene

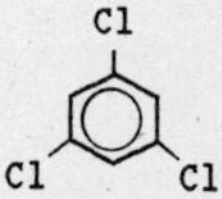

1,3,5-trichloro-benzene

b. dichlorobromobenzene

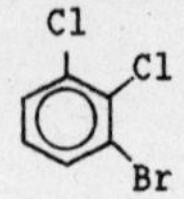

1,2-dichloro-3-bromobenzene

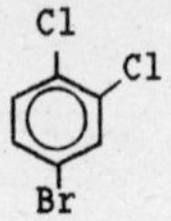

1,2-dichloro-4-bromobenzene

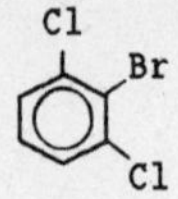

1,3-dichloro-2-bromobenzene

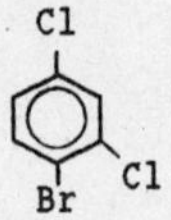

1,3-dichloro-4-bromobenzene

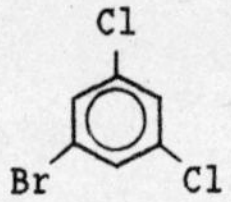

1,3-dichloro-5-bromobenzene

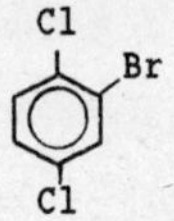

1,4-dichloro-2-bromobenzene

c. The benzene derivatives of formula C_8H_{10}:

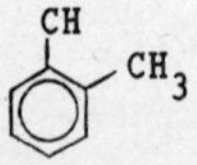

1,2-dimethylbenzene
or o-xylene

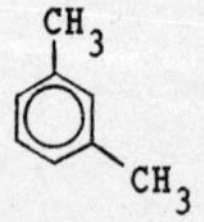

1,3-dimethylbenzene
or m-xylene

1,4-dimethylbenzene or
p-xylene

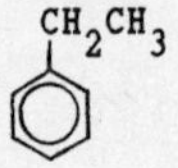

ethylbenzene

23. a. All isomers that can be written substituting a third chlorine atom on o-dichlorobenzene

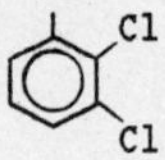

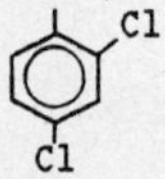

b. All isomers that can be written by substituting another chlorine in o-chlorobromobenzene

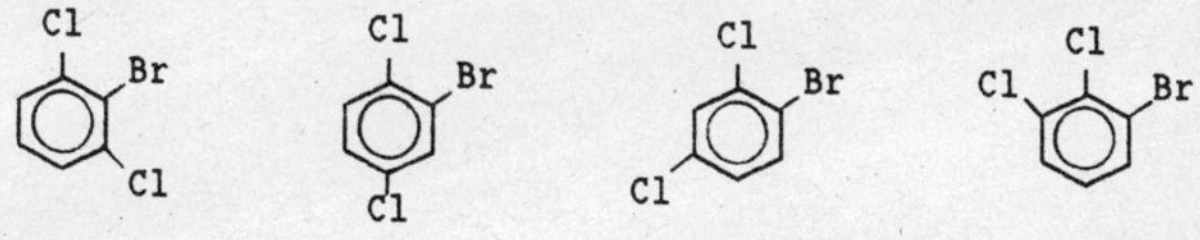

24. a. p-chloroethylbenzene or 4-chloro-1-ethylbenzene
b. styrene
c. propylbenzene
d. m-nitroaniline
e. 2,4-dibromobenzoic acid
f. m-nitroaniline
g. 2,4,6-tribromophenol
h. isopropylbenzene
i. triphenylmethane

25. Benzene does not undergo the typical reactions of an alkene. It does not decolorize bromine rapidly; it does not destroy the purple color of permanganate ions. The reactions of benzene are more like those of an alkane. Reaction of benzene with chlorine requires a catalyst. Benzene does not readily add the Cl_2 molecule to a double bond, but rather a hydrogen is replaced by a chlorine atom.

$$C_6H_6 + Cl_2 \xrightarrow{Fe} C_6H_5Cl + HCl$$

26. a.

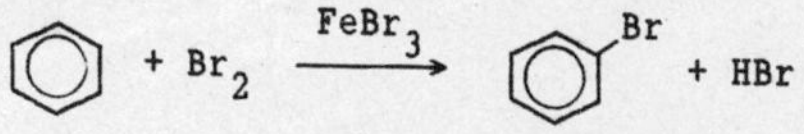

bromobenzene

b.

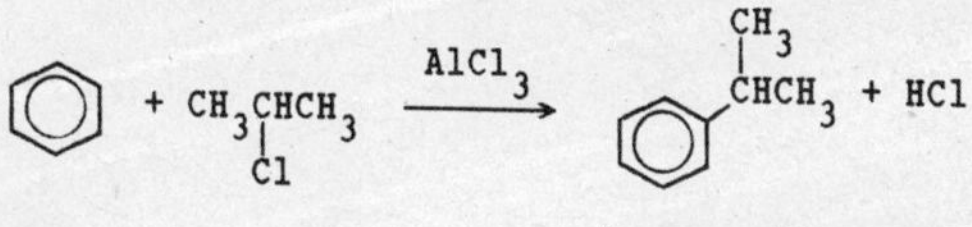

isopropylbenzene

c.

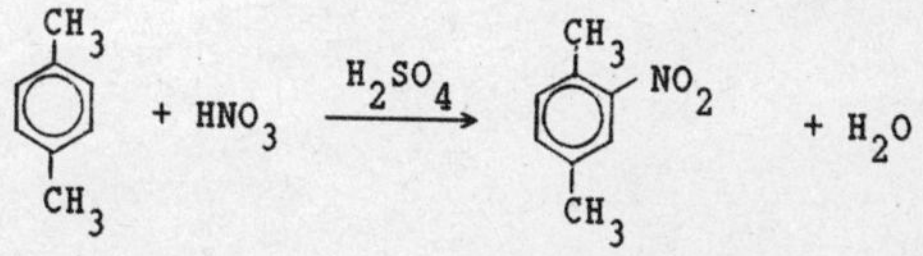

2-nitro-1,4-dimethylbenzene

d.

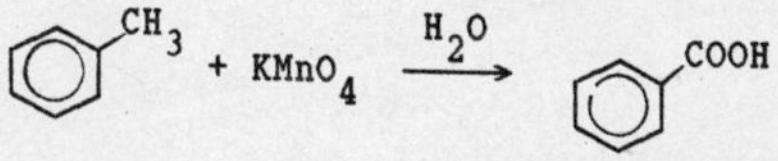

benzoic acid

27. The reaction mechanism by which benzene is brominated in the presence of $FeBr_3$:

$$FeBr_3 + Br_2 \longrightarrow FeBr_4^- + Br^+$$

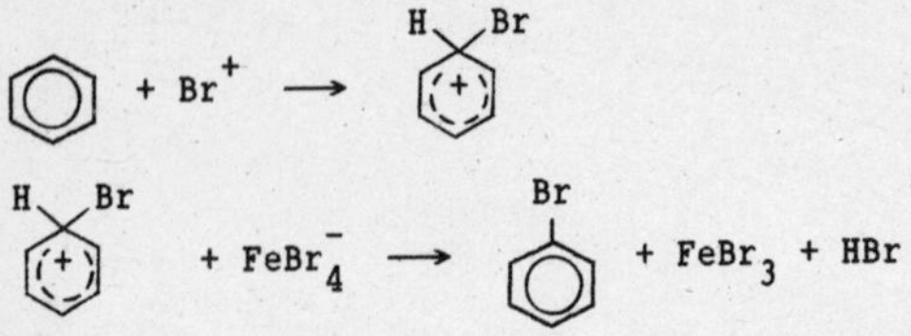

28. During the 10-year period from 1935 to 1945, the major source of aromatic hydrocarbons shifted from coal tar to petroleum due to the rapid growth of several industries which used aromatic hydrocarbons as raw material. These industries include drugs, dyes, detergents, explosives, insecticides, plastics, and synthetic rubber. Since the raw material needs far exceeded the aromatics available from coal tar, another source had to be found, and processes were developed to make aromatic compounds from alkanes in petroleum. World War II, which fell in this period, put high demands on many of these industries, particularly explosives.

29. The following statements are correct: a, f, g, h, i, j, k, m, n, o, q, t, w, x, and y.

B. 1. The question allows great freedom of choice. Those shown here are very simple examples of each type.

a. an alkyl halide

CH_3CH_2Cl

b. a phenol

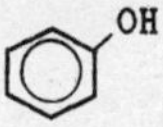

c. an ether

$CH_3CH_2OCH_2CH_3$

d. an aldehyde

$CH_3\overset{H}{C}{=}O$

e. a ketone

$CH_3\overset{O}{\overset{\|}{C}}CH_3$

f. a carboxylic acid

$CH_3C(=O)OH$

g. an ester

$CH_3\overset{O}{\overset{\|}{C}}{-}OCH_2CH_3$

2. Alkenes are almost never made from alcohols because the alcohols are almost always the higher value material. This is because recovering alkenes from hydrocarbon sources in an oil refinery (primarily catalytic cracking) is a relatively cheap process.

3. Isopropyl alcohol is usually used for rubbing alcohol, rather than n-propyl alcohol, because of price. The easy way to make the alcohol, from propene, yields isopropyl alcohol rather than n-propyl alcohol. Any process for producing n-propyl alcohol would be much more expensive.

4. Oxidation of primary alcohols yields aldehydes. Further oxidation yields carboxylic acids. Examples:

$$CH_3CH_2OH + [O] \longrightarrow CH_3C(H){=}O + H_2O$$

$$CH_3C(H){=}O + [O] \longrightarrow CH_3C({=}O)OH$$

5. The molecular weight of myricyl alcohol, an open chain saturated alcohol containing 30 carbon atoms:
Looking at the first alcohols CH_3OH C_2H_5OH C_3H_7OH

Leads to the formula: $C_{30}H_{61}OH$

Mol. Wt. = 30 × 12.0 + 62 × 1.0 + 1 × 16.0 = 438

6. Ethylene glycol is superior to methyl alcohol as an antifreeze because of its low volatility. Methyl alcohol is much more volatile than water. If the radiator leaks gas under pressure (normally

steam), it would primarily leak methanol vapor so you would soon have no antifreeze. Ethylene glycol has a lower volatility than water, so it does not present this problem.

7. Physiological effects of:

   a. methanol: It is a poisonous liquid capable of causing blindness or death if taken internally. Exposure to methanol vapors is also very dangerous.

   b. ethanol: It can act as a food, a drug, and a poison. The body can metabolize small amounts of ethanol to produce energy; thus it is a food. It depresses brain functions so that activities requiring skill and judgement are impaired; thus it is a drug. With very large consumption, the depression of brain function can lead to unconsciousness and death; thus it is a poison.

8. Ethanol (molecular weight 46) is a liquid whereas dimethyl ether (molecular weight 46) is a gas because ethanol has a significant amount of hydrogen bonding between molecules in the liquid state, and thus has a higher boiling point than would be predicted from molecular weight alone. Dimethyl ether is not capable of hydrogen bonding, so has low attraction between molecules.

9. a. methanol CH_3OH

   b. 2-butanol
   OH
   $CH_3CHCH_2CH_3$

   c. 3-methyl-1-hexanol
   CH_3
   $CH_3CH_2CH_2CHCH_2CH_2OH$

   d. 2-methyl-2-butanol
   OH
   $CH_3CCH_2CH_3$
   CH_3

   e. propylene glycol
   OH
   CH_3CHCH_2OH

   f. isopropyl alcohol
   OH
   CH_3CHCH_3

   g. 1-phenylethanol
   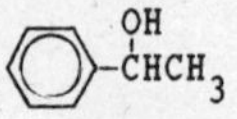

   h. cyclopentanol
   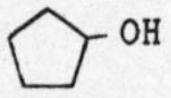

   i. 2,3-butanediol
   OH OH
   $CH_3CH\text{-}CHCH_3$

   j. sec-butyl alcohol
   OH
   $CH_3CH_2CHCH_3$

k. sodium propoxide — $CH_3CH_2CH_2ONa$

l. sodium isopropoxide —

$$\begin{array}{c} ONa \\ | \\ CH_3CHCH_3 \end{array}$$

10. a. C_3H_8O

Alcohols	Ethers
$CH_3CH_2CH_2OH$	$CH_3OCH_2CH_3$
$CH_3CH(OH)CH_3$	

b. C_4H_8O

Alcohols	Ethers
$CH_3CH_2CH_2CH_2OH$	$CH_3OCH_2CH_2CH_3$
$CH_3CH_2CH(OH)CH_3$	$CH_3OCH(CH_3)_2$
$(CH_3)_2CHCH_2OH$	$CH_3CH_2OCH_2CH_3$
$(CH_3)_3C\text{-}OH$	

11. Eight open chain isomeric alcohols with the formula $C_5H_{11}OH$:

1-pentanol

$CH_3CH_2CH_2CH_2CH_2OH$

2-pentanol

$$\begin{array}{r} OH \\ | \\ CH_3CH_2CH_2CHCH_3 \end{array}$$

3-pentanol

$$\begin{array}{c} OH \\ | \\ CH_3CH_2CHCH_2CH_3 \end{array}$$

3-methyl-1-butanol

$$\begin{array}{l} CH_3 \\ | \\ CH_3CHCH_2CH_2OH \end{array}$$

2-methyl-1-butanol

$$\begin{array}{c} CH_3 \\ | \\ CH_3CH_2CHCH_2OH \end{array}$$

2-methyl-2-butanol

$$\begin{array}{c} CH_3 \\ | \\ CH_3CH_2CCH_3 \\ | \\ OH \end{array}$$

3-methyl-2-butanol

$$\begin{array}{c} CH_3 \\ | \\ CH_3CHCHCH_3 \\ | \\ OH \end{array}$$

2,2-dimethyl-1-propanol

$$\begin{array}{c} CH_3 \\ | \\ CH_3CCH_2OH \\ | \\ CH_3 \end{array}$$

12. Of the eight compounds in question 11, the primary alcohols are: 1-pentanol; 3-methyl-1-butanol; 2-methyl-1-butanol; and 2,2-dimethyl-1-propanol. The secondary alcohols are: 2-pentanol; 3-pentanol; and 3-methyl-2-butanol. The tertiary alcohol is 2-methyl-2-butanol.

13. A glycol that is both a primary and a secondary alcohol:

$$\begin{array}{c} OH \\ | \\ CH_3CHCH_2OH \end{array}$$

14. The name of the compounds are:

 a. ethanol
 b. 2-propanol (isopropyl alcohol)
 c. 2-phenyl-1-ethanol
 d. 3-pentanol
 e. 3-methyl-3-pentanol (not 2-ethyl-2-butanol)
 f. 1-methylcyclopentanol
 g. 2-methyl-3-phenyl-1-propanol
 h. ethylene oxide (oxirane)
 i. 1,2-propanediol (propylene glycol)
 j. 2,3-dimethyl-1,4-butanediol

15. cis-cyclopentane-1,2-diol   trans-cyclopentane-1,2-diol

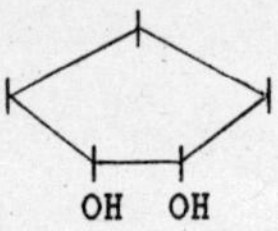

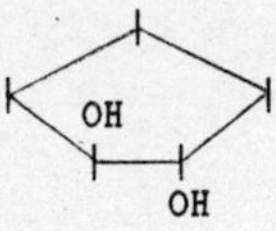

16. a. $CH_3CH(OH)CH(CH_3)CH_3 \xrightarrow[\Delta]{H_2SO_4} CH_3CH{=}C(CH_3)CH_3$ 2-methyl-2-butene

 b. $CH_3CH(OH)CH_2CH_2CH_3 \xrightarrow[\Delta]{H_2SO_4} CH_3CH{=}CHCH_2CH_3$ 2-pentene

 c. 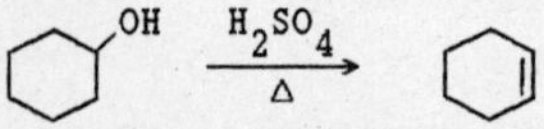 cyclohexene

17. $CH_3CH_2CH_2CH_2OH \xrightarrow{H_2SO_4} CH_3CH_2CH_2\overset{+}{C}H_2 \longrightarrow CH_3CH_2\overset{+}{C}HCH_3 \longrightarrow CH_3CH{=}CHCH_3$

18. a. $CH_3CHBrCH_3 + NaOH(aq) \longrightarrow CH_3CH(OH)CH_3 + NaBr(aq)$

 2-bromopropane   isopropyl alcohol (2-propanol)

 b. $CH_3CH(CH_3)CH_2CH_2Br + NaOH(aq) \longrightarrow CH_3CH(CH_3)CH_2CH_2OH + NaBr(aq)$

 3-methyl-1-bromobutane   3-methyl-1-butanol

 c. 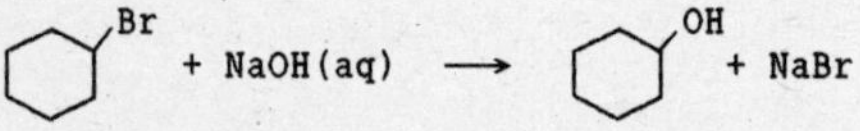

 bromocyclohexane   cyclohexanol

19. a. $CH_3CH{=}CH_2 + H_2O \xrightarrow{H+} CH_3CH(OH)CH_3$

b. $CH_3CH_2CH{=}CH_2 + H_2O \xrightarrow{H+} CH_3CH_2\underset{\substack{|\\OH}}{C}HCH_3$

c. $CH_3CH{=}CHCH_3 + H_2O \xrightarrow{H+} CH_3\underset{\substack{|\\OH}}{C}HCH_2CH_3$

d. $CH_3CH_2CH_2CH{=}CH_2 + H_2O \xrightarrow{H+} CH_3CH_2CH_2\underset{\substack{|\\OH}}{C}HCH_3$

e. $CH_3\underset{\substack{|\\CH_3}}{C}{=}CHCH_3 + H_2O \xrightarrow{H+} CH_3\overset{\substack{OH\\|}}{\underset{\substack{|\\CH_3}}{C}}CH_2CH_3$

20\. a. $CH_3\underset{\substack{|\\OH}}{C}HCH_3 \xrightarrow[H^+]{Cr_2O_7^{2-}} CH_3\overset{\substack{O\\\|}}{C}CH_3$

b. $CH_3CH_2CH_2CH{=}CH_2 \xrightarrow[H_2O]{H+} CH_3CH_2CH_2\underset{\substack{|\\OH}}{C}HCH_3$

c. $CH_3CH_2OH + Na(metal) \longrightarrow CH_3CH_2ONa + H_2$

d. $CH_3CH_2CH{=}CH_2 \xrightarrow[H_2O]{H+} CH_3CH_2\underset{\substack{|\\OH}}{C}HCH_3 \xrightarrow[H+]{Cr_2O_7^{2-}} CH_3CH_2\underset{\substack{\|\\O}}{C}CH_3$

e. $CH_3CH_2CH_2CH_2OH \xrightarrow[\Delta]{H_2SO_4} CH_3CH{=}CHCH_2 + CH_3CH_2CH{=}CH_2$

$\xrightarrow{HCl} CH_3CH_2\underset{\substack{|\\Cl}}{C}HCH_3$

f. $CH_3CH_2CH_2Cl \xrightarrow[H_2O]{NaOH} CH_3CH_2CH_2OH \xrightarrow[H+]{Cr_2O_7^{2-}} CH_3CH_2\underset{\substack{|\\H}}{C}{=}O$

21\. a. $CH_3CH_2OH + [O] \xrightarrow[\Delta]{Cu} CH_3\overset{\substack{H\\|}}{C}{=}O + H_2O$

acetaldehyde

b. $2\ CH_3CH_2OH + H_2SO_4 \xrightarrow{140°C} CH_3CH_2OCH_2CH_3 + H_2O$

diethyl ether

c. $CH_3CH_2CH_2OH + H_2SO_4 \xrightarrow{180°C} CH_3CH{=}CH_2 + H_2O$

propylene

d. $CH_3CH(OH)CH_2CH_3 \xrightarrow[H_2SO_4]{K_2Cr_2O_7} CH_3\overset{O}{\overset{\|}{C}}CH_2CH_3 + H_2O$

2-butanone

e. $CH_3CH_2\overset{O}{\overset{\|}{C}}\text{-}OCH_2CH_3 \xrightarrow[H_2O]{H+} CH_3CH_2COOH + CH_3CH_2OH$

propanoic acid ethanol

22. Two different organic compounds that can be obtained by oxidizing benzyl alcohol, $C_6H_5\text{-}CH_2OH$

$C_6H_5\text{-}\overset{H}{\overset{|}{C}}\text{=O}$ benzaldehyde

$C_6H_5\text{-}\overset{O}{\overset{\|}{C}}\text{-OH}$ benzoic acid

23. a. phenol
b. catechol
c. m-cresol
d. 2-bromo-4-nitrophenol
e. 4-hexylresorcinol
f. 2,4-dinitrophenol

24. a. o-cresol

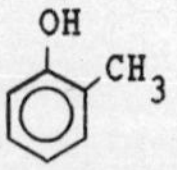

b. p-nitrophenol

c. resorcinol

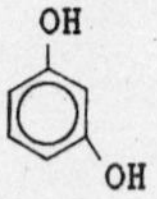

d. 2,6-dimethylphenol

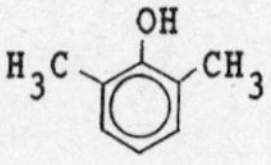

e. 4-hydroxy-3-methoxybenzaldehyde (vanillin)

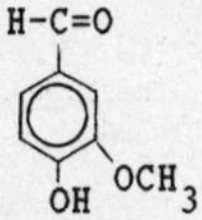

f. 2-isopropyl-5-methylphenol (thymol)

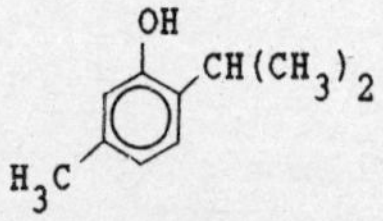

25. Phenols are solids, tend to have odors, are toxic to microorganisms, and some, particularly, are very poisonous to humans. They are weak acids, more acidic than alcohols, but weaker than acetic acid.

26. The cumene-hydroperoxide synthesis of phenol and acetone:

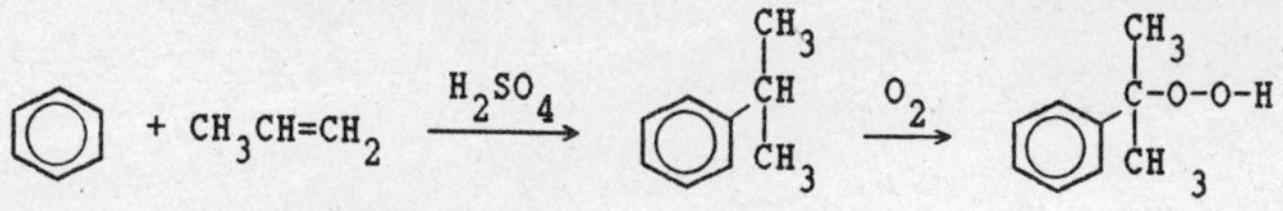

cumene cumene hydroperoxide

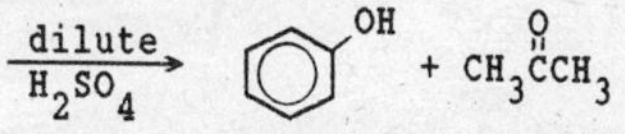

phenol acetone

27. Compounds (a) and (b) will react with sodium metal:

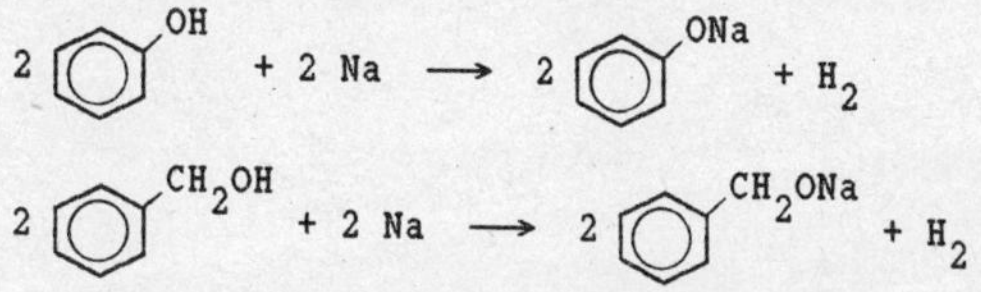

Compound (a) will also react with sodium hydroxide:

28. Preparation of ethyl-p-methylphenyl ether from p-cresol and ethane

$$CH_3CH_3 + Br_2 \xrightarrow{\text{light}} CH_3CH_2Br$$

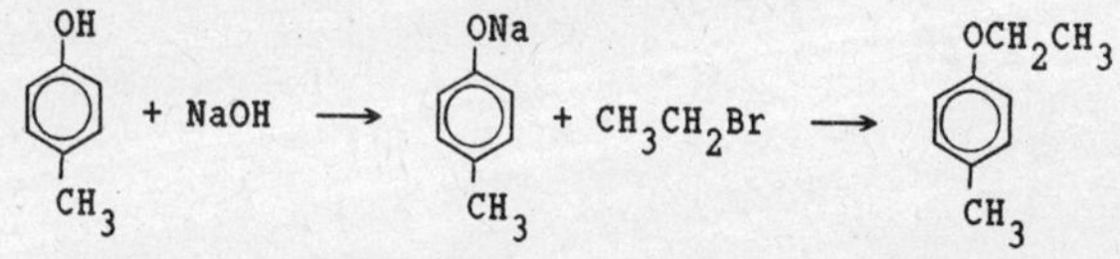

29. Low molecular weight ethers present two hazards. They are very volatile and their highly flammable vapors form explosive mixtures with air. They also slowly react with oxygen in the air to form unstable explosive peroxides.

30. Order of increasing solubility in water: (c, a, b, d)
c. $CH_3CH_2CH_2CH_2CH_3$; a. $CH_3CH_2OCH_2CH_2CH_3$;

b. $CH_3CH(OH)CH_2CH_2CH_3$; d. $CH_3CH(OH)CH(OH)CH_2CH_3$.

31. Increasing acidity

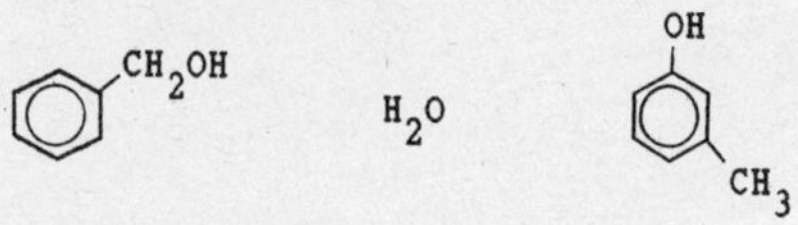

32. Six isomeric saturated ethers of formula $C_5H_{12}O$:

CH_3-O-$CH_2CH_2CH_2CH_3$

methyl butyl ether

CH_3CH_2-O-$CH_2CH_2CH_3$

ethyl propyl ether

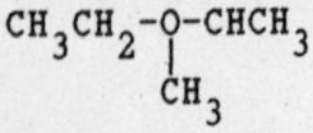

ethyl isopropyl ether

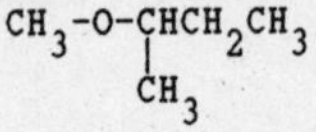

methyl sec-butyl ether

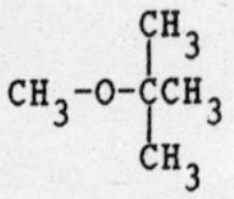

methyl tert-butyl ether

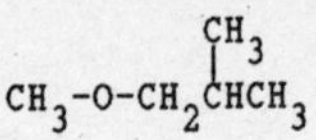

methyl isobutyl ether

33. The complete combustion of ethyl ether:

$$CH_3CH_2OCH_2CH_3 + 6\ O_2 \longrightarrow 4\ CO_2 + 5\ H_2O$$

34. Possible combination of reactants to make the following products by the Williamson synthesis:

a. $CH_3CH_2OCH_3$ — CH_3ONa + CH_3CH_2Cl

CH_3CH_2ONa + CH_3Cl

b. $CH_3CH_2CH_2OCH_2CH_2CH_3$ — $CH_3CH_2CH_2ONa$ + $CH_3CH_2CH_2Cl$

c. 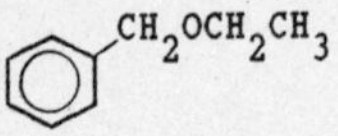

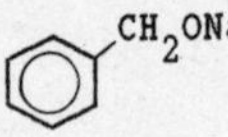 + CH_3CH_2Cl

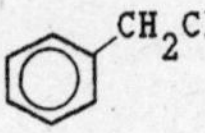 + CH_3CH_2ONa

d. 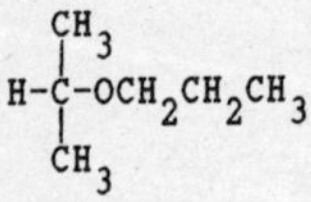

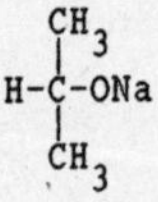 + $CH_3CH_2CH_2Cl$

$CH_3CH_2CH_2ONa$ + 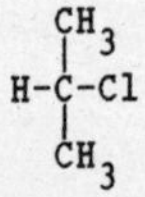

35. a. 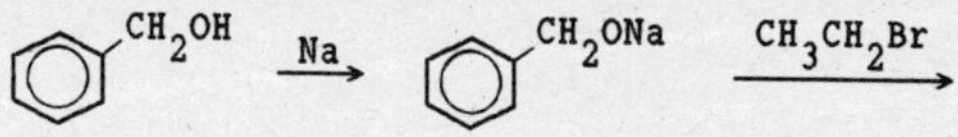

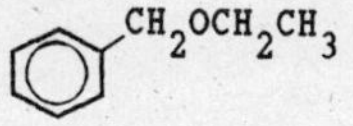 benzyl ethyl ether

b. 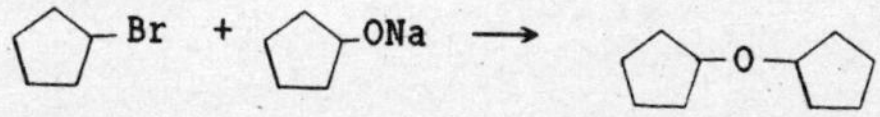

dicyclopentyl ether

36. A simple chemical test to distinguish between:

a. ethanol and dimethyl ether. Ethanol will react readily with potassium dichromate and sulfuric acid to make acetaldehyde. Visibly, the orange color of the dichromatic changes to green. Ethanol reacts with metallic sodium to produce hydrogen gas. Dimethyl ether does not react with either of these reagents.

b. 1-pentanol and 1-pentene. 1-pentene will rapidly decolorize bromine as it adds to the double bond. 1-pentanol does not react.

c. p-cresol and methyl phenyl ether. p-cresol will react with sodium hydroxide. Methyl phenyl will not react.

37. The correct statements are: a, b, e, g, h, i, j, m, o, and q.

B. 1. a. $R-\overset{H}{\overset{|}{C}}=O$

aldehyde

b. $R-\overset{O}{\overset{\|}{C}}-R$

Ketone

c. $H-\overset{O}{\overset{\|}{C}}-(CH_2)_n-\overset{O}{\overset{\|}{C}}-H$

a dialdehyde

d. $R-\underset{H}{\overset{OH}{C}}-OR'$

a hemiacetal

e. $R-\underset{R'}{\overset{OH}{C}}-OR''$

a hemiketal

f. $R-\underset{H}{\overset{OR'}{C}}-OR'$

an acetal

g. $R-\underset{R'}{\overset{OR''}{C}}-OR''$

a ketal

h. $R-\underset{H}{\overset{OH}{C}}-CN$

a cyanohydrin

i. RMgX

a Grignard reagent

2. Names of aldehydes:

a. $H_2C=O$ — methanal, formaldehyde

b. $CH_3CH_2CH_2\overset{H}{C}=O$ — butanal, butyraldehyde

c. $CH_3\underset{CH_3}{C}HCH_2\overset{H}{C}=O$ — 3-methylbutanal

d. 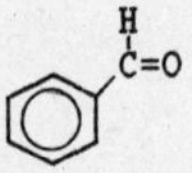 — benzaldehyde

e. $O=\overset{H}{C}-CH_2CH_2\overset{H}{C}=O$ — butanedial

f. 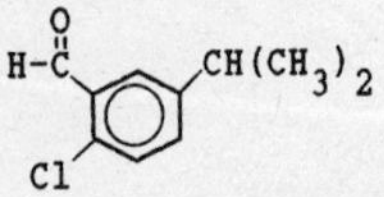 — 2-chloro-5-ispropylbenzaldehyde

g. 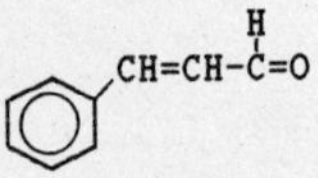 — 3-phenyl-2-propenal

h. $CH_3\underset{OH}{C}HCH_2\overset{H}{C}=O$ — 3-hydroxybutanal, β-hydroxybutyraldehyde

i. 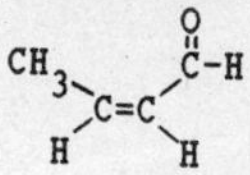 cis-2-butenal

3. Names of ketones:

a. $CH_3\overset{O}{\overset{\|}{C}}CH_3$ propanone, acetone, dimethyl ketone

b. $CH_3CH_2\overset{O}{\overset{\|}{C}}CH_3$ 2-butanone, methyl ethyl ketone

c. 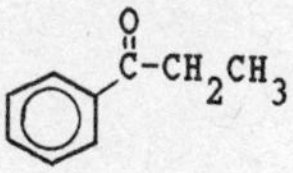 1-phenyl-1-propanone, phenyl ethyl ketone, propiophenone

d. 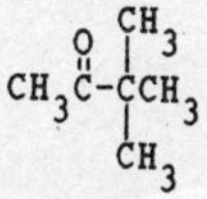 3,3-dimethyl-2-butanone, methyl t-butyl ketone

e. 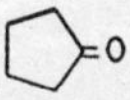 cyclopentanone

f. $CH_3\overset{O}{\overset{\|}{C}}CH_2CH_2\overset{O}{\overset{\|}{C}}CH_3$ 2,5-hexanedione

g. 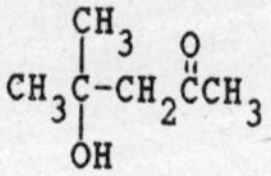 4-hydroxy-4-methyl-2-pentanone, diacetone alcohol

h. 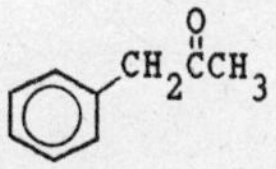 1-phenyl-2-propanone, methyl benzyl ketone

4. Structures for:

a. 1,3-dichloropropanone $ClCH_2\overset{O}{\overset{\|}{C}}CH_2Cl$

b. phenylacetaldehyde 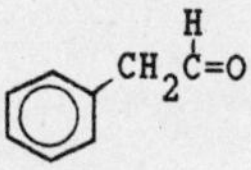

c. 3-butenal $CH_2{=}CHCH_2\underset{H}{\underset{|}{C}}{=}O$

d. 3-hydroxypropanal $HOCH_2CH_2\overset{H}{\overset{|}{C}}=O$

e. diisopropyl ketone $CH_3\overset{CH_3}{\overset{|}{C}H}-\underset{O}{\underset{\|}{C}}-\overset{CH_3}{\overset{|}{C}H}-CH_3$

f. 4-methyl-3-hexanone $CH_3CH_2-\overset{O}{\overset{\|}{C}}-\overset{CH_3}{\overset{|}{C}H}CH_2CH_3$

g. hexanal $CH_3CH_2CH_2CH_2CH_2\underset{H}{\underset{|}{C}}=O$

h. cyclohexanone

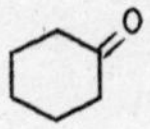

5. Propanal $CH_3CH_2\overset{H}{\overset{|}{C}}=O$ propanone $CH_3-\overset{O}{\overset{\|}{C}}-CH_3$

Each has the molecular formula C_3H_6O, so aldehydes and ketones appear to be isomeric with each other.

Butanal $CH_3CH_2CH_2\overset{H}{\overset{|}{C}}=O$ 2-butanone $CH_3-\overset{O}{\overset{\|}{C}}-CH_2CH_3$

Each has the molecular formula C_4H_8O, so the generalization seems to check out.

6. Aldehydes and ketones have lower boiling points than alcohols of similar molecular weight because they do not form hydrogen bonds, since they contain no -OH groups.

7. Higher boiling point

a. 1-pentanol b. pentanal c. benzyl alcohol
d. 2-pentanol e. butanone

8. Equations for the oxidation of:

a. 1-propanol

$$CH_3CH_2CH_2OH \xrightarrow[H_2SO_4]{K_2Cr_2O_7} CH_3CH_2\overset{H}{\overset{|}{C}}=O + H_2O$$

$$2\ CH_3CH_2CH_2OH + O_2 \xrightarrow[450°C]{Ag\ or\ Cu} 2_3CH_2CH\ \overset{H}{\overset{|}{C}}=O + 2\ H_2O$$

b. 3-pentanol

$$CH_3CH_2\overset{OH}{\overset{|}{C}}HCH_2CH_3 \xrightarrow[H_2SO_4]{K_2Cr_2O_7} CH_3CH_2\overset{O}{\overset{\|}{C}}CH_2CH_3$$

$$2\ CH_3CH_2\overset{OH}{\overset{|}{C}}HCH_2CH_3 + O_2 \xrightarrow[\Delta]{Ag\ or\ Cu} 2\ CH_3CH_2\overset{O}{\overset{\|}{C}}CH_2CH_3 + 2\ H_2O$$

c. 2,3-dimethyl-2-butanol

There is no reaction by either method because this is a tertiary alcohol, resistant to oxidations other than extreme measures which would result in the breaking of C-C bonds.

9. The alcohols which should be oxidized to give these ketones:

a. diethyl ketone: 3-pentanol $CH_3CH_2\overset{OH}{\overset{|}{C}}HCH_2CH_3$

b. diisopropyl ketone: 2,4-dimethyl-3-pentanol $CH_3\overset{CH_3}{\overset{|}{C}}H\underset{OH}{\underset{|}{C}}H\overset{CH_3}{\overset{|}{C}}HCH_3$

c. 4-phenyl-2-butanone: 4-phenyl-2-butanol

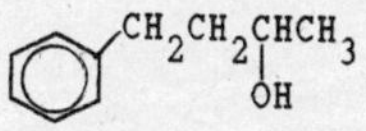

10. a. An aldehyde group, $-\overset{H}{\overset{|}{C}}=O$, must be present to give a positive Tollens test.

b. The visible evidence for a positive Tollens test is the formation of a silver mirror on the inner walls of a test tube.

c. $CH_3\overset{H}{\overset{|}{C}}=O + 2\ Ag^+ \xrightarrow[H_2O]{NH_3} CH_3COONH_4 + 2\ Ag(s)$ (silver mirror)

11. a. An aldehyde group, $-\overset{H}{\overset{|}{C}}=O$, must be present to give a positive Fehling test.

b. The visible evidence of a positive Fehling test is the formation of brick-red, Cu_2O, which precipitates during the reaction.

c. $CH_3\overset{H}{\overset{|}{C}}=O + Cu^{2+} \xrightarrow[H_2O]{NaOH} CH_3COONa + Cu_2O(s)$ (brick-red)

12. Give the major product when each of the following is treated with Tollens reagent:

a. butanal ⟶ butyric acid

b. benzaldehyde ⟶ benzoic acid

c. methyl ethyl ketone ⟶ no reaction

13. Equations for the aldol condensations for the following compounds:

a. acetaldehyde $2\ CH_3\overset{H}{\overset{|}{C}}=O \xrightarrow[NaOH]{dil} CH_3\overset{OH}{\overset{|}{C}}HCH_2\overset{H}{\overset{|}{C}}=O$

b. propanal $2\ CH_3CH_2CH{=}O \xrightarrow[NaOH]{dil} CH_3CH_2CH(OH)CH(CH_3)CH{=}O$

c. butanal $2\ CH_3CH_2CH_2CH{=}O \xrightarrow[NaOH]{dil} CH_3CH_2CH_2CH(OH)CH(CH_2CH_3)CH{=}O$

d. 3-pentanone $2\ CH_3CH_2C(=O)CH_2CH_3 \xrightarrow[NaOH]{dil} CH_3CH_2C(OH)(CH_2CH_3)\text{-}CH(CH_3)\text{-}C(=O)CH_2CH_3$

14. Four aldol condensation products are possible if a mixture of acetaldehyde (A) and propanal (P) are reacted:
A-A, P-P, A-P, P-A.

15. The completed equations are:

a. $CH_3CH_2CH{=}O + 2\ CH_3CH_2CH_2OH \xrightleftharpoons{\text{dry HCl}} CH_3CH_2\text{-}CH(OCH_2CH_2CH_3)\text{-}OCH_2CH_2CH_3 + H_2O$

b. $CH_3C(=O)CH_3 + HOCH_2CH_2OH \xrightleftharpoons{\text{dry HCl}} (CH_3)_2C(OCH_2)_2 + H_2O$

c. $CH_3CH_2CH{=}O + CH_3CH_2OH \xrightleftharpoons{H^+} CH_3CH_2CH(OH)\text{-}OCH_2CH_3$

d.

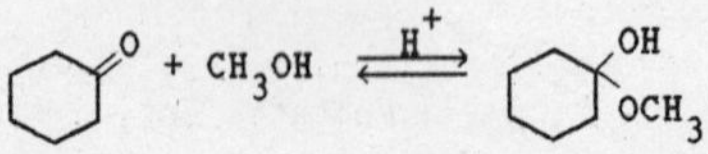

e. $CH_3CH_2CH_2CH(OCH_3)_2 \xrightarrow[H^+]{H_2O} CH_3CH_2CH_2CH{=}O + 2\ CH_3OH$

16.

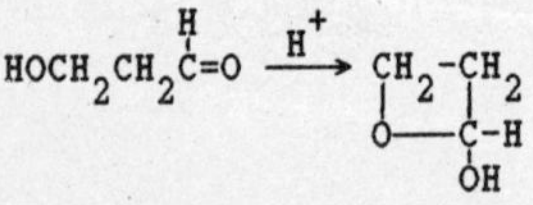

17. Sequence of reactions:

a.

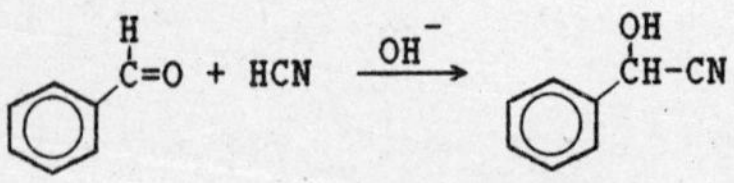

b. 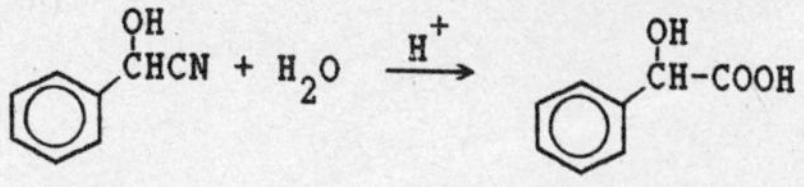

c. 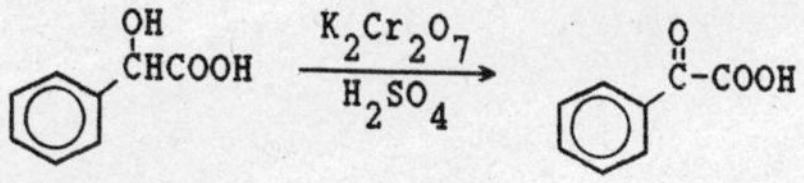

18. $CH_3\overset{H}{C}{=}O \xrightarrow{HCN} CH_3\overset{OH}{C}HCN \xrightarrow[H^+]{H_2O} CH_3CH(OH)COOH$
lactic acid

19. To give a positive iodoform test, a molecule must have, or be easily converted to, the methyl carbonyl structure:

$CH_3\overset{|}{C}{=}O$

20. Of the compounds shown, the following will give a positive iodoform test:

b. CH_3CH_2OH d. $CH_3\underset{OH}{CH}CH_3$ e. $CH_3\underset{H}{C}{=}O$

f. $CH_3CH_2\overset{CH_3}{C}{=}O$ h. $CH_3\underset{CH_3}{CH}\overset{O}{C}CH_3$ j. 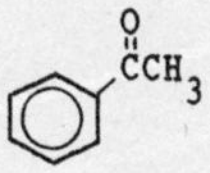

21. Of the four isomeric butyl alcohols only one, $CH_3CH_2\overset{OH}{C}HCH_3$, 2-butanol, will give a positive iodoform test.

22. a. $Mg + CH_3CH_2Br \xrightarrow{\text{dry ether}} CH_3CH_2MgBr$

b. (1) $CH_3CH_2MgBr + H\underset{H}{C}{=}O \xrightarrow{\text{dry ether}} CH_3CH_2CH_2OMgBr$

$\xrightarrow{H_2O} CH_3CH_2CH_2OH$ (1-propanol)

(2) $CH_3CH_2MgBr + CH_3CH_2CH_2\underset{H}{C}{=}O \xrightarrow{\text{dry ether}}$

$CH_3CH_2CH_2\overset{OMgBr}{C}HCH_2CH_3 \xrightarrow{H_2O} CH_3CH_2CH_2\overset{OH}{C}HCH_2CH_3$ (3-hexanol)

(3) $CH_3CH_2MgBr + CH_3CH_2\underset{O}{\overset{\|}{C}}CH_3 \xrightarrow{\text{dry ether}} CH_3CH_2\underset{CH_3}{\overset{OMgBr}{C}}CH_2CH_3$

$\xrightarrow{H_2O}$ $CH_3CH_2\overset{OH}{\underset{CH_3}{C}}CH_2CH_3$ (3-methyl-3-pentanol)

(4) $CH_3CH_2MgBr + CH_3\underset{CH_3}{C}H\overset{H}{C}{=}O \xrightarrow{\text{dry ether}} CH_3\underset{CH_3}{C}H\overset{OMgBr}{C}HCH_2CH_3$

$\xrightarrow{H_2O}$ $CH_3\underset{CH_3}{C}H\overset{OH}{C}HCH_2CH_3$ (2-methyl-3-pentanol)

(5) 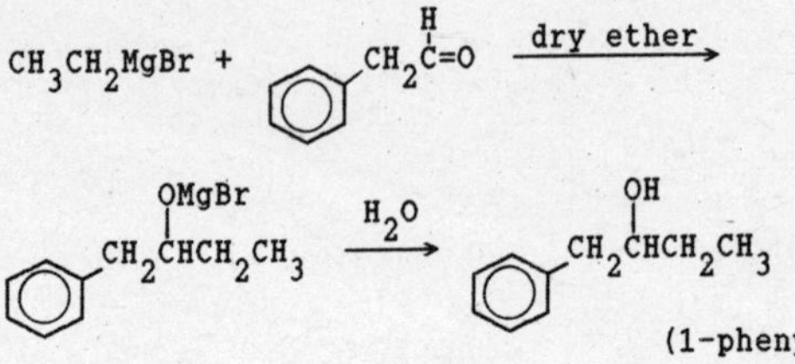

(1-phenyl-2-butanol)

23. What combination(s) of Grignard reagent and aldehyde or ketone are needed to prepare:

a. t-butyl alcohol — $CH_3\underset{O}{\underset{\|}{C}}CH_3$ and CH_3MgBr

b. 2-pentanol — $CH_3CH_2CH_2\underset{H}{C}{=}O$ and CH_3MgBr or

$CH_3\underset{H}{C}{=}O$ and $CH_3CH_2CH_2MgBr$

c. 1-hexanol — $H\underset{H}{C}{=}O$ and $CH_3CH_2CH_2CH_2CH_2MgBr$

d. benzyl alcohol — $H\underset{H}{C}{=}O$ and 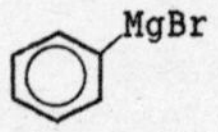

e. 1-phenylethanol — $CH_3\underset{H}{C}{=}O$ and 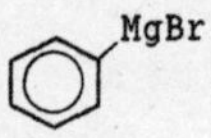

24. A test which will distinguish between the compounds in each pair:

a. $CH_3CH_2\overset{H}{C}{=}O$ and $CH_3\overset{O}{\overset{\|}{C}}CH_3$. The Tollens test or Fehling test will give positive results with the first compound (the aldehyde), but not with the ketone.

b. $CH_3\overset{H}{\overset{|}{C}}=O$ and $CH_3CH_2\overset{H}{\overset{|}{C}}=O$. The first compound will give a positive iodoform test.

c. $CH_3CH_2\overset{H}{\overset{|}{C}}=O$ and $CH_2=CH\overset{H}{\overset{|}{C}}=O$. Bromine will decolorize immediately with the second compound (the alkene).

d. 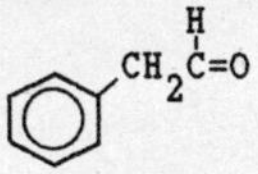 and 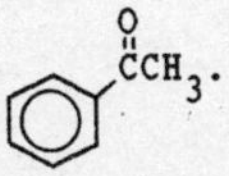. The first compound (an aldehyde) will give a positive Tollens test or Fehling test. Methyl phenyl ketone will give a positive iodoform test.

e. 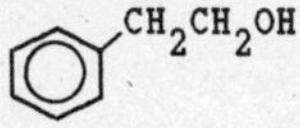 and 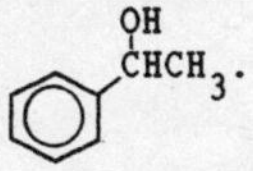. The second compound, 1-phenylethanol, will give a positive iodoform test.

25. The compounds A, B, C are butyl alcohols, C_4H_9OH.

a. The oxidation product of A is a aldehyde because A gives a positive Tollens test. Therefore, A must be a primary alcohol. A is $CH_3CH_2CH_2CH_2OH$.

b. The oxidation product of B has a methyl ketone group because it gives a positive iodoform test. B must then be a secondary alcohol. B is $CH_3CH_2CH(OH)CH_3$.

c. C is not oxidized (no reaction) so it must be a tertiary alcohol. C is $(CH_3)_3C\text{-}OH$.

26. $$\underset{A}{C_5H_{12}O} \xrightarrow[H_2SO_4]{K_2Cr_2O_7} \underset{B}{C_5H_{10}O}$$

Reaction with $K_2Cr_2O_7$ and H_2SO_4 indicates that A is oxidized to an aldehyde or a ketone. A negative Tollens test indicates that B is not an aldehyde. A negative iodoform test indicates that B does not have a methyl carbonyl ($CH_3\underset{|}{C}=O$) structure. Therefore, compounds A and B must be

$$\underset{A}{CH_3CH_2\overset{OH}{\overset{|}{C}}HCH_2CH_3} \qquad \underset{B}{CH_3CH_2\overset{O}{\overset{\|}{C}}CH_2CH_3}$$

27. $$\underset{\text{D}}{C_5H_{12}O} \xrightarrow[\text{Air}]{\text{Cu, }\Delta} \underset{\text{E}}{C_5H_{10}O}$$

Since E gives a positive Tollens and a negative iodoform test it must contain an aldehyde; compound D, therefore, must contain a primary alcohol group.

$$C_5H_{12}O \xrightarrow{-H_2O} \xrightarrow{\text{hydrogenation}} \underset{\text{F}}{CH_3\underset{\substack{|\\ CH_3}}{C}HCH_2CH_3}$$

The data establishes the carbon skeleton structure of compounds D and E. Compound D is a primary alcohol so the OH group can be on C-1 or C-4 of 2-methylbutane. Therefore,

compound D is $CH_3\overset{\substack{CH_3\\ |}}{C}HCH_2CH_2OH$ or $CH_3CH_2\overset{\substack{CH_3\\ |}}{C}HCH_2OH$ and

compound E is $CH_3\underset{\substack{|\\ CH_3}}{C}HCH_2\underset{\substack{|\\ H}}{C}{=}O$ or $CH_3CH_2\underset{\substack{|\\ CH_3}}{C}H\overset{\substack{H\\ |}}{C}{=}O$

28. The correct statements are: b, c, f, h, i, k, m, o, p, r, t, u, w, and y.

B. 1. a. CH_3COOH

b. 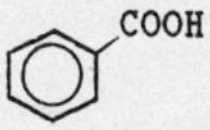

c. $CH_3CHCOOH$ with OH on the CH

d. $CH_3CHCOOH$ with NH_2 on the CH

e. CH_2CH_2COOH with Cl on the first CH_2

f. $COOH$–$COOH$

g. $CH_2{=}CHCOOH$

h. $CH_2C(=O){-}OCH_3$

i. CH_3CN

j. CH_3COONa

k. $CH_3C(=O){-}NH_2$

l. $CH_3C(=O){-}Cl$

m. $CH_2OC(=O)C_{17}H_{35}$
$CH{-}OC(=O)C_{17}H_{35}$
$CH_2OC(=O)C_{17}H_{35}$

n. CH_3NH_2 1° amine
$(CH_3)_2NH$ 2° amine
$(CH_3)_3N$ 3° amine

o. 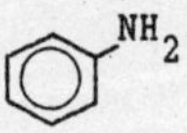

p. 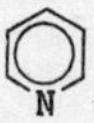

q. $(CH_3)_4N^+Cl^-$

2. Names:

a.	$CH_3(CH_2)_5COOH$	heptanoic acid
b.	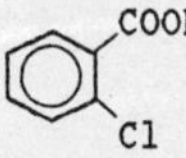	o-chlorobenzoic acid
c.	$CH_3CH{=}CHCOOH$	2-butenoic acid
d.	$CH_3(CH_2)_7CH{=}CH(CH_2)_7COOH$	oleic acid
e.	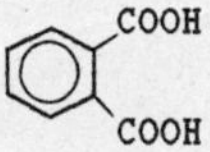	o-phthalic acid
f.	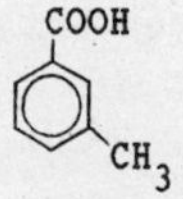	m-toluic acid
g.	$CH_3(CH_2)_{16}COOH$	stearic acid, octadecanoic acid
h.	$CH_3CH_2CH(OH)COOH$	2-hydroxybutanoic acid α-hydroxybutyric acid
i.	$CH_3CH_2COO^-Na^+$	sodium propanoate
j.	$CH_3CH_2COO^-NH_4^+$	ammonium propanoate

3. Structures:

   a. Caproic acid $CH_3(CH_2)_4COOH$

   b. Adipic acid $HOOC(CH_2)_4COOH$

   c. o-Toluic acid 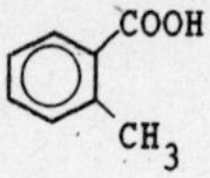

   d. Oxalic acid HOOCCOOH

   e. β-Hydroxybutyric $CH_3CH(OH)CH_2COOH$

   f. p-Phthalic acid 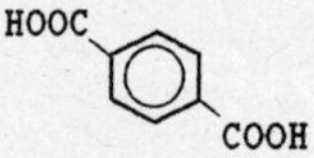

   g. Sodium benzoate 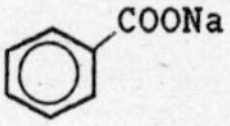

   h. Linoleic acid $CH_3(CH_2)_4CH{=}CHCH_2CH{=}CH(CH_2)_7COOH$

   i. 2-Chloropropanoic acid $CH_3CHClCOOH$

   j. Ammonium benzoate 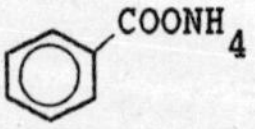

   k. p-Aminobenzoic acid 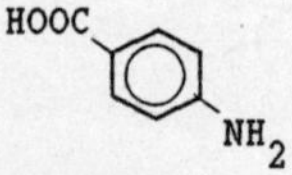

   l. o-Phthalic acid 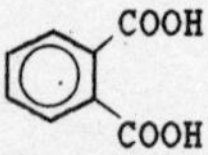

4. The butyric acid solution would be expected to have the more objectionable odor because salts normally exhibit little or no odor since salts are ionic and therefore have low volatility. For example, dilute solutions of acetic acid (vinegar) have considerable odor, but sodium acetate does not.

5. If a 2.0% solution of sodium stearate were acidified with a mineral acid, such as HCl, the insoluble stearic acid would precipitate. It could then be filtered and dried.

6. Arrangement in order of increasing pH for 0.01 M solutions:

   HCl, CH_3COOH, 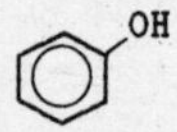 , NaCl, CH_3NH_2, NaOH

7. a. $H\overset{\overset{O}{\|}}{C}OCH_3$ — methyl formate or methyl methanoate

b. 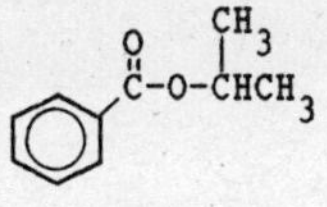 isopropyl benzoate or 2-propyl benzoate

c. 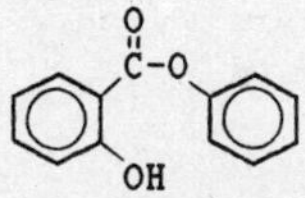 phenyl salicylate

d. $CH_3CH_2\overset{\overset{O}{\|}}{C}OCH_2CH_3$ — ethyl propionate, ethyl propanoate

e. $CH_3CH_2O\overset{\overset{O}{\|}}{C}-\overset{\overset{O}{\|}}{C}OCH_2CH_3$ — diethyl oxalate

f. $CH_2=CH\overset{\overset{O}{\|}}{C}-OCH_3$ — methyl-2-propenoate, methyl acrylate

8. a. methyl formate — $H-\overset{\overset{O}{\|}}{C}-OCH_3$

b. n-octyl acetate — $CH_3\overset{\overset{O}{\|}}{C}-O-(CH_2)_7CH_3$

c. n-propyl propanoate — $CH_3CH_2\overset{\overset{O}{\|}}{C}-O-CH_2CH_2CH_3$

d. ethyl benzoate — 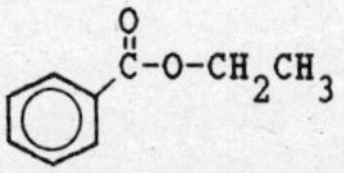

9. Write the structural formula and name of the principal organic product of each reaction.

a. $CH_3(CH_2)_7CH=CH(CH_2)_7\overset{\overset{O}{\|}}{C}OH \xrightarrow[H_2]{Ni} CH_3(CH_2)_{16}\overset{\overset{O}{\|}}{C}OH$ stearic acid

b. $CH_3CH_2\overset{\overset{O}{\|}}{C}OH + NH_3 \xrightarrow{\Delta} CH_3CH_2\overset{\overset{O}{\|}}{C}NH_2$ propanamide

c. 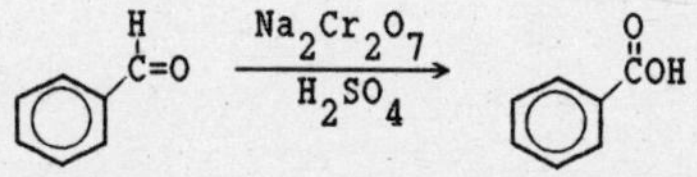 benzoic acid

d. 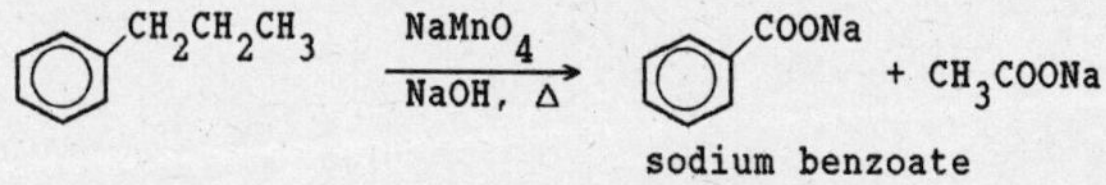

sodium benzoate

e. 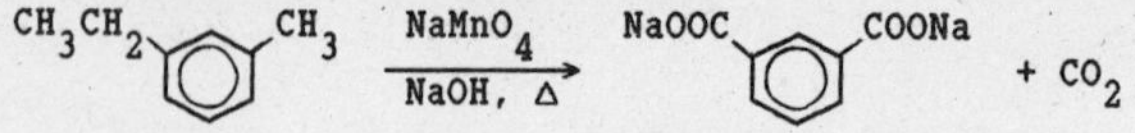

sodium isophthalate

f. $CH_3CH_2COOH + NaOH \longrightarrow CH_3CH_2COONa$ sodium propanoate

g. $CH_3CH_2CH_2CH_2CH_2\overset{O}{\overset{\|}{C}}OCH_2CH_2CH_3 \xrightarrow{NaOH} CH_3CH_2CH_2CH_2CH_2\overset{O}{\overset{\|}{C}}ONa$
sodium caproate

$+ CH_3CH_2CH_2OH$
1-propanol

10.

Acid	Amide: Common Name	Amide: IUPAC Names
formic acid	formamide	methanamide
acetic acid	acetamide	ethanamide
propionic acid	propionamide	propanamide
butyric	butyramide	butanamide
valeric acid	valeramide	pentanamide
caproic acid	caproamide	hexanamide
caprylic acid	caprylamide	octanamide
capric acid	capramide	decanamide

11. Write structural formulas for the products of the following reactions:

a. 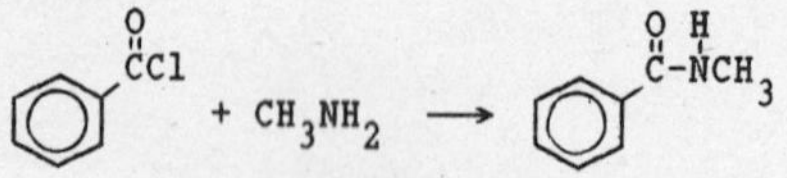

b. 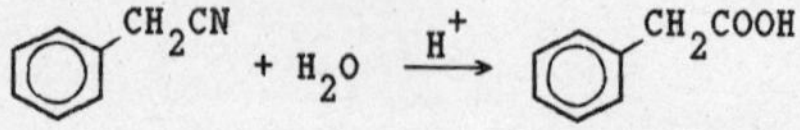

c. $CH_3CH_2CH_2Br + KCN \longrightarrow CH_3CH_2CH_2CN \xrightarrow{LiAlH_4} CH_3CH_2CH_2CH_2NH_2$

d. $CH_3CH_2\overset{O}{\overset{\|}{C}}NHCH_3 \xrightarrow{LiAlH_4} CH_3CH_2CH_2NHCH_3$

e. $CH_3CH_2CH_2CN \xrightarrow{H_2,\ Ni} CH_3CH_2CH_2CH_2NH_2$

12. Write structural formulas for the products of the following reactions:

a. $CH_3CH(COOH)_2 \xrightarrow{150°} CH_3CH_2COOH + CO_2$

b. 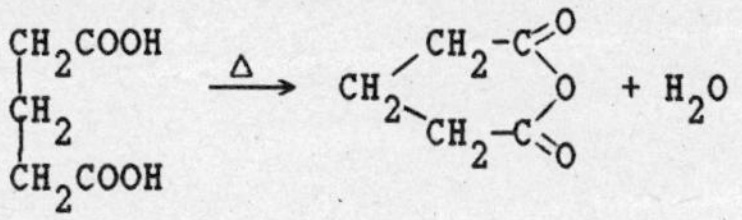

Glutaric anhydride

c. 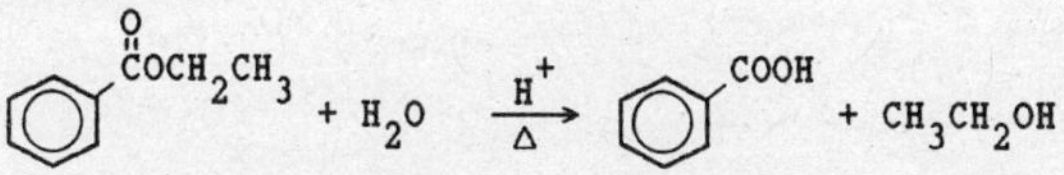

d. 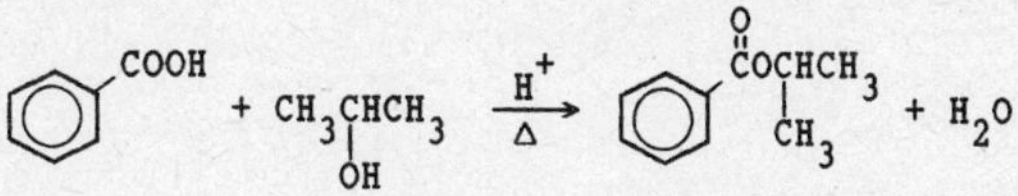

e. 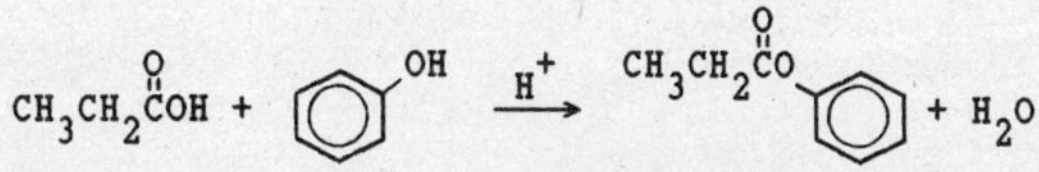

f. $$\begin{matrix}COOH\\ |\\ CH_2\\ |\\ COOH\end{matrix} + 2\ C_2H_5OH\ (2\ mol) \xrightarrow{H^+} \begin{matrix}COOC_2H_5\\ |\\ CH_2\\ |\\ COOC_2H_5\end{matrix} + 2\ H_2O$$

13. Write structural formulas for the organic products formed in the following reactions:

a. 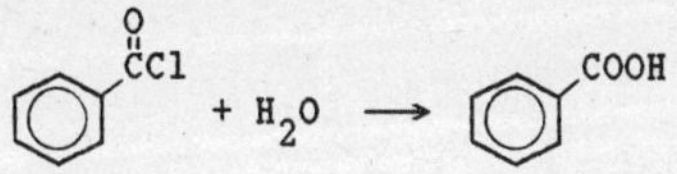

b. $CH{=}CHCOOH + Br_2 \longrightarrow CH_2BrCHBrCOOH$

c. $CH_3COOH + SOCl_2 \longrightarrow CH_3\overset{O}{\overset{\|}{C}}Cl$

d. $CH_3\overset{O}{\overset{\|}{C}}Cl + NH_3 \longrightarrow CH_3\overset{O}{\overset{\|}{C}}NH_2$

e. $CH_3\overset{O}{\overset{\|}{C}}Cl + CH_3CH_2OH \longrightarrow CH_3\overset{O}{\overset{\|}{C}}OCH_2CH_3$

f. 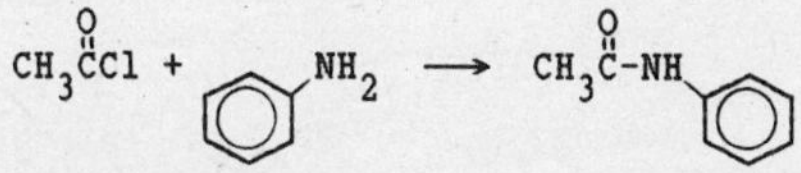

14. Write the structural formula for the ester that when hydrolyzed would yield:

a. methanol and acetic acid

$CH_3\overset{O}{\overset{\|}{C}}OCH_3$

b. ethanol and formic acid $\mathrm{H\overset{O}{\overset{\|}{C}}OCH_2CH_3}$

c. 1-octanol and acetic acid $\mathrm{CH_3\overset{O}{\overset{\|}{C}}O(CH_2)_7CH_3}$

d. 2-propanol and benzoic acid

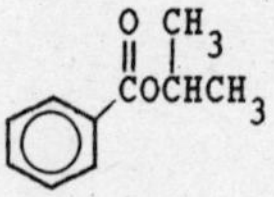

e. methanol and salicylic acid

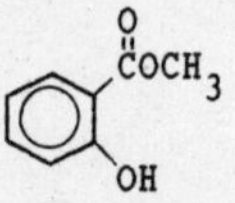

15. The amide formed from ammonia and acetic acid, acetamide, is molecular, both in the crystalline form and when dissolved in water. The ammonium salt formed from ammonia and acetic acid, ammonium acetate, is ionic, both in crystalline form and in solution. In solution, ammonium acetate dissociates, giving free ammonium ions and acetate ions.

16. Write structural formulas for:

a. isopropyl formate $\mathrm{H\overset{O}{\overset{\|}{C}}O\overset{CH_3}{\overset{|}{C}}HCH_3}$

b. methyl palmitate $\mathrm{CH_3(CH_2)_{14}\overset{O}{\overset{\|}{C}}OCH_3}$

c. diethyl adipate $\mathrm{CH_3CH_2O\overset{O}{\overset{\|}{C}}(CH_2)_4\overset{O}{\overset{\|}{C}}OCH_2CH_3}$

d. benzyl benzoate

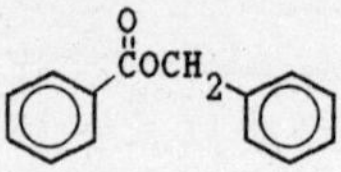

e. phenyl propionate

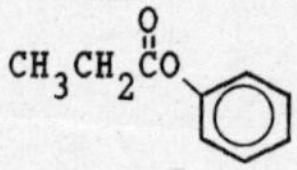

f. methyl-2-aminopentanoate $\mathrm{CH_3CH_2CH_2\underset{NH_2}{\underset{|}{C}}H\overset{O}{\overset{\|}{C}}OCH_3}$

17. Simple test to distinguish between:

a. Sodium benzoate is water soluble; benzoic acid is not.

b. Maleic acid and Malonic acid. Maleic acid has a carbon-carbon double bond, it will readily add and decolorize bromine; malonic acid will not.

18. a. cis-oleic acid

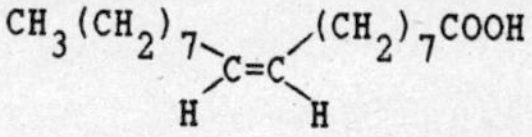

b. cis,cis-linoleic acid

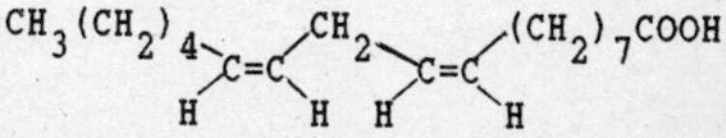

19.

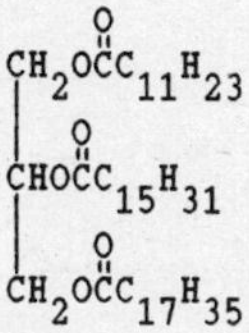

There would be two other triacylglycerols containing all three of these acids. Each of the three can be in the middle.

20. Names and formulas of products are:

a.

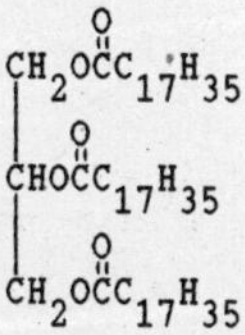

(glyceryl tristearate)

b. CH_2OH / $CHOH$ / CH_2OH (glycerol) and $CH_3(CH_2)_7CH{=}CH(CH_2)_7COOH$ (oleic acid)

c. CH_2OH / $CHOH$ / CH_2OH (glycerol) and $CH_3(CH_2)_7CH{=}CH(CH_2)_7\overset{O}{\overset{\|}{C}}{-}O^-K^+$ (potassium oleate)

d. CH_2OH / $CHOH$ / CH_2OH (glycerol) and $CH_3(CH_2)_{16}CH_2OH$ (stearyl alcohol) (1-octadecanol)

21. Which has the greater solubility in water?

a. Propanoic acid is more soluble than methyl propanoate.
b. Sodium palmitate is more soluble than palmitic acid.
c. Decylammonium chloride is more soluble than decylamine.
d. Sodium stearate is more soluble than barium stearate.
e. Sodium phenoxide is more soluble than phenol.

22. Structural formulas:

a. methyl propanoate $CH_3CH_2\overset{O}{\overset{\|}{C}}OCH_3$

propanoic acid $CH_3CH_2\overset{O}{\overset{\|}{C}}OH$

b. sodium palmitate $CH_3(CH_2)_{14}COONa$

palmitic acid $CH_3(CH_2)_{14}COOH$

c. decylamine $CH_3(CH_2)_9NH_2$

decylammonium chloride $CH_3(CH_2)_9\overset{+}{N}H_3Cl^-$

d. sodium stearate $CH_3(CH_2)_{16}COONa$

barium stearate $[CH_3(CH_2)_{16}COO]_2Ba$

e. phenol

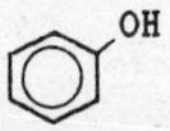

sodium phenoxide

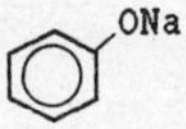

23. a. The major difference between fats and oils is that fats are solids at room temperature, oils are liquids. The fatty acids in the molecules are mostly saturated in fats; more unsaturated in oils.

b. Soaps are the sodium salts of high molecular weight fatty acids. Syndets are synthetic detergents and occur in several different forms. They have cleansing action similar to soaps, but have different structural and solubility characteristics, such as being soluble in hard water. Some syndets also contain long hydrocarbon chains.

c. Hydrolysis is the breaking apart of an ester in the presence of water to form an alcohol and a carboxylic acid. Mineral acids or digestive enzymes are used to speed the process. Saponification breaks the ester apart using sodium hydroxide to form alcohols and salts of carboxylic acids.

24. The principal advantage of synthetic detergents over soap is that the syndets do not form insoluble precipitates with the ions in hard water (Cu^{2+}, Mg^{2+}, Fe^{3+}).

25. $CH_3(CH_2)_{12}COONa$ would be more useful than $CH_3(CH_2)_{12}COOH$ as a cleaning agent in soft water. Both have a long hydrocarbon molecule which would dissolve in the fat, but the acid is not water soluble, whereas the sodium salt is soluble in water.

26. Sodium lauryl sulfate would be more effective as a detergent in hard water than sodium propyl sulfate because the hydrocarbon chain is only three carbons long in the latter, not long enough to dissolve grease well.

27. Like a soap, a detergent contains a grease soluble component and water soluble component. The grease soluble portion of many molecules dissolve in the grease film. The water soluble components attract water, causing small droplets of grease-bearing dirt to break loose and float away.

28. Only (a), hexadecyltrimethyl ammonium chloride would be a good detergent in water. It is cationic.

29. Monoglycerides function as emulsifying agents in making oleomargarine. A monoglyceride is able to emulsify water in oil because the hydrocarbon end of the molecule is soluble in oil or fat, and the glycerol end (hydroxy groups) is soluble in water.

30. Each molecule of triolein requires three molecules of H_2 (one for each double bond).

a. $1.00\ \text{kg triolein} \times \frac{10^3\ \text{g}}{\text{kg}} \times \frac{1\ \text{mol triolein}}{886\ \text{g triolein}} \times \frac{22.4\ \text{L}}{1\ \text{mol}\ H_2}$
$= 75.8\ \text{L}\ H_2$

b. $1.00\ \text{kg triolein} \times \frac{890\ \text{g tristearin}}{884\ \text{g triolein}} = 1.01\ \text{kg tristearin}$

31. The trimethylamine in 1.0 N NaOH would have the more objectionable odor because in the acid it would form a salt which would have less odor or no odor.

32. a. Structures of amines with formula C_3H_9N.

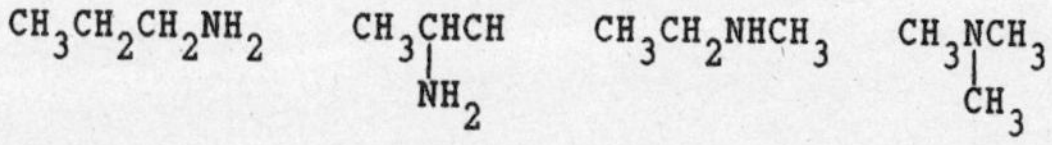

b. Structures of amines with formula $C_4H_{11}N$.

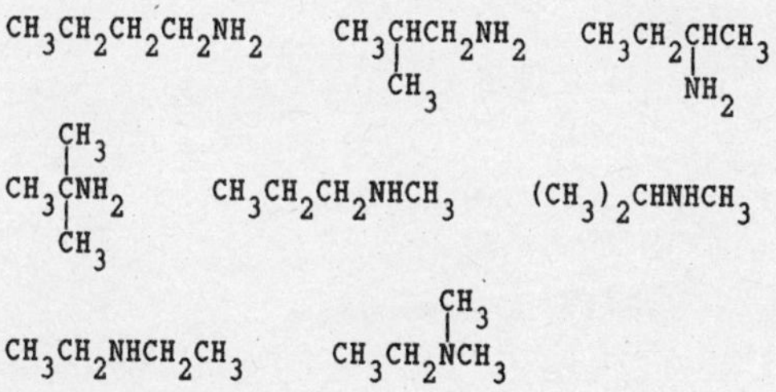

c. Structures of quaternary salts with formula $C_4H_{12}N^+Cl^-$

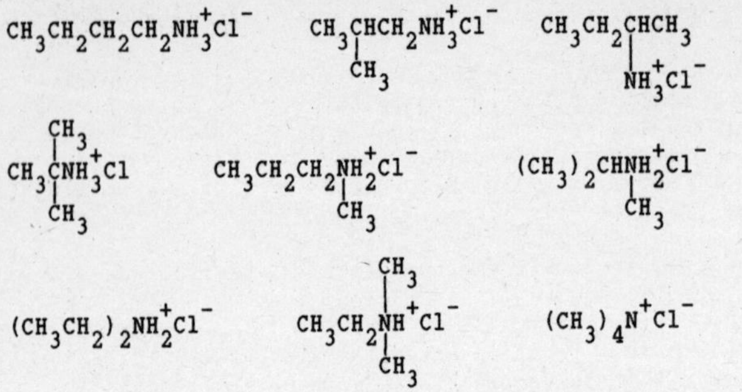

33. Name the compounds:

a. 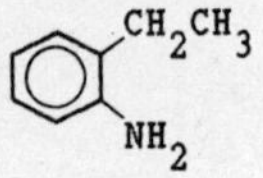 o-ethylaniline

b. $CH_3CH_2NH_3^+Br^-$ ethylammonium bromide

c. 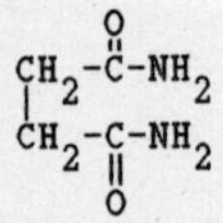 succinamide

d. 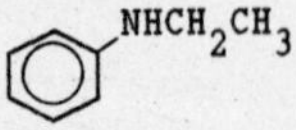 N-ethylaniline

e.  pyridine

f. $CH_3\overset{O}{\overset{\|}{C}}\text{-}NHC_2H_5$ N-ethylacetamide

g. 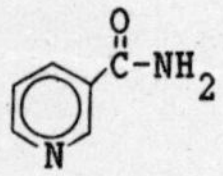 nicotinamide

h. 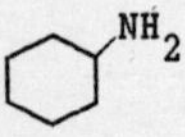 cyclohexylamine
$C_6H_{11}NH_2$

i. 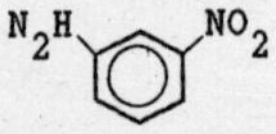 m-nitroaniline

j. $(C_2H_5)_4N^+I^-$ tetraethylammonium iodide

k. 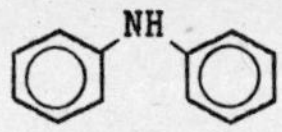 diphenylamine

34. Write structural formulas for the following:

a. methylethylamine $CH_3NHCH_2CH_3$

b. tributylamine $(CH_3CH_2CH_2CH_2)_3N$

c. aniline 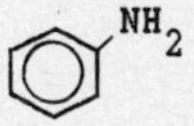

d. ethylammonium chloride $CH_3CH_2NH_3^+Cl^-$

e. N-methylanilinium chloride 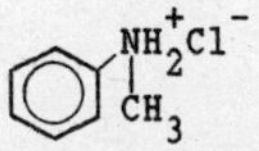

f. 1,4-butanediamine $H_2NCH_2CH_2CH_2CH_2NH_2$

g. N,N-dimethylaniline 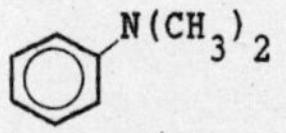

h. ethylisopropylmethylamine $CH_3CH_2NCH(CH_3)_2$ (with CH_3 on N)

i. pyridine 

j. 2-amino-1-pentanol $CH_3CH_2CH_2CHCH_2OH$ (with NH_2 on C-2)

35. a. $CH_3CH_2CH_2Br + NH_3 \longrightarrow CH_3CH_2CH_2NH_2$
(excess)

b. $CH_3CH_2CH_2Br + KCN \longrightarrow CH_3CH_2CH_2CN \xrightarrow[Ni]{H_2} CH_3CH_2CH_2CH_2NH_2$

c. 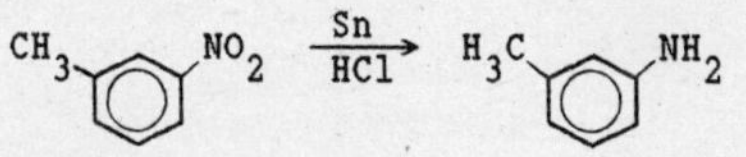

36. a. Heterocyclic compounds are those in which all the atoms in the ring are not alike.
b. The number of heterocyclic rings in each of the compounds is: (1) LSD, 2; (2) ampicillin, 2; (3) methadone, 0; (4) nicotine, 2.

37. One gram of acetic acid will require more base for neutralization than one gram of propanoic acid because it contains more moles,

having the smaller molecular weight, and both acids are monoprotic, requiring one mole of base per mole of acid.

38. Synthesis of:

a. acetic acid

$$CH_3CH_2OH \xrightarrow[H_2SO_4]{K_2Cr_2O_7} CH_3COOH$$

b. ethyl acetate

$$CH_3COOH + CH_3CH_2OH \overset{H^+}{\rightleftharpoons} CH_3\overset{O}{\overset{\|}{C}}OCH_2CH_3$$

c. N-ethylacetamide

$$CH_3COOH + NH_3 \xrightarrow{\Delta} CH_3\overset{O}{\overset{\|}{C}}\text{-}NH_2 \xrightarrow[H_2]{Ni} CH_3CH_2NH_2$$

$$CH_3COOH + SOCl_2 \longrightarrow CH_3\underset{\underset{O}{\|}}{C}Cl$$

$$CH_3\overset{O}{\overset{\|}{C}}Cl + CH_3CH_2NH_2 \longrightarrow CH_3\overset{O}{\overset{\|}{C}}\text{-}NHCH_2CH_3$$

d. β-hydroxybutyric acid

$$CH_3CH_2OH + \text{air} \xrightarrow{\text{Cu tube}} CH_3\overset{H}{\overset{|}{C}}=O$$

$$2\ CH_3\overset{H}{\overset{|}{C}}=O \xrightarrow{\text{dil NaOH}} CH_3\underset{\underset{OH}{|}}{C}HCH_2\overset{H}{\overset{|}{C}}=O \xrightarrow[\substack{NH_3 \\ \text{Tollens} \\ \text{reagent}}]{Ag_2O} CH_3\underset{\underset{OH}{|}}{C}HCH_2COOH$$

39. The ester would be propyl propanoate $CH_3CH_2\underset{\underset{O}{\|}}{C}OCH_2CH_2CH_3$.

Compound A is an acid; B is an alcohol. If B is oxidized to an acid which is the same as A, then A and B both have the same carbon structure, three carbon atoms each. The acid A must be propanoic acid, CH_3CH_2COOH. The alcohol can be 1-propanol or 2-propanal.

Only 1-propanol can be oxidized to propanoic acid which is the same as compound A.

40. The correct statements are: a, b, f, g, h, i, j, m, n, o, p, r, t, w, y, and z.

B. 1. An addition polymer is one that is produced by the successive addition of repeating monomer molecules. A condensation polymer is one that is formed from monomer molecules in a reaction which splits out water or some other simple molecule. Condensation polymerization usually involves two different monomers.

2. Those polymers which soften on reheating are thermoplastic polymers; those which set to an infusible solid and do not soften on reheating are thermosetting polymers.

3. The free radical mechanism for the polymerization of propylene to polypropylene:

   a. Free radical formation, using an organic peroxide

   RO:OR $\longrightarrow$ 2 RO· (free radical)

   b. Propagation of polymeric chain

   $RO\cdot + CH_3CH{=}CH_2 \longrightarrow ROCH(CH_3)CH_2\cdot$

   $ROCH(CH_3)CH_2\cdot + CH_3CH{=}CH_2 \longrightarrow ROCH(CH_3)CH_2CH(CH_3)\cdot$, etc.

4. How many ethylene units are in a polyethylene that has a molecular weight of approximately 25,000?

   Polyethylene $-(CH_2-CH_2)_n-$

   Molecular weight of one unit = 2(12) + 4(1) = 28

   $25{,}000 \times \frac{\text{ethylene unit}}{28} = 8.9 \times 10^2$ units

5. Structural formulas of the polymers that can be formed from:

   a. ethylene $-(CH_2-CH_2)_n-$

   b. propylene $-(CH_2-CH(CH_3))_n-$

   c. 1-butene $-(CH_2-CH(CH_2CH_3))_n-$

   d. 2-butene $-(CH(CH_3)-CH(CH_3))_n-$

6. Structural formulas for:

   a. Saran $-(CH_2CCl_2)_n-$

   b. Orlon $-(CH_2CH(CN))_n-$

   c. Teflon $-(CF_2CF_2)_n-$

   d. Polystyrene

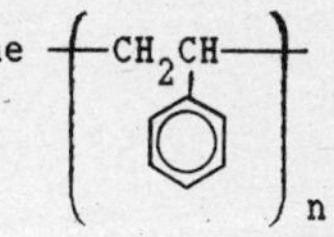

e. Lucite $\left(CH_2-\overset{\displaystyle CH_3}{\underset{\displaystyle \underset{\displaystyle \overset{\|}{O}}{C}OCH_3}{C}} \right)_n$

7. Two possible ways in which vinyl chloride can polymerize to form polyvinyl chloride:

a. $\left(CH_2\underset{Cl}{C}HCH_2\underset{Cl}{C}HCH_2\underset{Cl}{C}HCH_2\underset{Cl}{C}H \right)_n$

b. $\left(CH_2\underset{Cl}{C}H-\underset{Cl}{C}HCH_2CH_2\underset{Cl}{C}H-\underset{Cl}{C}HCH_2 \right)_n$ (other possibilities also)

8. a. From 2,3-dimethyl-1,3-butadiene, $CH_2=\underset{CH_3}{C}H-\underset{CH_3}{C}H=CH_2$,

this polymer can be made $\left(CH_2\underset{H_3C}{C}=\underset{CH_3}{C}CH_2 \right)_n$

b. If produced by the free-radical mechanism, a random mixture of cis and trans connections are made. It is possible, using catalysts, for the reaction to proceed by an ionic mechanism which will give a stereochemically controlled polymer.

9. Chemical structures for the monomers of:

a. natural rubber $CH_2=\underset{CH_3}{C}CH=CH_2$

b. synthetic rubber 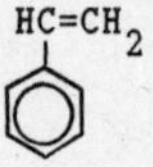 and $CH_2=CHCH=CH_2$

c. synthetic natural rubber $CH_2=\underset{CH_3}{C}CH=CH_2$

10. Smoggy atmospheres affect natural rubber and many synthetic rubbers by the attack of ozone at the sites of the double bonds, causing "age hardening" and cracking.

11. Vulcanization is the process of heating raw rubber with sulfur. Sulfur atoms are introduced as cross links between the polymeric chains.

12\. a. Natural rubber (all cis)

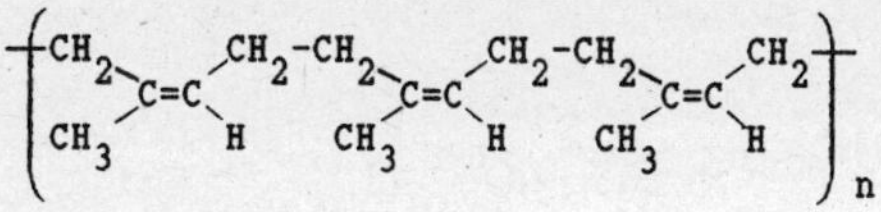

b. Gutta percha (all trans)

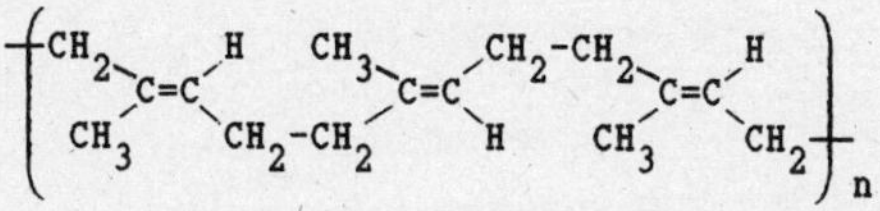

13\. Nitrile rubber (Buna N)

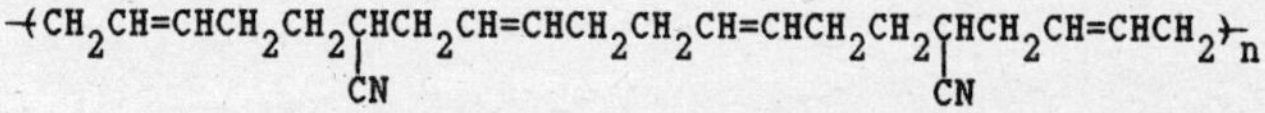

14\. a. Isotactic polypropylene

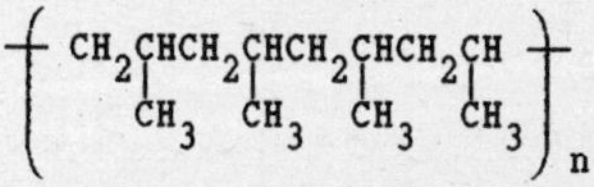

b. Another form of polypropylene

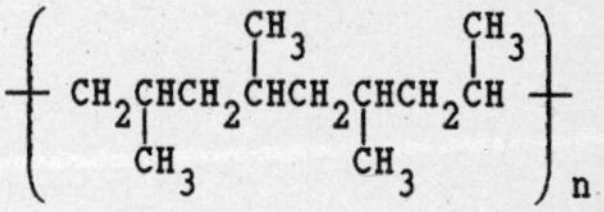

15\. a. Dacron

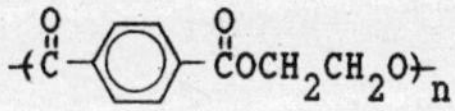

b. Nylon 66 $\left(\overset{O}{\overset{\|}{C}}(CH_2)_4\overset{O}{\overset{\|}{C}}NH(CH_2)_6NH\right)_n$

c. Bakelite

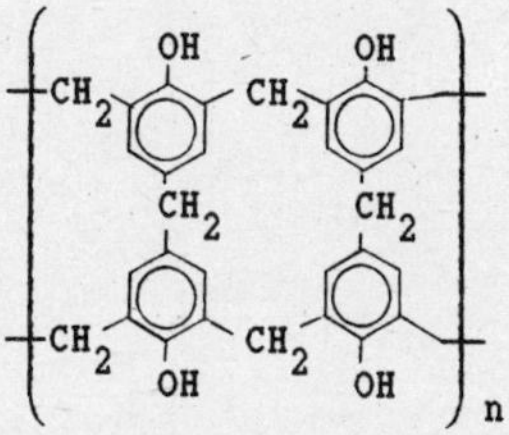

d. polyurethane

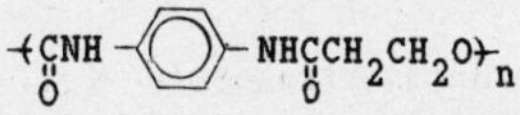

16. In phenol, three positions, ortho, ortho, and para to the OH group are used in the reaction to form a thermosetting polymer. However, in p-cresol the para position is occupied by a methyl group and cannot react with formaldehyde. This leaves the p-cresol molecule as a bifunctional monomer, which results in a more linear, thermoplastic polymer.

17. Monomers of the given polymers:

a. $CH_2=CH-C(=O)-OC_2H_5$ — a polyvinyl

b. 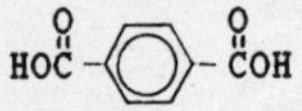 — a polyester

and

c. 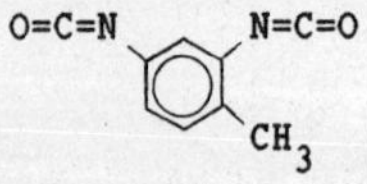 — a polyurethane

and

$HO-CH_2CH_2-OH$

d. $HOC(=O)(CH_2)_8C(=O)OH$ — a polyamide

and

$H_2N(CH_2)_6NH_2$

e. $CH_2=C(CH_3)Cl$ — a polyvinyl

f. 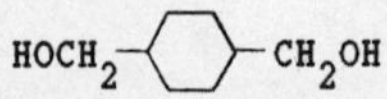 — a polyester

and

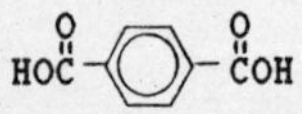

18. In order to form a thermosetting polyester, there must be cross linking of polymer chains. This requires that some monomer molecules be trifunctional. If all molecules are only bifunctional, then only long chains can result, and they will not be thermosetting.

19. A glyptol polyester would more likely be thermosetting, because the glycerol is trifunctional, and would thus allow cross linking between chains in forming ester linkages.

20. Foam or sponge rubber materials achieve the foam by incorporating chemicals that release a gas within the material during the polymerization or the molding process.

21. Silicone polymers are more resistant to high temperatures than the usual organic polymers because they contain silicon-oxygen chains which, like quartz, are stable at high temperature.

22. The correct statements are: a, b, d, f, i, and k.

B. 1. The objects which are chiral: (a) a wood screw; (b) a pair of pliers; (d) the letter G; (e) a coiled spring; (f) your ear.

2. An asymmetric carbon atom is one to which four different atoms or groups are attached and is a center of dissymmetry in a molecule. In the following three compounds, the asymmetric carbon atoms are marked with an asterisk. (These are merely three examples; there are an infinite number of compounds which contain one asymmetric carbon atom.)

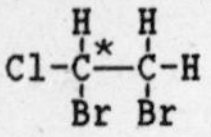

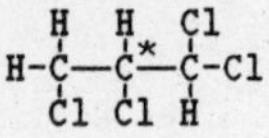

$CH_3\overset{*}{C}H(OH)CH_2OH$

3. When the axes of two pieces of polaroid film are parallel, you have maximum brightness of the light passing through both. When one piece has been rotated by 90° the polaroid appears black, indicating very little light passing through.

4. How many asymmetric carbon atoms are present in each of the following: (In addition to the number, each asymmetric carbon atom has been marked with an asterisk.)

a. 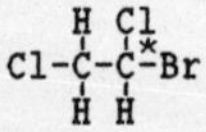 one

b. 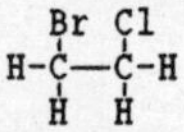 none

c. $CH_3CH_2CH_2\overset{*}{C}HClCH_3$ one

d. 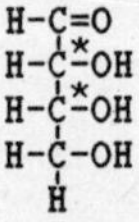 two

e. 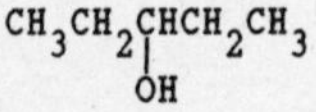 none

f. 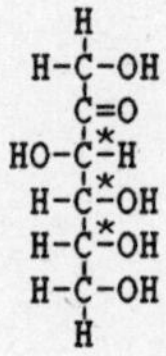 three

g. 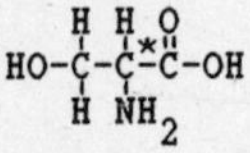 one

5. Using projection formulas decide which of the following compounds will show optical activity.

a. 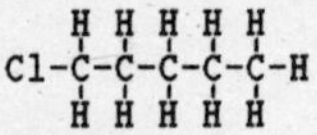 no

b. 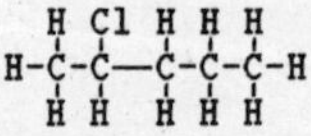 yes

c. 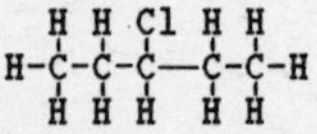 no

d. 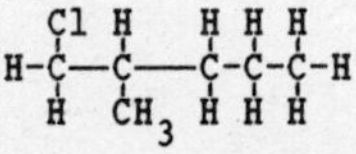 yes

e. 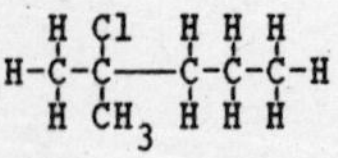 no

f. 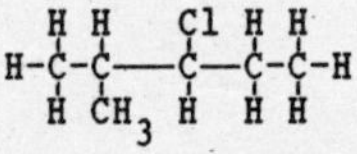 yes

g. 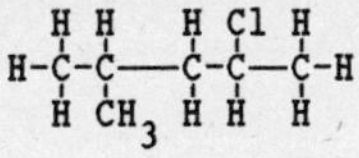 yes

h. 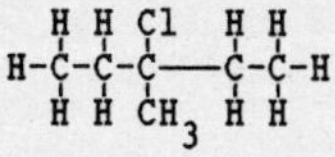 no

6. If four different groups were attached to a central carbon atom in a planar arrangement, it would not rotate polarized light because there is a plane of symmetry. No such plane of symmetry is possible when the four different groups are arranged in a tetrahedral structure.

7. A necessary and sufficient condition for a compound to show enantiomerism is that the compound not be superimposable on its mirror image.

8. Glucose, which has four asymmetric carbon atoms, would have 16 theoretical stereoisomers. This can be obtained from 2^n. $2^4 = 16$.

9. The two projection formulas (a) and (b) are the same compound, for it takes two changes to make (b) identical to (a).

(a) (b) 1 change 2 changes

10. (-)-lactic acid is 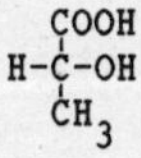

a. 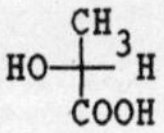 e. 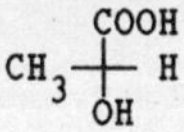 f. 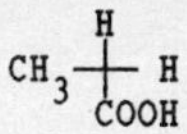

(+)-lactic acid is 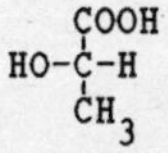

b. 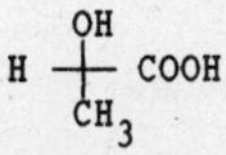 c. 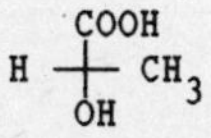 d. 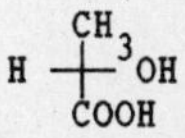

11. All possible stereoisomers of the following compounds, with enantiomers and meso compounds labeled:

a. 1,2-dibromopropane 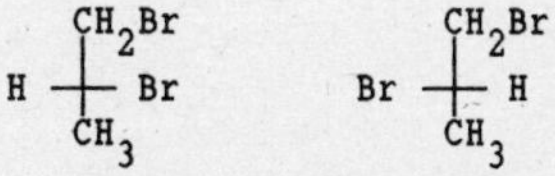

enantiomers

b. 2,3-dichlorobutane

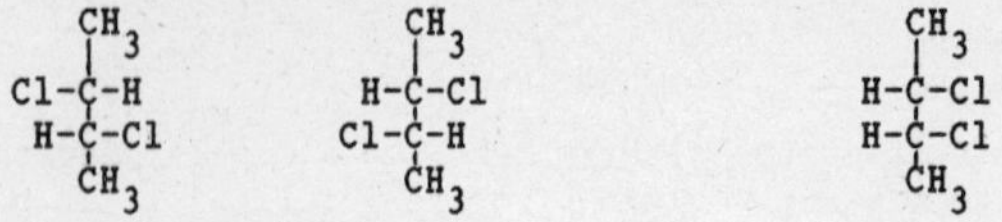

enantiomers meso

c. 2-butanol

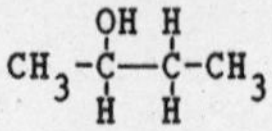 enantiomers 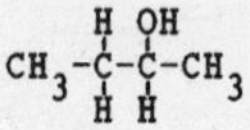

d. 2,4-dibromopentane

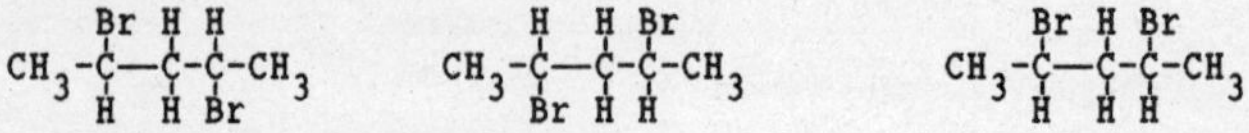

enantiomers meso

e. 3-chlorohexane

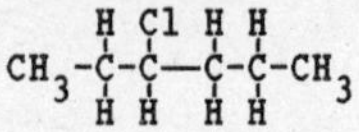 enantiomers 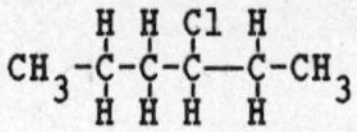

12. All the steroisomers of 1,2,3-trihydroxybutane:

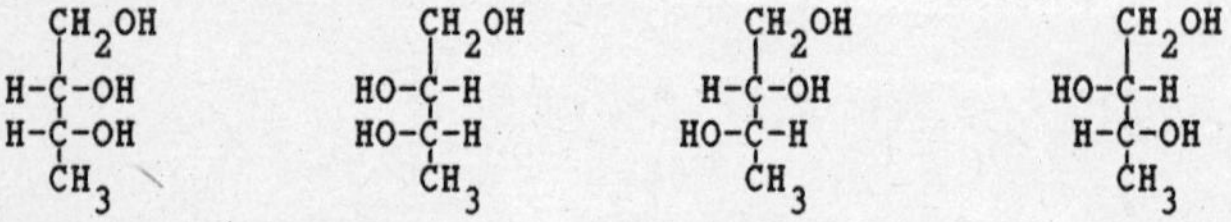

A B C D

Compounds A and B, and C and D are enantiomer pairs. There are no meso compounds. Pairs of diastereomers are A and C, A and D, B and C, and B and D.

13. The four stereoisomers of 3-pentene-2-ol.

cis trans cis trans

14. c. is chiral

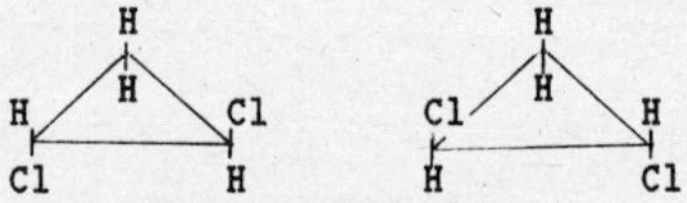

enantiomers

15. Physical properties of a pair of enantiomers

	(+)2-methyl-1-butanol	(-)2-methyl-1-butanol
specific rotation	+5.76°	-5.76°
boiling point	129°C	129°C
density	0.819 g/ml	0.819 g/ml

16. All stereoisomers of 2,3,4-tribromopentane:

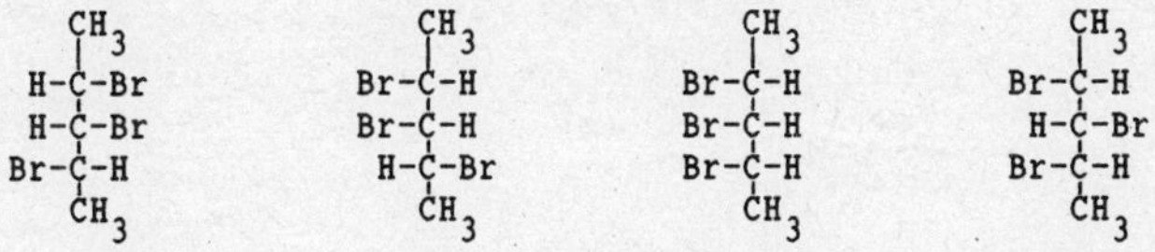

A B C D

Compounds A and B are a pair of enantiomers. Compounds C and D are both meso compounds.

17. a. $CH_3CH_2CH_2CHCl_2$ (a) $CH_3CH_2CHClCH_2Cl$ (b) $CH_3CHClCH_2CH_2Cl$ (c)

$CH_2ClCH_2CH_2CH_2Cl$ (d) $CH_3CH_2CCl_2CH_3$ (e) $CH_3CHClCHClCH_3$ (f)

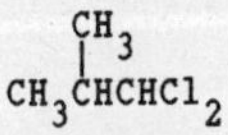

(g)

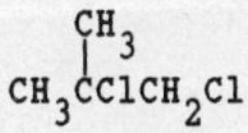

(h)

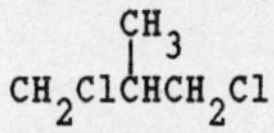

(i)

b. (b) is chiral

enantiomers

(c) is chiral

enantiomers

f. can be both chiral and meso

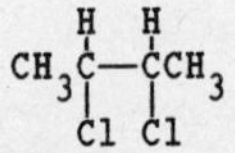

meso

enantiomers

a, d, e, g, h, and i are achiral.

18. Assume (+)-2-chlorobutane is

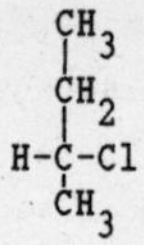

All possible isomers formed when (+)2-chlorobutane is further chlorinated to dichlorobutane:

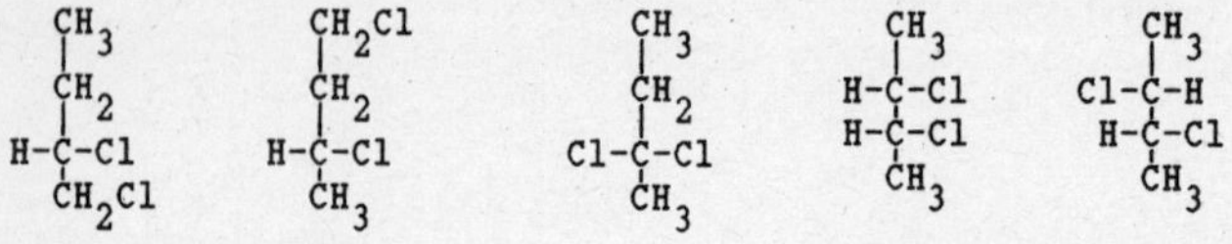

A B C D E

Compounds A, B, and E would be optically active.

19. If 1-chlorobutane and 2-chlorobutane were obtained by chlorinating butane, and then distilled, they would be separated into the two fractions, because their boiling points are different. 1-chlorobutane has no asymmetric carbon, so would not have an enantiomer. 2-chlorobutane would exist as a racemic mixture (equal quantities of enantiomers) because substitution of Cl for H on carbon-2 gives equal amounts of the two enantiomers. Distillation would not separate the enantiomers, because their boiling points are identical. The optical rotation of the two enantiomers of 2-chlorobutane would exactly cancel, and thus would not show optical activity.

20. a. Enantiomers have identical physical properties except effect on polarized light, so cannot be separated by ordinary chemical and physical means.

b. Diastereoisomers do not have identical physical properties, so the differences form a basis for chemical or physical separation. Differences of boiling point, freezing point, and solubilities would be most commonly used.

21. $$[\alpha] = \frac{\text{observed rotation in degrees}}{\text{(length of sample tube in dm)(sample conc. in g/ml)}}$$

$$[\alpha] = \frac{12.5^\circ}{2.00\text{ dm} \times 0.100\text{ g/ml}} = 62.5^\circ$$

22. Compounds (a) and (d) are meso.

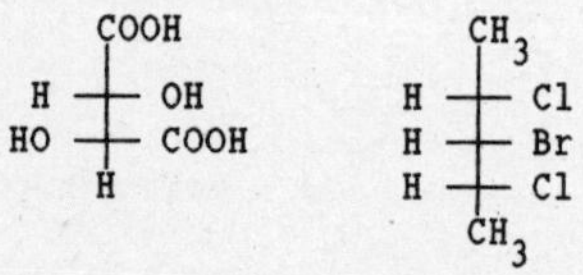

23. If a compound $R_1R_2R_3R_4N^+X^-$ can rotate polarized light, then the orientation of the bonds around the nitrogen atom must be tetrahedral.

24. a. A chiral primary alcohol of formula $C_5H_{12}O$.

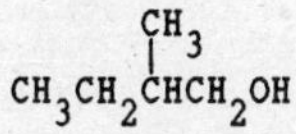

b. A compound with three primary alcohol groups is chiral, and has the formula $C_6H_{14}O_3$.

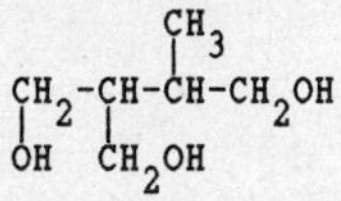

25. A compound of formula $C_3H_8O_2$:

a. is chiral; contains two OH groups

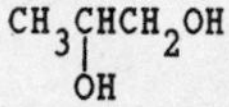

b. is chiral; contains one OH group

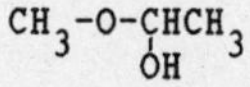

c. is achiral; contains two OH groups

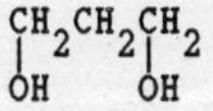

CH_3-O-CH_2-O-CH_3

CH_3-O-CH_2CH_2OH

26. The correct statements are: a, f, g, i, k, and l.

B. 1. In general, the carbohydrate carbon oxidation state determines the carbon's metabolic energy content. The more oxidized a carbon is, the less energy it may provide in biological systems.

2. The notations D and L in the name of a carbohydrate specify the configuration on the last asymmetric carbon. If that configuration is the same as D-glyceraldehyde, then the carbohydrate is designated as D. If it is the same as L-glyceraldehyde, it is designated as L.

3. The notations (+) and (-) in the name of a carbohydrate specify whether the compound rotates the plane of polarized light to the right (+) or to the left (-).

4. Carbohydrates form five membered or six membered heterocyclic rings (one oxygen atom, the rest carbon atoms). If it forms a five-membered ring, it is termed a furanose, after the compound furan. If it forms a six-membered ring it is termed a pyranose, after the compound pyran.

Furan

C_4H_4O

Pyran

C_5H_6O

5. Galactosemia is the inability of infants to metabolize galactose. The galactose concentration increases markedly in the blood and also appears in the urine. Galactosemia causes vomiting, diarrhea, enlargement of the liver, and often mental retardation. If not recognized a few days after birth it can lead to death.

6. There are four pairs of epimers among the D-aldohexoxes in figure 27.1. They are: allose and altrose; glucose and mannose; gulose and idose; and galactose and talose.

7. Cyclic structures. Either the Fischer projection formulas or perspective formulas are satisfactory.

a. α-D-glucopyranose

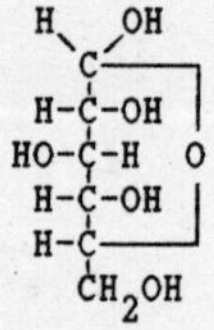

b. β-D-galactopyranose

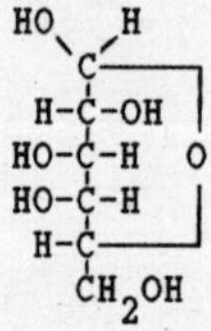

c. α-D-mannopyranose

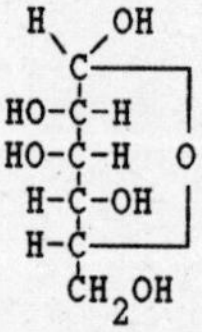

α-D-gluco-
pyranose

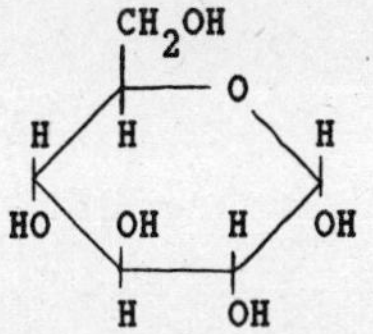

β-D-galacto-
pyranose

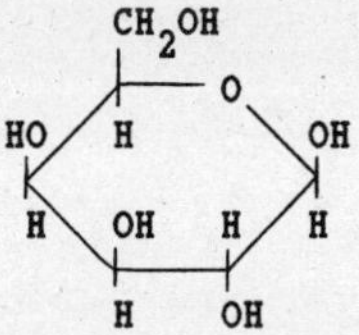

α-D-manno-
pyranose

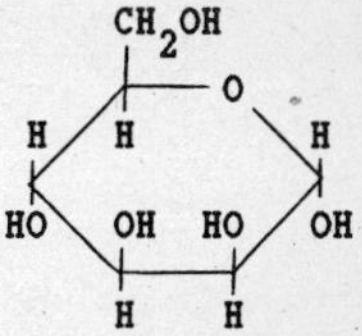

8. α-D-glucopyranose and β-D-glucopyranose differ in the configuration at the number 1 carbon in the cyclic structure. In the open-chain structure, carbon 1 is an aldehyde group, and not asymmetric. In the cyclic structure that carbon contains a hemiacetal structure, which is asymmetric. When the ring forms, it can form in two ways at number 1 carbon, leading to the two structures, called α and β.

9. The structure of L-glyceraldehyde is

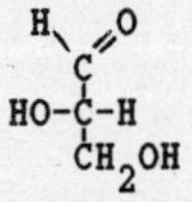

1st carbon oxidation number = +1
2nd carbon oxidation number = 0
3rd carbon oxidation number = -1

Carbon number 3 has the most negative oxidation number and is the most reduced.

10. D-mannose and D-galactose should provide about equal amounts of metabolic energy. The carbons in each carbohydrate have equivalent oxidation numbers and, thus, will release equivalent amounts of energy when they are oxidized.

11. Kiliani-Fischer synthesis of D-glucose from the proper D-tetrose (D-erythrose)

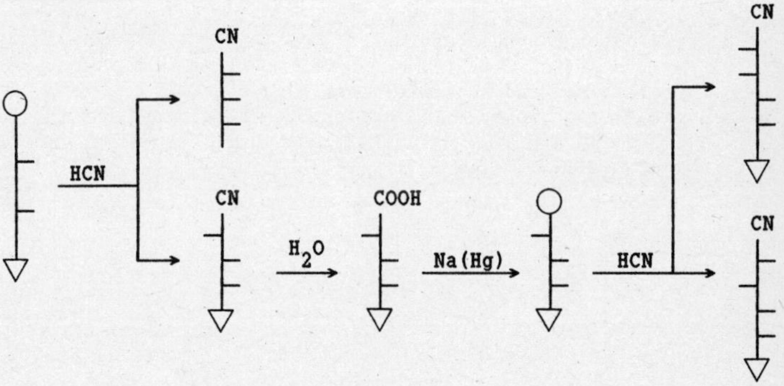

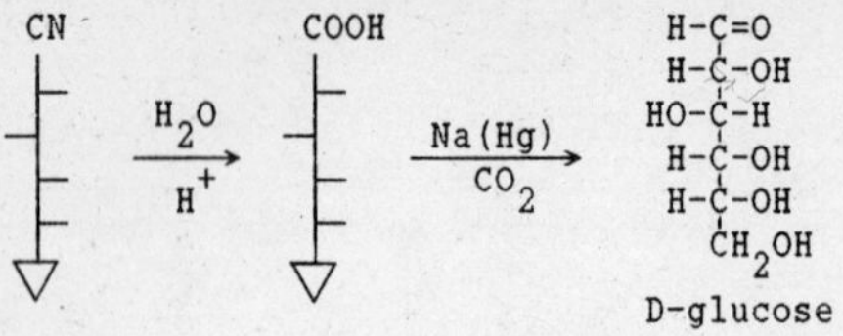

12. The cyclic forms of monosaccharides are hemiacetals, because the number one carbon has an ether and an alcohol group; whereas a glycoside is an acetal, two ether linkages.

13. Mutarotation is the phenomenon by which the α or β form of a sugar, when in solution, will undergo change to reach an equilibrium mixture, not necessarily 50%-50%, of the two forms. To achieve this equilibrium, the chain must open up and then reclose. On closing, it has the possibility of closing in either the α or β form as the equilibrium mixture is achieved.

14. Yes, D-2-deoxyarabinose is the same as D-2-deoxyribose. It was at the number 2 carbon that the structures of D-ribose and D-arabinose were different, so if that OH group has been changed to just H, the difference has been removed.

15. Major sources:

    a. sucrose: sugar beets and sugar cane
    b. lactose: milk
    c. maltose: sprouting grain and partially hydrolzed starch

16. The monosaccharide composition of:

    a. sucrose: one glucose and one fructose unit
    b. maltose: two glucose units
    c. lactose: one glucose and one galactose unit
    d. glycogen: many glucose units
    e. amylose: many glucose units

17. Both cellobiose and isomaltose are disaccharides composed of two glucose units. However, in cellobiose the glucose monosaccharides are linked by a β(1 → 4) acetal bond while for isomaltose the linkage is α(1 → 6).

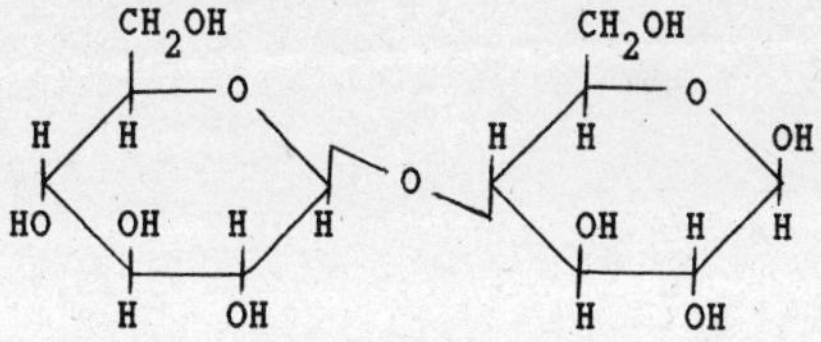

cellobiose

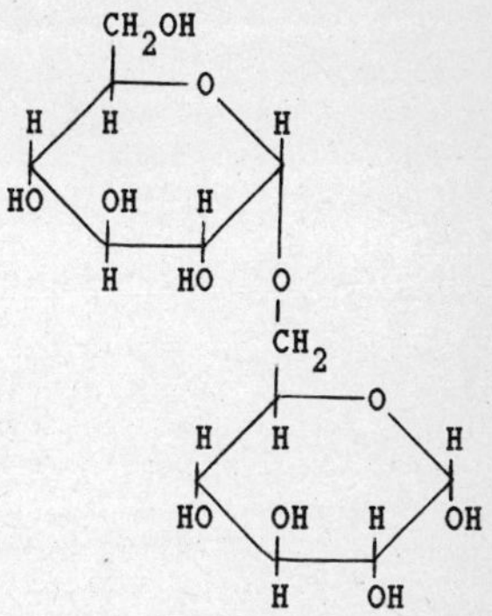

isomaltose

18. Aspartame supplies many fewer calories than sucrose. In addition, oral bacteria cannot use aspartame as efficiently as sucrose and will form fewer dental carries.

19. Invert sugar is sweeter than sucrose because it is a 50-50 mixture of fructose and glucose. Glucose is somewhat less sweet than sucrose, but fructose is much sweeter, so the mixture is sweeter.

20. Maltose and lactose will show mutarotation; sucrose will not. Maltose and lactose have a hemiacetal structure in one of their units which will open for mutarotation; sucrose forms an acetal between C-1 of glucose and C-2 of fructose and thus does not have a hemiacetal unit so will not open, and therefore, does not undergo mutarotation.

21.

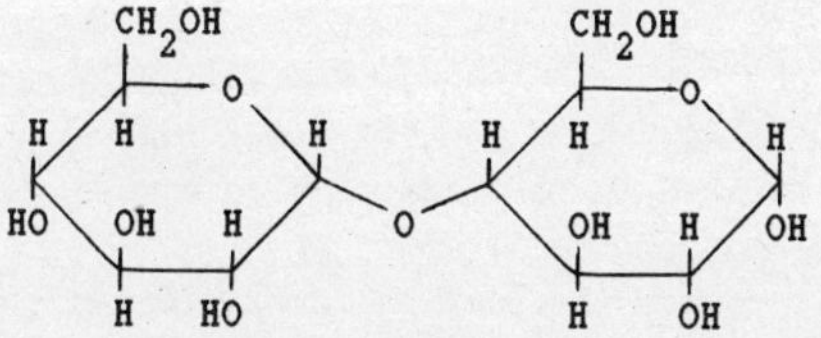

maltose

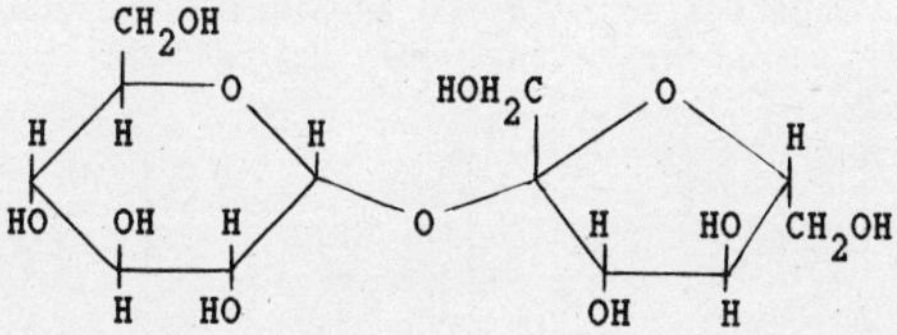

sucrose

Maltose, with a hemiacetal structure which opens to form an aldehyde group, is a reducing sugar. Sucrose, lacking the hemiacetal structure, is not a reducing sugar.

22. The systematic name for isomaltose is α-D-glucopyranosyl-(1-6)-α-D-glucopyranose. The systematic name for lactose is β-D-galactopyranosyl-(1-4)-α-D-glucopyranose.

23. In Benedict's test, both concentrated and dilute glucose solutions will give a precipitate. However, the concentrated glucose solution will produce a redder precipitate while the dilute glucose solution will appear more greenish-yellow.

24. D-ribose and D-xylose are the two aldopentoses in Figure 27.1 which would be optically inactive when reduced; because they would be meso forms. When reduced they would be:

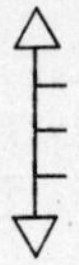

and

25. a. D-glyceraldehyde

H-C=O
H-C-OH
CH_2OH

L-glyceraldehyde

H-C=O
HO-C-H
CH_2OH

b. If glyceraldehyde is reacted with hydrogen in the presence of a platimun catalyst, the product will be:

CH_2OH
H-C-OH
CH_2OH

26. The following parts all related to the eight D-aldohexoses shown in the text (Figure 27.1):

a. Four pairs will each produce the same osazone. They are allose and altrose; glucose and mannose; gulose and idose; and galactose and talose.

b. If each of the aldohexoses is oxidized by nitric acid to dicarboxylic acids, allose and galactose would become meso forms.

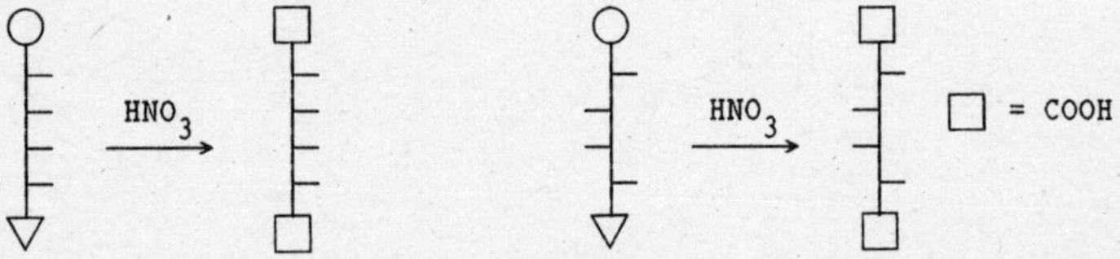

allose galactose

c. Names and structures of enantiomers of D(+)-mannose and D(-)-iodose:

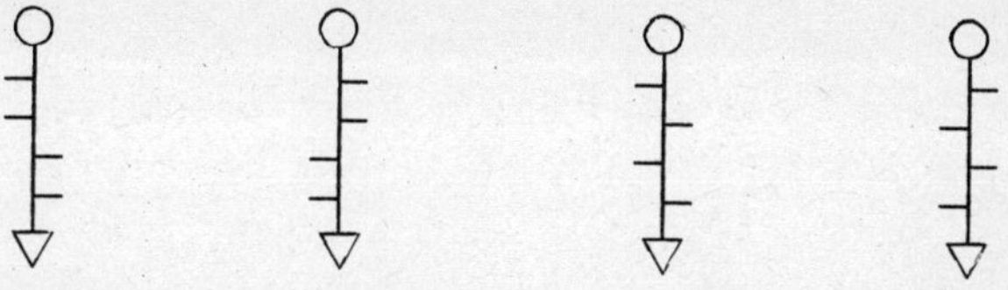

D-(+)-mannose L-(-)-mannose D-(-)-idose L-(+)-idose

27. The principal differences and similarities between the members of the following pairs:

a. D-glucose and D-fructose. Glucose is an aldose; fructose is a ketose. Glucose often forms a pyranose ring structure; fructose commonly is found in a furanose structure. Both are hexoses, and both are reducing sugars.

b. D-ribose and 2-deoxy-D-ribose. The 2-deoxy-D-ribose has no OH group on the number 2 carbon, only 2 hydrogen atoms. Both are five carbon aldoses.

c. Maltose and sucrose. Maltose is composed of two glucose units; sucrose is composed of one glucose and one fructose unit. Maltose is a reducing sugar; sucrose is not. Both are common disaccharides, with formulas $C_{12}H_{22}O_{11}$.

d. Amylose and amylopectin. Amylose is a straight chain polysaccharide; amylopectin has branched chains, and more monomer units per molecule. Both are large polysaccharides composed of α-D-glucose units.

e. Cellulose and glycogen. Cellulose is composed of glucose units linked by β-1,4-glycosidic linkages; glycogen is composed of glucose units linked by α-1,4-glucosidic linkages. Glycogen is much more readily hydrolyzed or digested than cellulose. Both are large polymers of glucose.

f. D(-)-ribose and D(+)-glucose. The ribose is a pentose that rotates polarized light to the left; the glucose is a hexose that rotates polarized light to the right. Both are aldoses; both give positive tests with Tollens' reagent and Fehling's solution.

28. 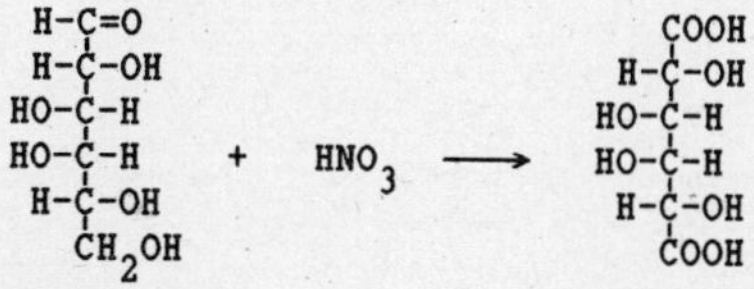

galactose mucic acid (galactaric acid)

29. The four L-aldopentoses

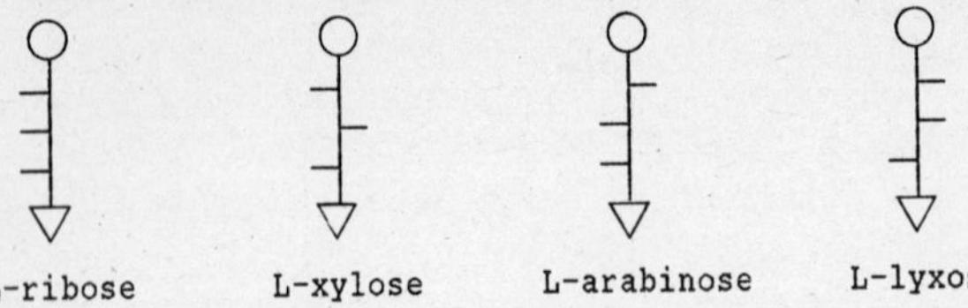

L-ribose and L-arabinose will yield identical osazones. These two sugars differ only at C-2, a difference which is lost upon reaction with phenylhydrazine.

30. a.

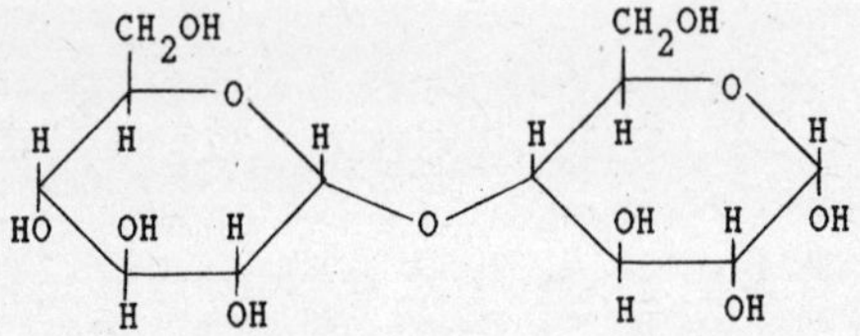

b.

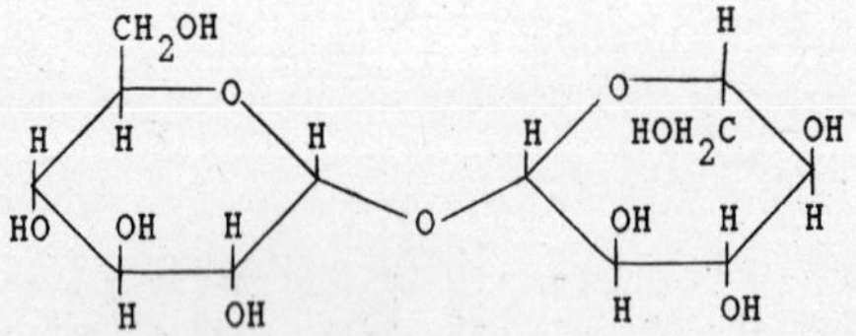

a 1,1-glycosidic linkage

31. Starch consists of glucose units joined by α-1,4 and α-1,6-glycosidic linkages, whereas cellulose consists of glucose units joined by β-1,4-glycosidic linkages. Starch often contains molecules with a branching structure while cellulose molecules are linear.

32. The two main components of starch are amylose and amylopectin. They are both composed of glucose units joined by α-1,4-glycosidic linkages. The difference is that amylopectin also has branching which occurs through α-1,6-glycosidic linkages about every 25 glucose units.

33. Cellobiose, from the hydrolysis of cellulose, would be:

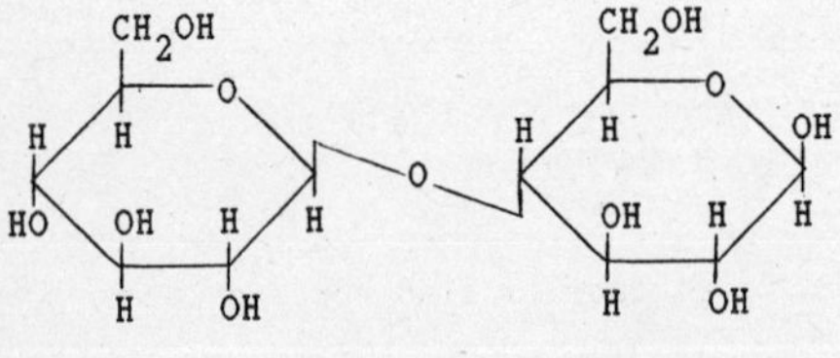

34. The Haworth formula for β-D-glucopyranosyl-(1-4)-α-D-galactopyranose is

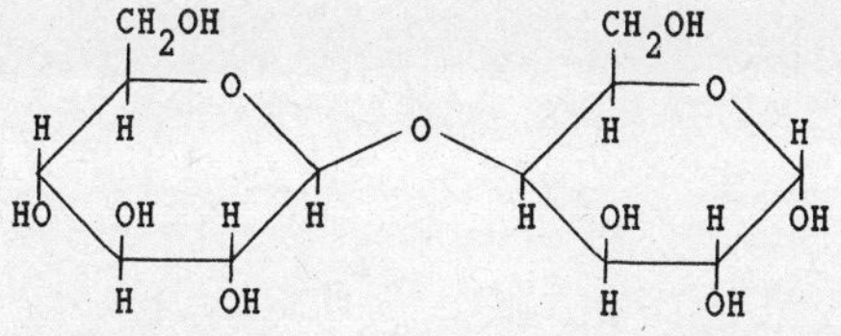

The Haworth formula for β-D-galactopyranosyl-(1-6)-β-D-mannopyranose is

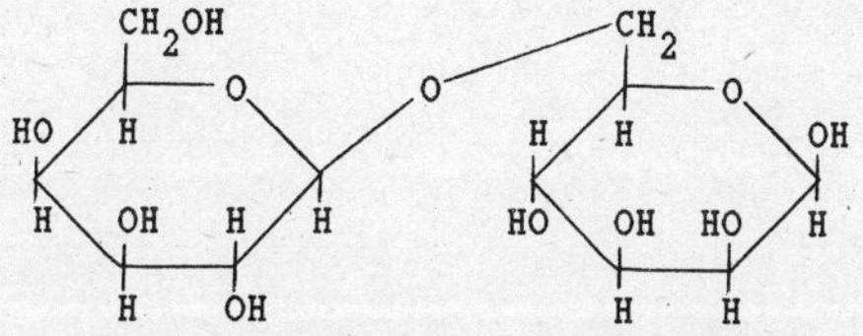

35. In a polysaccharide, one water molecule has been removed for each glucose unit, so one monomer unit is $C_6H_{12}O_6 - H_2O = 180 - 18 = 162$.

$$325{,}000 \text{ amu} \times \frac{\text{monomer}}{162 \text{ amu}} = 2000 \text{ monomer units}$$

36. Cellulose acetate:

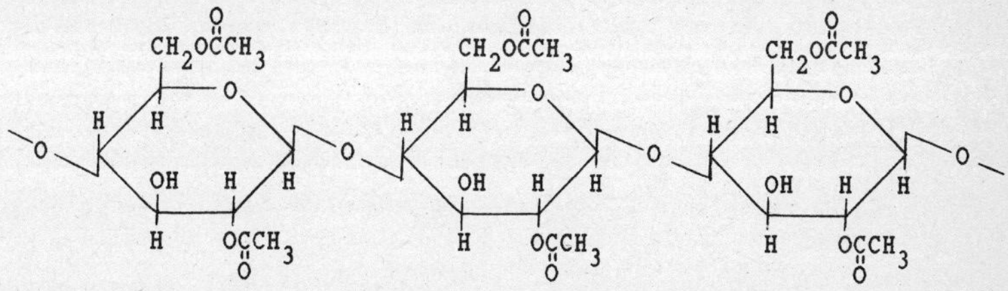

Approximately 2/3 of the OH groups have been esterified with acetic acid.

37. Three units of the polymeric structure of inulin:

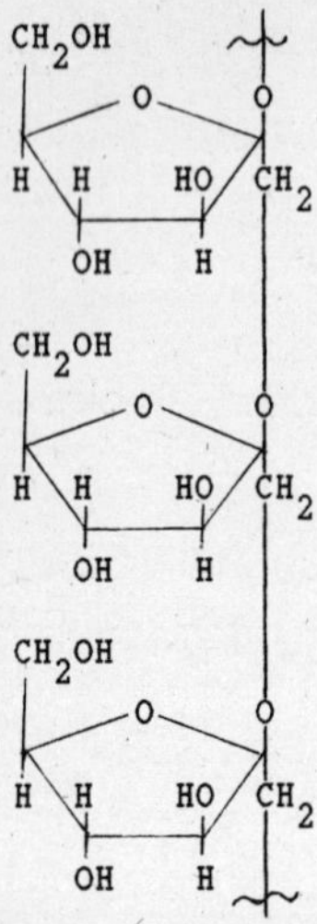

38. The correct statements are: c, e, g, h, i, l, n, p, q, r, s, v, w. The others are not correct for these reasons:

a. α-D-glucopyranose and β-D-glucopyranose differ only at the anomeric carbon. Enantiomers (mirror-images) must differ at all chiral centers.

b. The carbon of a secondary alcohol has an oxidation number of 0, while the carbon of a primary alcohol has an oxidation number of -1. Thus, the primary alcohol carbon is more reduced.

d. To be epimers, D-threose and L-threose would have to be identical at carbon numbers 3 and 4, but they are opposites, not identical, thus are not epimers.

f. Osazones eliminate any differences at carbons one and two, but D-glucose and L-glucose would be opposites at carbons three, four and five after forming the osazones.

j. D-mannitol is a sugar alcohol and is obtained from D-mannose by reduction, not oxidation.

k. Methyl glucosides are acetals, are stable in alkaline solutions, and thus the ring does not open to form an aldehyde with reducing ability.

m. Fructose is a hexose, a monosaccharide, a ketose, but not an aldose.

o. The disaccharide found in mammalian milk is lactose. Galactose is one of the two monosaccharides in lactose.

t. Invert sugar is composed of both glucose and fructose and is not as sweet as pure fructose.

u. Sucrose is not a reducing sugar because it lacks the functional groups which will react in Benedict's test, for example, (aldehydes, ketones adjacent to hydroxyl groups, hemiacetals, hemiketals).

B. 1. The lipids, which are dissimilar substances, are arbitrarily classified as a group on the basis of their solubility in fat solvents and their insolubility in water.

2. Although caproic acid, $CH_3(CH_2)_4COOH$, has the same number of polar bonds as stearic acid, $CH_3(CH_2)_{16}COOH$, caproic acid has a shorter, nonpolar hydrocarbon chain and, therefore, is more water soluble.

3. A triacylglycerol containing one unit each of palmitic, stearic, and oleic acids:

$CH_2-O-\overset{O}{\overset{\|}{C}}-(CH_2)_{14}CH_3$ palmitic acid

$CH-O-\overset{O}{\overset{\|}{C}}-(CH_2)_{16}CH_3$ stearic acid

$CH_2-O-\overset{O}{\overset{\|}{C}}-(CH_2)_7CH=CH(CH_2)_7CH_3$ oleic acid

There would be two other triacylglycerols possible from these same components. Since the top and bottom attachments are equivalent, it only matters which acid is attached to the middle of the glycerol.

4. Hydrolysis of a triacylglycerol will yield three fatty acids and glycerol. The products will be:

CH_2OH
$CHOH$ glycerol
CH_2OH

$CH_3(CH_2)_{14}COOH$ palmitic acid

$CH_3(CH_2)_7CH=CH(CH_2)_7COOH$ oleic acid

$CH_3(CH_2)_4CH=CHCH_2CH=CH(CH_2)_7COOH$ linoleic acid

5. The three essential fatty acids are linoleic, linolenic, and arachidonic acids. Diets lacking these fatty acids lead to impaired growth and reproduction, and skin disorders such as eczema and dermatitis.

6. Arachidonic acid is used to synthesize two types of hormone-like substances, the prostaglandins and the leukotrienes.

7. Fats contain more biochemical energy than carbohydrates because (a) the carbons of fat are more reduced than those found in carbohydrates and (b) there are more reduced carbons per gram of fat than for the same weight of carbohydrate.

8. Aspirin relieves inflammation by blocking the conversion of arachidonic acid to prostaglandins.

9. Waxes serve as a protective coating because they are very hydrophobic. They do not dissolve in water and many compounds cannot pass through a wax coating.

10. A membrane lipid must be (a) partially hydrophobic, to act as a barrier to water and (b) partially hydrophilic, so that the membrane can interact with water along its surface.

11. Phospholipids are mainly produced in the liver.

12. Lecithins and cephalins are both derivatives of phosphatidic acids, which are glycerol esters of fatty acids and phosphoric acids. Lecithins have choline attached to the phosphoric acid; cephalins have ethanolamines attached to the phosphoric acid. The unique parts are shown here:

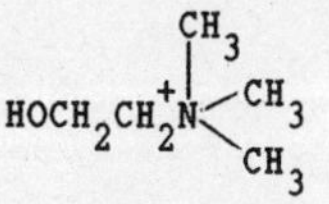

choline

$HOCH_2CH_2NH_2$

ethanolamine

13. The structure of phosphatidyl serine is

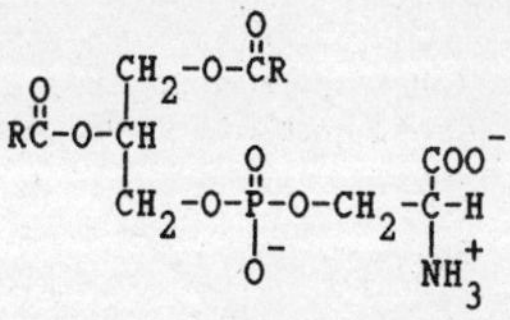

14.

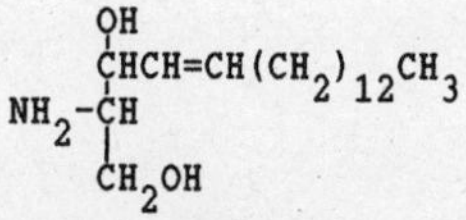

sphingosine

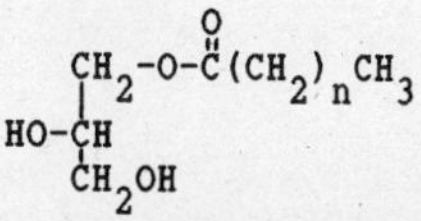

monoacylglycerol

Sphingosine is similar to the monoacylglycerol in that (a) both molecules contain a long hydrophobic chain, (b) the sphingosine amino group reacts with a fatty acid as does the secondary alcohol of the monoacylgycerol and, (c) for both compounds, the primary alcohol will react further with either acids (to form esters) or sugars (to form acetals).

15. The structure of a glucocerebroside is

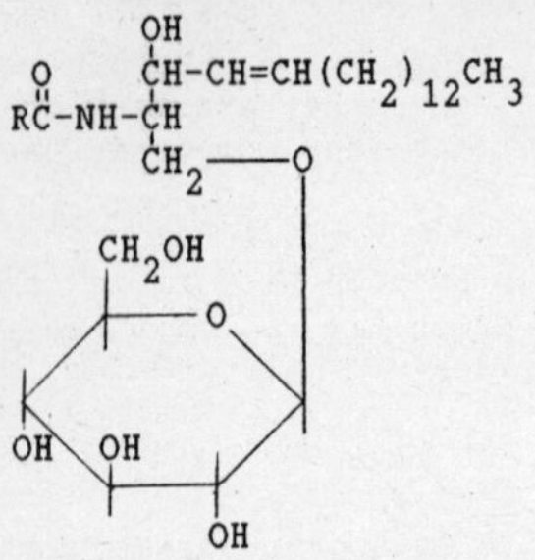

16. In general, a membrane lipid will have both hydrophilic and hydrophobic parts. Sphingomyelin can serve as a membrane lipid because its phosphate and choline groups are hydrophilic and its two long carbon chains are hydrophobic.

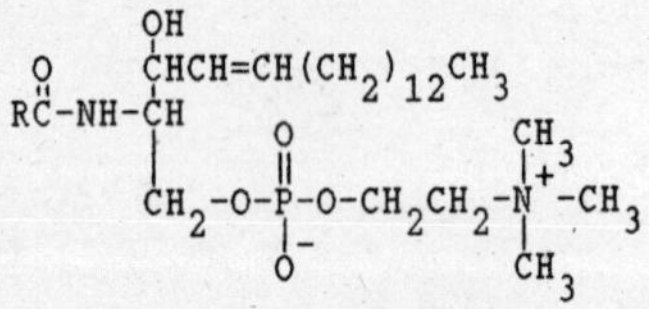

sphingomyelin

17. All steroids possess a 17-carbon unit structure containing four fused rings known as the steroid nucleus. Many steroids could be shown; these are two common ones:

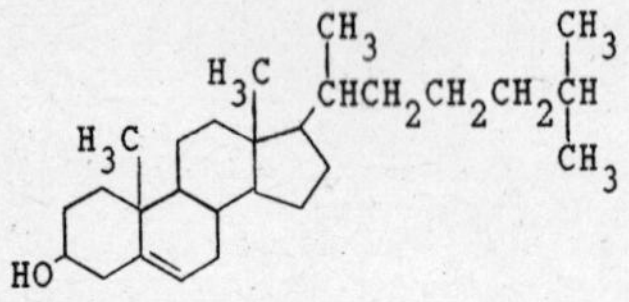

cholesterol

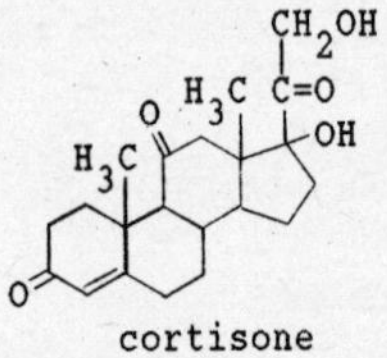

cortisone

18. Cholesterol is found in the body primarily in the brain, the spinal chord, the liver, and under the skin. It is mainly produced in the liver, with some also produced in the adrenal glands and the reproductive organs.

19. Atherosclerosis is the deposition of cholesterol and other lipids on the inner walls of the large arteries. These deposits, called plaque, accumulate, making the arterial passages narrower and narrower. Blood pressure increases as the heart works to pump sufficient blood through the restricted passages. This may lead to a heart attack, or the rough surface can lead to coronary thrombosis.

20. Dietary cholesterol is transported first to the liver where it is bound to other lipids and proteins to form the very low density lipoprotein (VLDL). This aggregate moves through the blood stream delivering lipids to various tissues. As lipids are removed, the VLDL is converted to an intermediate density lipoprotein (IDL) and, then, to a low density lipoprotein (LDL). Cells needing cholesterol can absorb LDL.

21. Because the interior of a lipid bilayer is very hydrophobic, molecules with hydrophilic character only cross the lipid bilayer with difficulty. A lipid bilayer acts as a barrier to water-soluble compounds.

22. Many different membrane lipids could be drawn. In general, their structures will have both hydrophilic and hydrophobic parts. One common example is phosphatidyl choline.

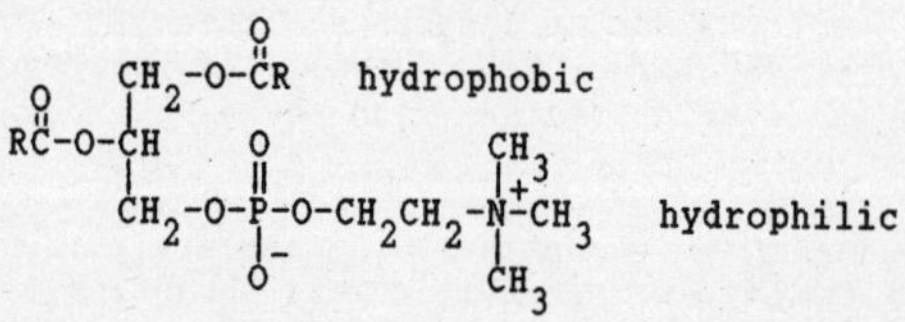

23. The myelin sheath is an insulator because the sheath lipid serves as a barrier to the passage of water-soluble ions.

24. Intrinsic proteins are found in the oily interior of membranes. These proteins must be hydrophobic in order to remain in this nonaqueous environment.

25. Both facilitated diffusion and active transport catalyze movement of compounds through membranes. Active transport requires an input of energy while facilitated diffusion does not.

26. The correct statements are: d, e, g, and j. The others are not correct for these reasons:

   a. In general, a lipid is hydrophobic and does not interact well with water.

   b. As a class, lipids are relatively large molecules (containing more than about 10 carbons per molecule).

   c. A triacylglycerol is high in biochemical energy because this molecule contains many reduced carbons.

   f. A sphingolipid may contain carbohydrate but does not always have this component.

   h. Potassium ion crosses the nerve cell membrane from a low outside concentration to high inside potassium concentration. This process opposes diffusion and requires energy. It is an example of active transport.

   i. A lipid bilayer has a hydrophobic interior.

B. 1. Foods which are major sources of protein (any four of these): fish, beans, nuts, cheese, eggs, poultry, and meat of all kinds.

2. The amino acids of proteins are called alpha amino acids because the amine group is always attached to the alpha carbon atom, that is, the carbon atom next to the carboxyl group, COOH.

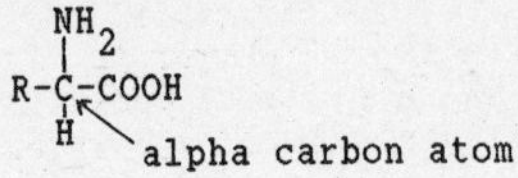

3. All amino acids and proteins contain carbon, hydrogen, oxygen, and nitrogen. Sulfur is contained in some of the amino acids, and thus in most proteins.

4. Proteins from some foods are of greater nutritional value than others because they are "complete", which means they contain all eight essential amino acids, those which the human body cannot synthesize.

5. The amino acids which are essential to humans are isoleucine, leucine, lysine, methionine, phenylalanine, threonine, tryptophan, and valine.

6. There are two general methods for improving the nutritional value of corn protein. They are:
   a. Synthesize lysine and add it to the corn product.
   b. Use maize varieties with greatly improved lysine content.

7. Those amino acids which contain a heterocyclic ring are histidine, proline, and tryptophan.

8. D-serine

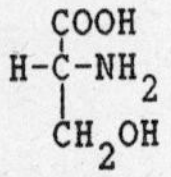

L-serine (form commonly found in proteins)

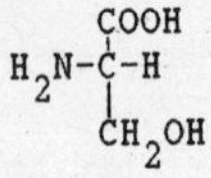

9. Amino acids are amphoteric because the carboxyl group can react with a base to form a salt, or the amine group can react with an acid to form a salt. They are optically active because the alpha carbon is asymmetric, except for glycine. They commonly have the L configuration at carbon two, as in L-serine.

10. The structural formula for threonine at its isoelectric point is:

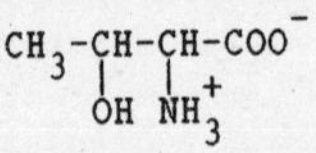

11. For phenylalanine:

a. molecular formula

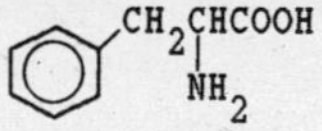

b. "zwitterion" formula

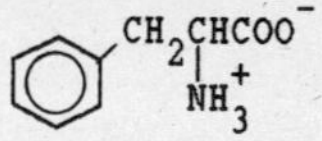

c. formula in 0.1 M H_2SO_4

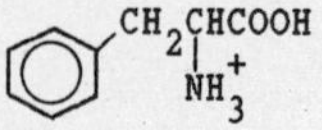

d. formula in 0.1 M NaOH

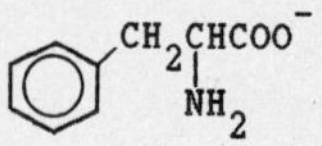

12. Ionic equations showing how alanine acts as a buffer towards:

a. H^+ ion

$$CH_3\underset{\underset{NH_3^+}{|}}{C}HCOO^- + H^+ \longrightarrow CH_3\underset{\underset{NH_3^+}{|}}{C}HCOOH$$

b. OH^- ion

$$CH_3\underset{\underset{NH_3^+}{|}}{C}HCOO^- + OH^- \longrightarrow CH_3\underset{\underset{NH_2}{|}}{C}HCOO^- + H_2O$$

13. a. Arginine will not migrate to either electrode in an electrolytic cell at a pH of 10.8.

b. Arginine will migrate towards the positive electrode at pH greater than 10.8, that is, more basic.

14. At its isoelectric point, a protein molecule must have an equal number of positive and negative charges.

15. The pH range for the isoelectric points of these amino acids:

a. Histidine, with three nitrogens to only one carboxyl group, would be basic, therefore, in the range of 7.8 to 10.8.

b. Phenylalanine, with one nitrogen and one carboxyl, would be neutral, therefore, in the range of 4.8 to 6.8.

c. Glutamic acid, with two carboxyl groups to one nitrogen, would be acidic, therefore, in the range of 2.8 to 3.3. Since it appears in Table 28.2 we know its isoelectric point is 3.2.

16. The two dipeptides containing glycine and phenylalanine:

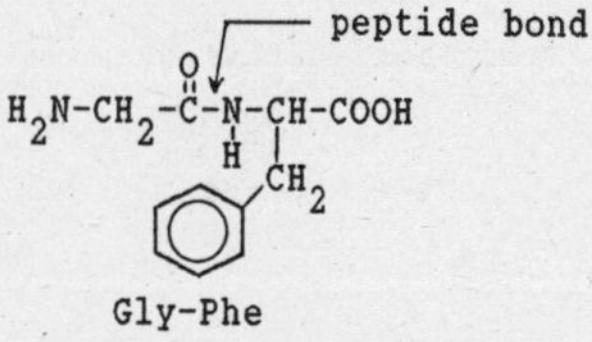

Gly-Phe

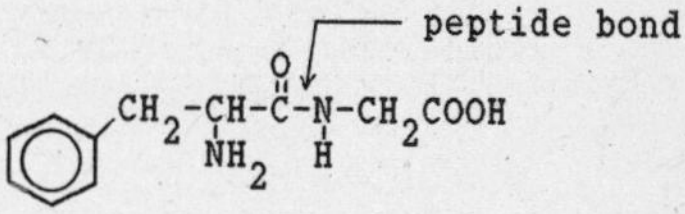

Phe-Gly

17. a. glycylglycine

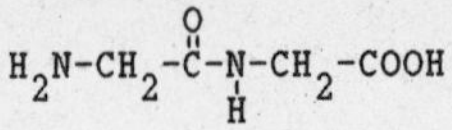

b. glycylglycylalanine

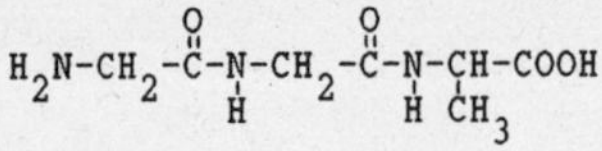

c. leucylmethionylglycylserine

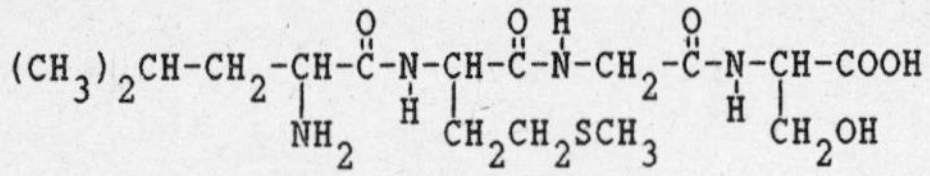

18. All the possible tripeptides containing one unit each of glycine, phenylalanine, and leucine:

Gly-Phe-Leu	Gly-Leu-Phe	Phe-Gly-Leu
Phe-Leu-Gly	Leu-Gly-Phe	Leu-Phe-Gly

19. Nineteen dipeptides can be written with glycine on the N-terminal side. Another nineteen are possible with glycine on the C-terminal end. Finally, one dipeptide can be written with two glycines giving a total of thirty-nine.

20. a. Primary structure. The number, kind, and sequence of amino acid units comprising the polypeptide chain making up a molecule.

b. Secondary structure. The helix and pleated sheet structures of Pauling and Corey due to the hydrogen bonding between the O of C=O groups and the H of the N-H groups in the polypeptide chains.

c. Tertiary structure. The distinctive and characteristic conformation or shape of a protein molecule.

21. The sulfur-containing amino acid, cysteine, has the special role in protein structure of creating disulfide bonding between polypeptide chains which helps control the shape of the molecule.

22. The major structural difference between hemoglobin and myoglobin is that hemoglobin is composed of four sub-units while myoglobin only contains one. Hemoglobin's quaternary structure allows a more effective control of oxygen transport than is possible with myoglobin.

23. Enzymes lower the energy barrier (activation energy) which controls the rate of biochemical reactions.

24. Enzymes are proteins. As noted in the answer to question 23, they serve as catalysts for chemical reactions in the body.

25. A coenzyme is the nonprotein part of a conjugated enzyme. An apoenzyme is the protein part of a conjugated enzyme.

26. Sucrase catalyzes the hydrolysis of sucrose; lactase catalyzes the hydrolysis of lactose; and maltase catalyzes the hydrolysis of maltose.

27. Enzymes function as catalysts by locking onto the substrate, forming a substrate-enzyme complex unit. While locked together, the substrate, activated by the enzyme in the complex, reacts to form the products, and breaks away, regenerating the enzyme. Each enzyme has a specific structure which allows it to catalyze a specific reaction. In general, a different structure is required for each different reaction.

28. The six general classes of enzymes are: (a) oxidoreductases, (b) transferases, (c) hydrolyases, (d) lyases, (e) isomerases, and (f) ligases.

29. A K_m is the substrate concentration necessary to achieve half the maximum velocity in an enzyme-catalyzed reaction. As such, the K_m is a measure of an enzyme's attraction for substrate.

30. The V_{max} is a measure of an enzyme's maximum catalytic ability. Enzyme B is a more effective catalyst than enzyme A because it shows a larger V_{max} when the two enzymes are studied under identical conditions.

31. An enzyme inhibitor binds to the enzyme and blocks a catalyzed reaction while an enzyme substrate binds to the enzyme and undergoes a catalyzed reaction.

32. Proteins tend to be easily denatured because their three dimensional structure is bonded together by many, weak noncovalent bonds.

33. Hydrolysis breaks the peptide bonds, thus disrupting the primary structure of the protein. Denaturation involves alteration or disruption of the secondary, tertiary, or quaternary but not of the primary structure of proteins.

34. When a protein is hydrolyzed to amino acids, one water molecule is added for each peptide bond that is broken. A hydrogen atom attaches to the nitrogen to complete the amino group, and the OH group attaches to the carboxyl carbon.

35. When proteins are partially hydrolyzed, small groupings of amino acids are formed--dipeptides, tripeptides, etc.

36. Trypsin and chymotrypsin are both enzymes involved in the metabolism of protein. The enzyme trypsin hydrolyzes the polypeptide chain on the carboxyl side of arginine and lysine. Chymotrypsin hydrolyzes the polypeptide chain on the carboxyl side of tryrosine, phenylalanine, and tryptophan.

37. The visible evidence observed in the:

    a. Xanthoproteic reaction is a yellow-colored reaction product.

    b. Biuret test is a violet color.

    c. Ninhydrin test is a blue solution with all amino acids except proline and hydroxyproline, both of which produce a yellow solution.

38. The expected results if Thr-Ala-Gly is reacted with:

    a. Sanger's reagent: would show that the terminal amino group was on Thr, threonine.

    b. Concentrated HNO_3: would not produce the yellow color because none of these amino acids contains a benzene ring.

    c. Ninhydrin: there would be a reaction, giving a blue solution.

39. Amino acids containing a benzene ring give a positive xanthoproteic test (formation of yellow-colored reaction products). Among the common amino acids, these would include phenylalanine, tryptophan, and tyrosine.

40. a. Thin layer chromatography is a way of separating substances based on a differential distribution between two phases, the liquid phase and the solid phase.

    b. A strip of sheet is prepared with a thin coating of alumina or other adsorbent. A tiny spot of solution containing a mixture of amino acids is placed near the bottom of the strip. After the spot dries, the bottom edge of the strip is placed in a suitable solvent. The solvent rises in the strip, carrying the different amino acids upwards at different rates. When the solvent front nears the top, the strip is removed from the solvent and dried.

    c. Ninhydrin is the reagent used to locate the different amino acids on the strip.

41. Protein column chromatography uses a column packed with polymer beads (solid phase) through which a protein solution (liquid phase) is passed. Proteins separate based on differences in how they interact with the solid phase. The proteins move through the column at different rates and can be collected separately.

42. Given the decapeptide Ala-Tyr-Gly-Gln-Lys-Phe-Gly-Arg-Ser-Gly, these fragments will be obtained by hydrolysis with:

    a. pepsin: attacks the amino end of Tyr and Phe, so gives fragments of Ala; Tyr-Gly-Gln-Lys; and Phe-Gly-Arg-Ser-Gly.

    b. trypsin: hydrolyzes on the carboxyl side of Arg and Lys, so gives fragments of Ala-Tyr-Gly-Gln-Lys; Phe-Gly-Arg; and Ser-Gly.

c. chymotrypsin: hydrolyzes the polypeptide chain on the carboxyl side of tyrosine, phenylalanine, and tryptophan, so gives fragments of Ala-Tyr; Gly-Gln-Lys-Phe; and Gly-Arg-Ser-Gly.

d. carboxypeptidase: hydrolyzes C-terminal amino acids so gives fragments of Gly and Ala-Tyr-Gly-Gln-Lys-Phe-Gly-Arg-Ser.

e. aminopeptidase: hydrolyzes N-terminal amino acids, so gives fragments of Ala and Tyr-Gly-Gln-Lys-Phe-Gly-Arg-Ser-Gly.

f. Sanger's reagent: ties up the N-terminal amino acid and hydrolyzes it from the remainder, so gives a fragment of Tyr-Gly-Gln-Lys-Phe-Gly-Arg-Ser-Gly.

g. Edman's reagent: splits up the chain one acid at a time, so given sufficient time, would give fragments of the eight amino acids in the molecule (Gly appears three times).

43. The steroisomers of threonine:

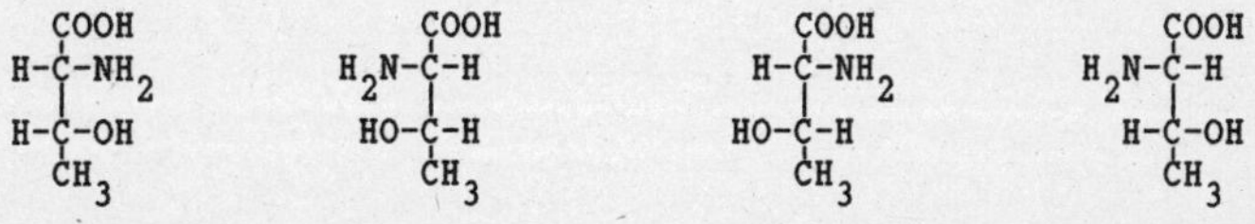

44. A single Edman degradation of the tripeptide, Ala-Leu-Gly would follow the sequence:

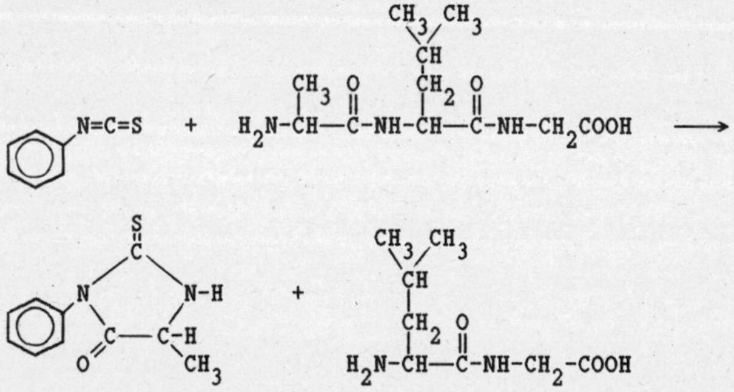

45. The sequence of amino acids in these two heptapeptides is:

a. Phe-Ala-Gly-Phe-Leu-Ala-Tyr
b. Phe-Ala-Ala-Leu-Phe-Gly-Tyr

46. The amino acid sequence in this nonapeptide is:
Arg-Pro-Pro-Gly-Phe-Ser-Pro-Phe-Arg

47. $$6.0 \text{ g nitrogen} \times \frac{100 \text{ g protein}}{16 \text{ g nitrogen}} \times \frac{1}{100 \text{ g food product}} \times 100\%$$

= 38% protein

48. $\frac{4 \text{ Fe atoms}}{\text{molecule}} \times \frac{55.8 \text{ amu}}{\text{Fe atom}} \times \frac{100 \text{ amu hemoglobin}}{0.33 \text{ amu Fe}} = 6.8 \times 10^4$

Thus, the molecular weight of hemoglobin is 6.8×10^4

49. The correct statements are: b, c, d, f, g, i, m, n, p, and t. The others are incorrect for these reasons:

a. Proteins are needed for growth and/or maintenance of body tissue.

e. Glycine does not have an asymmetric carbon atom.

h. A compound with three amino acids is known as a tripeptide. The name is not based on the number of peptide linkages, but on the number of amino acids linked.

j. The primary structure of a protein is the number, kind, and sequence of amino acids making up the protein.

k. A protein's primary structure is the protein's sequence of amino acids and helps to determine the protein's tertiary structure (the overall three dimensional arrangement of the polypeptide chain).

l. The amino acid residues in a peptide chain are numbered beginning with the N-terminal amino acid.

o. Enzymes decrease the activation energy.

q. The substance acted on by the enzyme is called the substrate.

r. Denaturation involves the destruction of a protein's native three dimensional structure but does not cause the peptide bonds to break.

s. Denaturation often involves irreversible coagulation or precipitation of proteins, but this does not always occur in denaturation.

B. 1. The five nitrogen bases found in nucleotides:

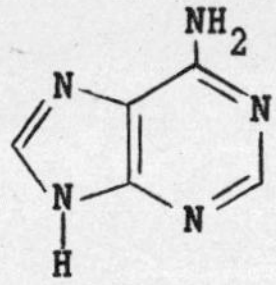

Adenine

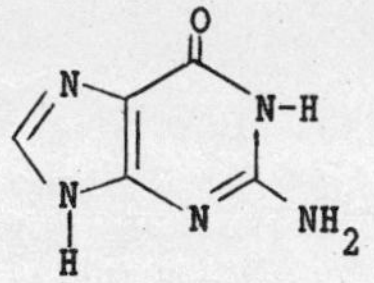

Guanine

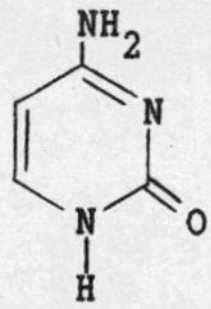

Cytosine

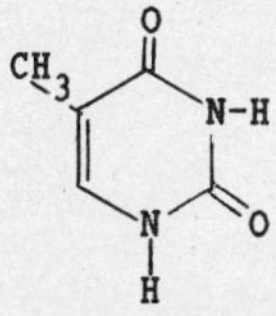

Thymine

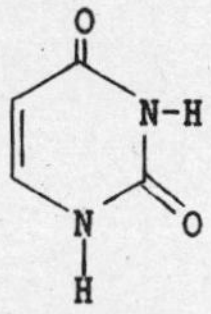

Uracil

2. A nucleoside is a purine or pyrimidine base linked to a sugar molecule, usually ribose or deoxyribose. A nucleotide is a purine or pyrimidine base linked to a ribose or deoxyribose sugar which in turn is linked to a phosphate group.

3. Identify the compound represented by the following letters:

   a. A, adenosine; AMP, adenosine-5'-monophosphate; ADP, adenosine-5'-diphosphate; ATP, adenosine-5'-triphosphate.

   b. G, Guanosine; GMP, guanosine-5'-monophosphate; GDP, guanosine-5'-diphosphate; GTP, guanosine-5'-triphosphate.

4. The three units making up a nucleotide are one molecule each of phosphoric acid, a sugar (ribose or deoxyribose) and one of the five nitrogen bases.

5. Structural formulas

   a. A

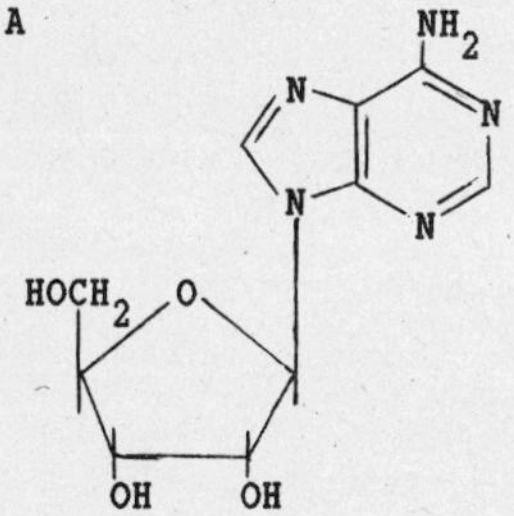

   b. AMP

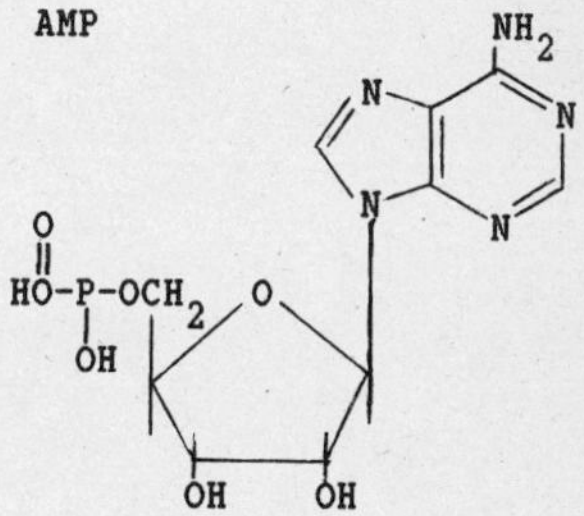

c. ADP

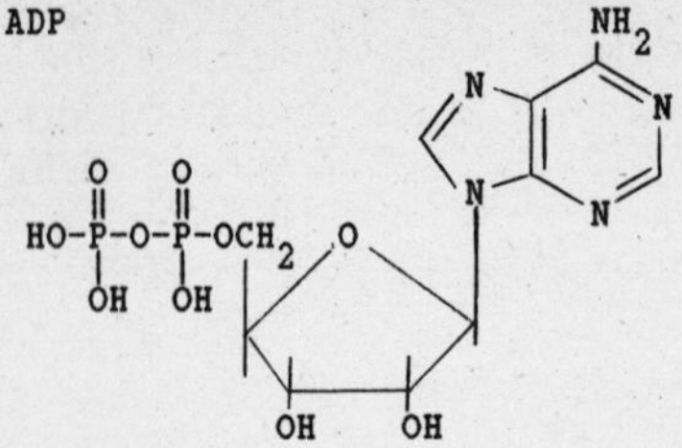

d. ATP

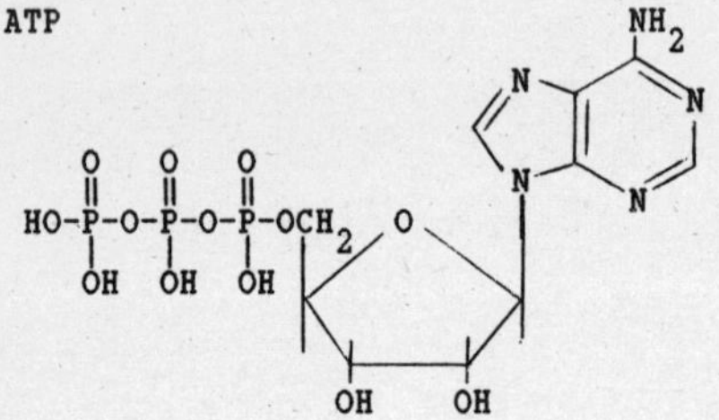

e. dGTP

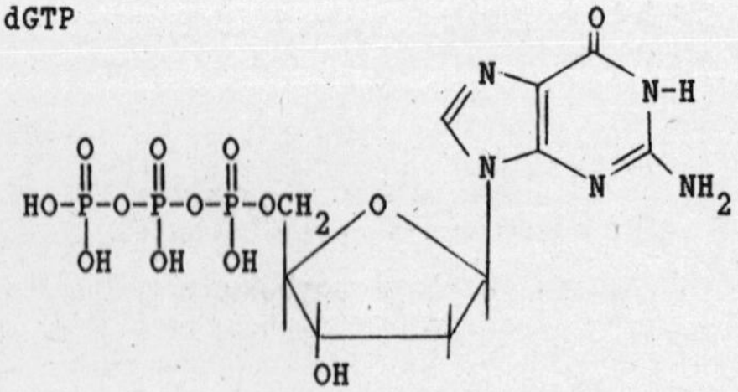

f. CDP

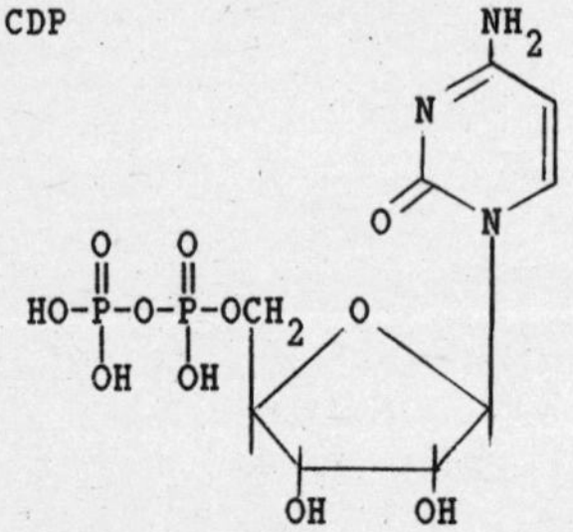

6. The major function of ATP in the body is to store chemical energy, and to release it when called upon to carry out many of the complex reactions that are essential to most of our life processes. An equilibrium exists with ADP.

$$\text{ATP} + H_2O \underset{\text{energy storage}}{\overset{\text{energy utilization}}{\rightleftharpoons}} \text{ADP} + P_i + 7.3\ \text{kcal}$$

7. There are three structural differences between DNA and RNA.

   a. In RNA the sugar molecule is always ribose. In DNA, the sugar molecule is always deoxyribose, which has H instead of OH at carbon number two.

   b. Both molecules use a mixture of four nitrogen bases. Both use cytosine, adenine, and guanine. In DNA, the fourth base is thymine. In RNA, the fourth base is uracil.

   c. DNA exists as a double helix whereas RNA is a single strand of nucleotides.

8. A three-nucleotide segment of RNA.

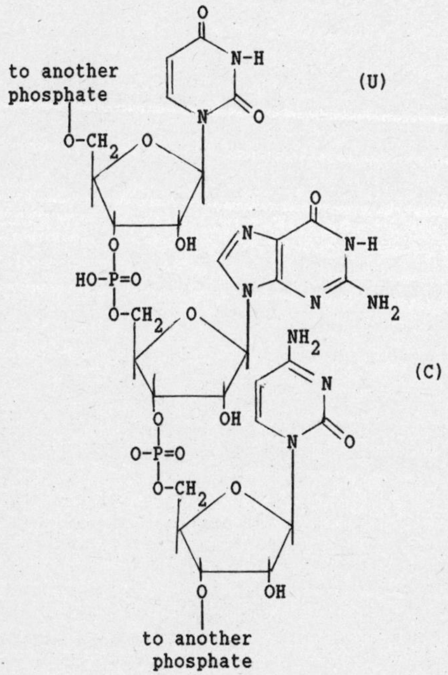

9. The hydrogen bonding between adenine and uracil.

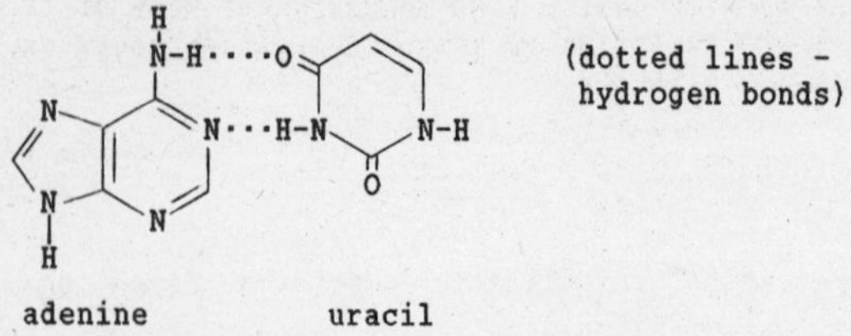

10. The structure of DNA as proposed by Watson and Crick is a double-stranded helix. Each strand has a backbone of alternating phosphate and deoxyribose molecules. Each deoxyribose molecule has one of the four nitrogen bases attached, but coming off the backbone, not part of the backbone. These nitrogen bases thus link to their complementary nitrogen base on the other strand of the double helix.

11. Complementary bases are the pairs that "fit" to hydrogen bond to each other between the two helixes of DNA. For DNA the complementary pairs are thymine with adenine, and cytosine with guanine, or T-A and C-G. For RNA it is U-A and C-G.

12. Thymine and adenine are complementary bases in DNA; therefore each time one appears, its complement appears. There is no fixed relationship between the amount of thymine and the amount of guanine, since they are not complementary. Table 30.2 shows that the one set of bases is often far different from the other set, varying by as much as 3:2 ratio. The complementary bases "fit" with each other with respect to H-bonding between them; also with respect to the size of the molecules.

13. DNA is considered to be the genetic substance of life, because it contains the sequence of bases that carries the code for genetic characteristics.

14. In the replication process of DNA: (a) The double helix unwinds, separating at the point of the hydrogen bonds. (b) Each strand then serves as a template to combine with the proper nucleotides (each nitrogen base can only pair with the proper nitrogen base) to duplicate the missing half of the original double helix.

15. The chromosomes, which are in the nucleus of each cell, contain strings of genes, or DNA molecules. The sequence of nucleotides in the DNA molecules is the code for reproducing the species, and the replication process assures that as cells divide, each new cell will contain an exact copy of the original genes.

16. The genetic code is the sequence of nitrogen bases in a strand of DNA. The code determines the sequence of amino acids in protein molecules which will be assembled according to the code. It takes three nucleotides to make a codon, which determines one amino acid in a sequence.

17. Since there are only four different bases to make up the code, one nucleotide could only specify four possible amino acids; two nucleotides could specify 16 amino acids; 3 nucleotides could specify 64 amino acids. Since there are at least 20 amino acids needed, three nucleotides are required.

18. The three kinds of RNA are:

    a. Ribosomal RNA (rRNA) is found in the ribosomes where it is associated with protein roughly in the proportion 62% protein, 33% rRNA.

    b. Messenger RNA (mRNA) carries genetic information from DNA to the ribosomes. It is a template made from DNA and carries the codons that direct the synthesis of proteins.

    c. Transfer RNA (tRNA) is used to bring amino acids to the ribosomes for incorporation into protein molecules.

19. An oncogene codes for a protein which can cause a cell to become cancerous when this gene is transcribed too often.

20. A codon is a triplet of three nucleotides, and each codon specifies one amino acid. The cloverleaf model of transfer RNA has an anticodon loop consisting of seven unpaired nucleotides. Three of these make up the anticodon, which is complementary to, and hydrogen-bonds to the codon on mRNA.

21. When the tRNA molecule brings its correct amino acid to the ribosome site, the anticodon of the tRNA links up to the complementary codon on the mRNA molecule attached to the ribosome. Thus each amino acid has been brought to the ribosome in the proper order.

22. If there are 146 amino acid residues in the beta chain of hemoglobin, then the number of required nucleotides in mRNA is $3 \times 146 = 438$.

23. In RNA the guanine content does not have to be equal to the cytosine content, because RNA is a single strand. Its complements are on the DNA template, not on the RNA. They have to be equal in DNA, because it is a double helix, with the complements together.

24. A brief outline of the biosynthesis of proteins:

    a. A DNA strand produces a complementary mRNA strand which leaves the nucleus and travels to the cytoplasm where it becomes associated with a cluster of ribosomes, binding to five or more ribosomes.

    b. With the aid of an enzyme, the proper amino acid attaches to a tRNA molecule by an ester linkage.

    c. The amino acids are brought to the synthesis site by tRNA.

d. The beginning is always an mRNA codon AUG, which ties to the tRNA anticodon UAC. This code brings N-formylmethionine for procaryotic cells. The amino group is blocked by the formyl group, leaving the -C=O group available to react with the amino group of the next amino acid.

e. The next tRNA, bringing an amino acid, comes in to the mRNA and links up, anticodon to codon. The peptide linkage is then made between amino acids.

f. The first tRNA is ejected, and a third enters the ribosome.

g. The polypeptide chain terminates at a nonsense or termination-codon. The protein molecule breaks free.

25. The role of N-formylmethionine in procaryotic protein synthesis is to start the polypeptide chain so it goes in the right direction. It can only build from the carboxyl end. After the synthesis it breaks loose from the protein.

26. A segment of DNA strand consists of GCTTAGACCTGA.

a. The order in the complementary mRNA will be: CGAAUCUGGACU.

b. The anticodon order in tRNA will be: GCUUAGACCUGA.

c. The sequence of amino acids coded by the DNA is Arg-Ile-Trp-Thr.

27. What will the anticodon be in tRNA if the codon in mRNA is: (a) GUC, CAG; (b) ACC, UGG; (c) CGA, GCU: (d) UUU, AAA; (e) CCA, GGU.

28. Complete hydrolysis of RNA would yield ribose, phosphoric acid, uracil, adenine, guanine, and cytosine.

29. From time to time a new trait appears in an individual that is not present in either parents or ancestors. These traits which are generally the result of genetic or chromosomal changes are called mutations.

30. Mutations occur because of mutagens, agents capable of causing mutations. These might be chemical agents, or radiation such as X-rays, cosmic rays, and ultraviolet rays. A mutagen may alter genetic material by causing a chromosome or chromosome fragment to be added or removed. A common type of alteration is the substitution of one purine or pyrimidine for another, thus causing misinformation in the genetic code.

31. The Ames Test uses a special strain of bacteria to check the mutagenicity of various substances. By quickly identifying mutagens, this test provides scientists with an early warning of potential carcinogens.

32. DNA damage does not always result in a mutation. Each normal cell contains enzymes which function to excise and replace damaged DNA. These repair processes occur continuously in normal cells.

33. Often, the basic steps in genetic engineering follow this sequence: (a) the double stranded DNA is split at specific locations by restriction endonucleases; (b) a new segment of DNA is inserted and covalently attached to the original DNA; (c) the newly modified DNA is introduced into living cells.

34. The correct statements are: a, b, d, g, h, j, m, p, and r. The others are incorrect for these reasons:

c. Only DNA is responsible for transmitting genetic information from parent to daughter cells.

e. Codons are combinations of base units in an mRNA molecule.

f. The double helical structure of DNA is held together by hydrogen bonds between nitrogen bases.

i. The letters ATP stand for adenosine triphosphate.

k. Messenger RNA is complementary to a transcribed section of DNA.

l. These are not the pairs of bases complementary to each other. See statement 34.b. for the correct pairings.

n. Humans have 23 pairs, or 46 chromosomes.

o. The sperm and egg cells are produced by meiosis so that they each have half of the normal chromosome number and, after they join, the fertilized cell again will have 46 chromosomes.

q. Messenger RNA carries the code for the synthesis of proteins.

s. DNA contains deoxyribose, not ribose.

t. RNA's are single stranded but they do not always form a single helix conformation. For example, ribosomal RNA assumes a shape which is determined by the structure of the ribosome.

B. 1. Milk contains (a) lipids, (b) minerals (calcium), (c) water and (d) carbohydrates.

2. The three classes of nutrients which do not commonly supply energy for the cells are (a) vitamins, (b) minerals and (c) water.

3. The "energy allowance" represents the recommended dietary energy supply needed to maintain health.

4. Marasmus is a chronic calorie deficiency while kwashiorkor represents a protein deficiency condition.

5. 9 kg of body fat represents about 70,000 kcal of energy. Complete starvation for about 30 days would cause a net loss of 9 kg of fat. Thus, this new diet is unreasonable.

6. Candy contains primarily sucrose, a source of calories, but little or no other nutrients. Thus, candy is said to provide "empty calories".

7. Cellulose is important as dietary fiber. Although it is not digested, it provides dietary bulk which helps maintain a healthy digestive tract.

8. Fat (9 kcal/g) contains more than twice as much energy as compared with carbohydrate (4 kcal/g). Thus, a tablespoon of butter will greatly increase the calorie content of a medium baked potatoe.

9. Essential fatty acids are required for normal growth and development. On a fat-free diet many deleterious physiological changes occur, including skin lesions, kidney damage, poor growth and, impaired fertility.

10. The essential fatty acids have 18 or more carbon atoms per molecule and are polyunsaturated.

11. There are eight essential amino acids. They are isoleucine, leucine, lysine, methionine, phenylalanine, threonine, tryptophan and, valine.

12. Animal proteins generally are a source of all twenty common amino acids while specific vegetable proteins may be deficient in one or several of the amino acids.

13. The fat-soluble vitamins include A, D, E and K. The water-soluble vitamins are C and the B-complex.

14. The B-complex vitamins are water-soluble and can be leached from vegetables upon cooking.

15. Many vitamins act as coenzymes. In addition, vitamin A is an important component of vision and vitamin D acts to control calcium metabolism.

16. Major elements are needed in relatively large quantities by the body while only small amounts of the trace elements are required.

17. Calcium is a major constituent of bones and teeth and is also important in nerve transmission among other functions.

18. Both vitamins and trace elements are needed in only small amounts by the body and cannot be synthesized within the body's cells. Vitamins are organic molecules and are often modified by the body. Trace elements are inorganic and, in general, are not modified before use.

19. Solid foods provide about 30-40% of the average daily water consumption while liquids yield about another 50% of the total.

20. When a food label contains a nutrition claim (or the food contains an added nutrient) the following information must be included on the label; serving or portion size; servings or portions per container; calorie content per serving; protein, grams per serving; carbohydrate, grams per serving; fat, grams per serving; proteins, vitamins, and minerals as percentages of the USRDA.

21. Some common categories of food additives are: (1) nutrients; (2) preservatives; (3) anticaking agents; (4) emulsifiers; (5) thickeners; (6) flavor enhancers; (7) colors; (8) nonsugar sweeteners.

22. The U.S. Food and Drug Administration is responsible for regulating the use of food additives.

23. The foods of higher animals must be digested before they can be utilized because foods are primarily large molecules. Foods must be broken down to much smaller molecules in order to pass through the intestinal walls into the blood and lymph systems where they can be utilized in the metabolic process of animals.

24. The five principal digestive juices are: saliva, gastric juice, pancreatic juice, bile, and intestinal juice.

25. Chyme is the material of liquid consistency found in the stomach consisting of food particles reduced to small size and mixed with gastric juices.

26. The pH values of the digestive juices are: (a) saliva, 6.6 to 6.8; (b) gastric juice, 1.5 to 2.5; (c) pancreatic juice, 7.5 to 8.0; (d) bile, 7.1 to 7.7.

27. Enzymes present in the digestive juices are: saliva: salivary amylase; gastric juice: pepsin, and gastric lipase; pancreatic juice, trypsinogen, chymotrypsinogen, procarboxypeptidase, amylopsin, steapsin; bile: no enzymes; intestinal juice: sucrase, maltase, lactase, aminopeptidase, dipeptidase, nucleases, phosphatase, and intestinal lipase.

28. a. Carbohydrates are digested in the mouth and small intestine.
    b. Fats are digested in the small intestine.
    c. Proteins are digested in the stomach and small intestine.

29. The absorption of food occurs only in the small intestine.

30. The end products of carbohydrate digestion are monosaccharides. They are commonly glucose, fructose, and galactose.

31. The digestive function of the liver is the production of bile, one of the important digestive juices. The bile is stored in the gall bladder.

32. The principal digestive enzymes that act on proteins are pepsin, trypsin, chymotrypsin, carboxypeptidase, and aminopeptidase.

33. Intestinal mucosal cells produce many enzymes needed to complete digestion, i.e. disaccharidases, aminopeptidases, and dipeptidases.

34. The liver takes up foreign compounds which are then oxidized by enzymes on the endoplasmic reticulum.

35. The correct statements are: a, d, e, g, i, j, l, and m. The others are not correct for these reasons:

    b. Specific vegetable proteins may be deficient in one or several amino acids.

    c. Oleic acid, which contains only one double bond, is not an essential fatty acid.

    f. Vitamins are organic nutrients in contrast to trace elements which are inorganic.

    h. Calcium is an important major element.

    k. Most digestion occurs in the small intestine.

    n. Bile acids are needed to emulsify fats not carbohydrates.

    o. The liver oxidizes toxic substances, it does not reduce them.

B. 1. As more carbon dioxide is formed, it will react with water to form carbonic acid.

$$H_2O + CO_2 \longrightarrow H_2CO_3$$

Carbonic acid will create a more acidic environment for the muscle tissue.

$$H_2CO_3 \longrightarrow H^+ + HCO_3^-$$

2. Rapid breathing expels more carbon dioxide causing more bicarbonate to react with hydrogen ions

$$HCO_3^- + H^+ \longrightarrow H_2CO_3$$

$$H_2CO_3 \longrightarrow H_2O + CO_2(g)$$

Thus, the blood will become less acidic and the pH will rise.

3. Hemoglobin contains amine groups which can react with carbon dioxide to form carbamino ions,

$$R\text{-}NH_2 + CO_2 \longrightarrow R\text{-}NH\text{-}CO_2^- + H^+$$

4. A catabolic pathway is destructive causing large molecules to be broken down, commonly with oxidation, to produce energy. An anabolic pathway synthesizes complex substances from simple substances.

5. Fats and carbohydrates are good sources of cellular energy because they contain many reduced carbon atoms.

6. The mitochondria oxidize larger molecules to carbon dioxide and produce energy for the cell.

7. The mitochondria carry out much of the oxidation-reduction reactions for the cell. Because these reactions involve electron movement, they are more easily controlled in a non-aqueous environment. Thus, mitochondria contain 90% membrane by weight.

8. Higher plant cells need mitochondria as well as chloroplasts in order to oxidize energy storage molecules to provide for the cellular energy needs.

9. The most common high energy phosphate bond in the cell is the phosphate anhydride bond.

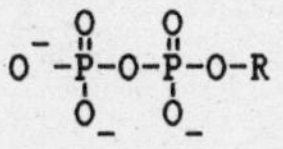

It is generally associated with adenosine triphosphate (ATP).

10. ATP is known as the "common energy currency" of the cell because energy from many different catabolic processes is funneled into this molecule. In turn, most anabolic pathways draw energy from this common pool.

11. For each mole of ATP converted to ADP, 35 kJ of energy ar released. Thus, a conversion of 3 moles of ATP will produce 3 × 35 kJ or 105 kJ.

12. An oxidation-reduction coenzyme is a reusable organic compound which helps an enzyme carry out an oxidation-reduction reaction.

13. The nicotinamide ring of NAD^+ becomes reduced.

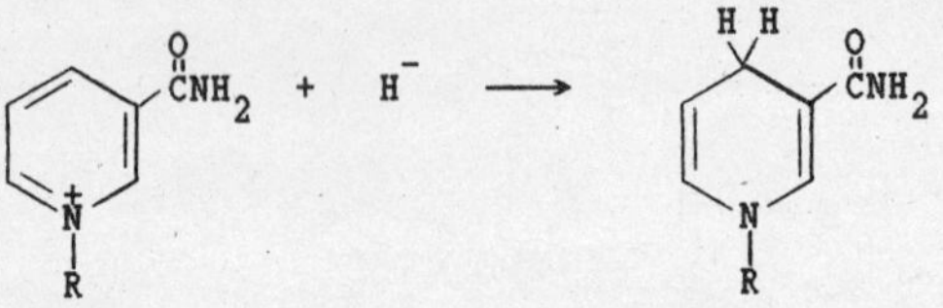

NAD^+ NADH

14. Oxidative phosphorylation uses the mitochondrial electron transport system to directly produce ATP from oxidation-reduction reactions. Substrate-level phosphorylation does not use the electron transport system but, instead, involves transfer of a phosphate group from a substrate to ADP.

15. Oxidative phosphorylation takes place in the mitochondria.

16. During electron transport and oxidative phosphorylation, each NADH will give rise to three ATPs while each $FADH_2$ will allow formation of two ATPs. Thus, 2NADH and $3FADH_3$ will provide for formation of 12 ATPs.

17. Since phosphoenolpyruvate contains only one high energy phosphate bond, substrate-level phosphorylation can produce only one ATP.

18. Both chloroplasts and mitochondria are organelles which are bounded by two membranes. In both cases they contain much, folded internal membrane (which facilitates oxidation-reduction reactions).

19. The chloroplast electron transport causes the reduction of coenzymes with formation of oxygen gas. The mitochondrial electron transport consumes oxygen gas while oxidizing coenzymes.

20. Chloroplast pigments trap light to provide energy for photosynthesis.

21. The overall photosynthetic reaction in higher plants is as follows:

$$6\ CO_2 + 6\ H_2O + 2820\ kJ\ (673\ kcal) \longrightarrow C_6H_{12}O_6 + 6\ O_2$$

22. Photosynthesis starts by removing electrons from water. These electrons are finally used to reduce $NADP^+$.

23. The correct statements are: a, c, e, f, h, i, j, k and l. The others are not correct for these reasons:

    b. Anabolic processes form complex biological substances from simpler compounds.

    d. As a typical carbohydrate carbon is less reduced than a typical fatty acid carbon, the carbohydrate carbon will supply less energy.

    g. Carbon dioxide contains highly oxidized carbons and will supply no cellular energy.

    m. $NADP^+$ is reduced during photosynthesis.

B. 1. Carbohydrates are considered energy storage molecules because they contain biologically-usable, reduced carbons.

2. Glucose catabolism breaks down glucose to the smaller carbon dioxide in an oxidative process. This produces energy for the cell and is catabolic. Photosynthesis produces glucose from carbon dioxide via a reductive path. Light energy is required and this is an anabolic process.

3. Each mole of glucose will yield 2820 kJ of energy when oxidized. Thus, 3 moles of glucose will provide 3 moles × 2820 kJ/mole or 8460 kJ.

4. Enzymes are needed so that biochemical reactions will proceed fast enough to keep the cell alive. Metabolism is often controlled by controlling enzyme activity.

5. Glycogen is the major storage carbohydrate which is found in the liver.

6. Glycogenolysis breaks down glycogen to form glucose.

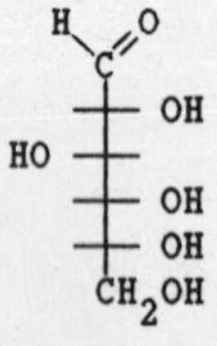

D-glucose

7. The Embden-Meyerhof pathway uses substrate-level phosphorylation to produce ATP. That is, following carbon oxidation a phosphate group is transferred from a substrate to ADP, forming ATP.

8. Although glucose is oxidized, molecular oxygen (O_2) is not used in the Embden-Meyerhof pathway. Thus, this path is anaerobic.

9. The end products of anaerobic glucose catabolism are lactic acid in the muscle tissue and ethanol plus carbon dioxide in yeast cells.

10. The final reactions in anaerobic glucose metabolism convert NADH back to NAD^+. This coenzyme can then be reused to oxidize more glucose.

11. The Embden-Meyerhof pathway is defined as catabolic because (a) glucose is broken down to smaller compounds, (b) the carbons of glucose are oxidized and, (c) this pathway causes production of ATP.

12. Each glucose yields a net 2 ATPs directly through the Embden-Meyerhof pathway. Thus, three glucose molecules will provide six ATPs.

13. $CH_3CH(OH)COOH$

lactic acid

All three carbons of lactic acid are more reduced than the carbon in CO_2. Thus, lactic acid can provide more cellular energy upon further oxidation of the carbons.

14. One high energy phosphate bond (in GTP) is formed directly in the citric acid cycle. In addition, one $FADH_2$ and three NADHs are produced.

15. Electron transport and oxidative phosphorylation are needed to convert the reduced coenzyme products of the citric acid cycle into oxidized coenzymes with ATP formation.

16. The anaerobic metabolic sequence produces two molecules of ATP per molecule of glucose. The aerobic sequence produces 32 molecules of ATP per molecule of glucose.

17. The citric acid cycle requires free oxygen. Large amounts of free oxygen were probably not available on earth until photosynthetic organisms had evolved.

18. Acetyl-CoA is an acetyl group bonded to a coenzyme A group. Coenzyme A contains adenine, ribose, diphosphate, pantothenic acid, and thioethanolamine. In the citric acid cycle, the acetyl group enters the cycle, leaving the coenzyme. In the process of the cycle, the acetyl group is oxidized to two molecules of carbon dioxide. In the process, energy is transferred, creating 12 ATP molecules. Acetyl-CoA serves to tie metabolism of carbohydrates, fats and certain amino acids to the citric acid cycle.

19. Hormones are chemical substances that act as control or regulatory agents in the body. Hormones are secreted by the endocrine, or ductless glands directly into the bloodstream and are transported to various parts of the body to exert specific control functions.

20. Hormones function as chemical messengers. They are chemical substances that act as control or regulatory agents in the body. Enzymes are catalysts for specific reactions, allowing these reactions to occur faster and under milder conditions than would otherwise be possible.

21. (a) The glucose concentration in blood under normal fasting conditions is about 70 to 90 milligrams/100 mL of blood.

    (b) The blood glucose concentrations considered as (1) hyperglycemic is 90 to 140 mg/100 mL of blood; and (2) hypoglycemic is 50 to 70 mg/100 mL of blood.

22. The renal threshold is the concentration of a substance in the blood above which the kidneys begin to excrete that substance.

23. Normal fasting blood sugar level is the sugar content of the blood when the body has had no nourishment for several hours, as in the morning before breakfast.

24. The presence of glucose in the urine does not establish that the condition of diabetes mellitus is present, though it strongly indicates that it may be present. If food particularly high in carbohydrate has been eaten recently, it is possible for the glucose level in the blood to exceed the renal level before the control mechanism can transform the excess glucose into glycogen.

25. The body maintains blood glucose concentrations within certain definite limits by the control of three hormones. If blood glucose concentration is high, insulin stimulates glycogenesis, the formation of glycogen from glucose. If blood glucose concentration is low, adrenalin (epinephrine) and glucagon stimulate glycogenolysis, the formation of glucose from glycogen.

26. Adrenalin is sometimes called the emergency or crisis hormone because it stimulates glycogenolysis, which raises the concentration of glucose in the blood, giving the body additional energy for an emergency or crisis situation.

27. When blood glucose concentration rises above acceptable levels, insulin stimulates glycogenesis, the production of glycogen from glucose. When blood glucose concentration drops below an acceptable level, adrenalin stimulates glycogenolysis, the hydrolysis of glycogen to glucose.

28. If a large overdose of insulin was taken by accident, the blood glucose concentration would drop to a low level, probably in the hypoglycemic range. This could result in fainting, convulsions, and unconsciousness.

29. Insulin is not effective when given orally because it is a protein, and would be hydrolyzed to amino acids in the gastro-intestinal tract.

30. The glucose tolerance test: After not eating for at least 12 hours, a fasting blood and urine specimen is taken for a beginning reference level. The patient then drinks a solution containing 100 grams of glucose. Blood and urine specimens are then collected at 0.5-, 1-, 2-, and 3-hour intervals and tested for glucose content. In a normal person, the blood glucose has returned to normal in three hours. Persons with mild diabetes show a slower drop in glucose levels. In a severe diabetic, the level stays high for the entire three hours.

31. The correct statements are: a, c, h, i, j, m, and n. The others are not correct for these reasons:

   b. Almost no biological reaction will proceed at a sufficient rate to maintain life in the absence of an enzyme (catalyst).

   d. Glycolysis refers to the anaerobic oxidation of glucose to produce lactic acid.

   e. The Embden-Meyerhof pathway uses substrate-level phosphorylation to form ATP.

   f. Lactic acid is used in gluconeogenesis and is formed by glycolysis.

   g. ATP is produced when the citric acid cycle, electron transport, and oxidative phosphorylation operate in concert.

   k. Hormones are secreted through the blood stream to control metabolism.

   l. A person with diabetes mellitus suffers from hyperglycemia or high blood sugar.

B. 1. A fatty acid contains many reduced carbons which can be oxidized to supply cellular energy.

2. It is possible to become obese despite a low fat diet because the body is capable of synthesizing fats from nonfat starting materials.

3. Knoop prepared a homologous series of straight chain fatty acids with a phenyl group at one end and a carboxyl group at the other end. He fed each of these acids to test animals and analyzed the urine of the animals for metabolic products of the acid. Those animals which had been fed acids with an even number of carbons excreted phenylaceturic acid. Those animals which had been fed acids with an odd number of carbon atoms excreted hippuric acid. The experiment showed that the metabolism of the acid must occur by shortening the chain two carbon atoms at a time.

4. In the oxidation of fatty acids, the carbon atom beta to the carboxyl group is oxidized forming a β-keto acid which is then cleaved between the α and β carbon atoms.

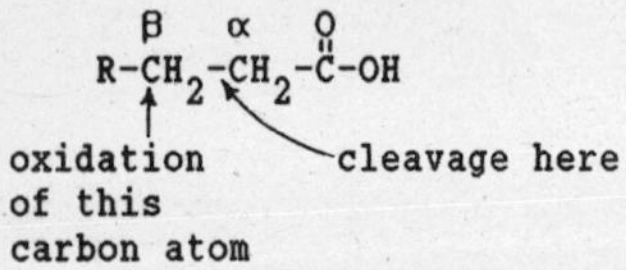

5. The oxidation of caproic acid to butyric acid:

a. Activation: formation of thioester with CoA

$$CH_3CH_2CH_2CH_2CH_2COOH + CoASH + ATP \longrightarrow$$

$$CH_3CH_2CH_2CH_2CH_2\overset{O}{\overset{\|}{C}}\text{-S-CoA} + AMP + 2\ P_i$$

b. Oxidation: dehydrogenation at C-2 and C-3 (α and β carbons)

$$CH_3CH_2CH_2CH_2CH_2\overset{O}{\overset{\|}{C}}\text{-S-CoA} \xrightarrow[FAD]{-2\ H} CH_3CH_2CH_2CH{=}CH\text{-}\overset{O}{\overset{\|}{C}}\text{-S-CoA}$$

c. Hydration: conversion to secondary alcohol

$$CH_3CH_2CH_2CH{=}CH\overset{O}{\overset{\|}{C}}\text{-S-CoA} \xrightarrow{H_2O} CH_3CH_2CH_2\overset{OH}{\overset{|}{C}}H\text{-}CH_2\text{-}\overset{O}{\overset{\|}{C}}\text{-S-CoA}$$

d. Oxidation: dehydrogenation of C-3 (β carbon) to keto group

$$CH_3CH_2CH_2\overset{OH}{\overset{|}{C}}H\text{-}CH_2\text{-}\overset{O}{\overset{\|}{C}}\text{-S-CoA} \xrightarrow[NAD]{-2\ H} CH_3CH_2CH_2\overset{O}{\overset{\|}{C}}\text{-}CH_2\text{-}\overset{O}{\overset{\|}{C}}\text{-S-CoA}$$

e. Carbon chain cleavage (β-cleavage)

$$CH_3CH_2CH_2\overset{O}{\overset{\|}{C}}-CH_2-\overset{O}{\overset{\|}{C}}-S-CoA \quad + \quad CoASH \longrightarrow$$

$$CH_3CH_2CH_2\overset{O}{\overset{\|}{C}}-S-CoA \quad + \quad CH_3\overset{O}{\overset{\|}{C}}-S-CoA$$

6. The acyl carrier protein (ACP) serves to hold the growing fatty acid chain until the fatty acid synthesis is complete.

7. Three unsaturated fatty acids, linoleic, linolenic, and arachidonic acids are essential in our diet. Fats are not required in our diet except as a source of these three fatty acids because fatty acids and body fat are continuously metabolized and synthesized in the body.

8. β-oxidation is a pathway with relatively few reactions, requiring FAD, NAD^+, ATP and, coenzyme A. This process forms no ATP directly but depends upon mitochondrial electron transport and oxidative phosphorylation to make use of the reduced coenzymes, NADH and $FADH_2$, and form ATP.

9. β-hydroxybutyric acid is a ketone body which does not contain a ketone group.

$$CH_3\underset{\underset{OH}{|}}{C}HCH_2COOH$$

10. Ketone bodies are relatively soluble in the blood stream and can be transported to energy-deficient cells in time of need. Ketone bodies supply energy via β-oxidation and the citric acid cycle.

11. Ketosis is the accelerated production of ketone bodies leading to high blood levels of these compounds with a corresponding increase in blood acidity.

12. β-oxidation occurs in the mitochondria while fatty acids are synthesized in the cytoplasm. In fatty acid synthesis, the growing molecule is linked to an acyl carrier protein while in β-oxidation the fatty acid being broken down is linked to a coenzyme A. Finally, fatty acid synthesis uses some new reactions and a new coenzyme ($NADP^+$) which are not found in β-oxidation.

13. In addition to being a food reserve, fat has two principal functions in the body. It acts as a cushion or shock absorber around internal organs, and it acts as an insulating blanket.

14. In the enzymatic oxidation of fatty acids from fats, the fatty acids are oxidized, ultimately forming acetyl-CoA. Acetyl-CoA enters the citric acid cycle to produce energy and/or ATP.

15. One molecule of glucose is converted to two molecules of pyruvic acid by the Embden-Meyerhof pathway. This pyruvic acid can then enter the citric acid cycle to form two molecules of acetyl-CoA.

    A six-carbon fatty acid will produce three molecules of acetyl-CoA and therefore will produce more ATP in the citric acid cycle than is produced by glucose.

16. The biosynthesis of fatty acids begins with a conversion of acetyl-CoA and malonyl-CoA to acetyl-ACP and malonyl-ACP. These molecules will combine to form a four carbon compound linked to ACP. After four more steps, this results in butyryl-ACP. This process can then be repeated again, reacting the butyryl-ACP with malonyl-ACP, which will add two more carbon atoms to the chain, etc.

17. Acetyl-CoA is considered to be an important, central intermediate in metabolism for the following reasons:
    a. Complete catabolism of most energy containing nutrients is achieved by conversion to acetyl-CoA and oxidation via the citric acid cycle.
    b. Fats and amino acids can be synthesized from acetyl-CoA.
    c. Carbohydrates, fats and amino acids can be converted to acetyl-CoA.
    d. Acetyl-CoA is the central intermediate in converting (1) carbohydrates to fats or amino acids, (2) fats to amino acids or, (3) amino acids to fats.

18. The conversion of starch to body fat:

$$\text{starch} \xrightarrow{\text{hydrolysis}} \text{glucose} \xrightarrow{\text{Embden-Meyerhof}} \text{acetyl-CoA}$$

$$\text{fats} \longleftarrow \text{fatty acids} \longleftarrow \text{(acetyl-CoA)}$$

19. Soybeans enrich the soil because their roots serve as host to the symbiotic, nitrogen-fixing bacteria.

20. The possible metabolic fates of amino acids are:
    a. incorporation into a protein;
    b. utilization in the synthesis of other nitrogenous compounds such as nucleic acids;
    c. deamination to a keto acid which can be (1) utilized to synthesize other compounds, or (2) oxidized for energy to carbon dioxide and water.

21. A negative nitrogen balance means more nitrogen is excreted than ingested. A low protein diet will further decrease the amount of nitrogen ingested and should cause the nitrogen balance to become more negative.

22. Transamination is the reaction which is central in metabolic nitrogen transfer. Each transamination involves transfer of an amino group from an α-amino acid to an α-keto acid. One common example of transamination is as follows:

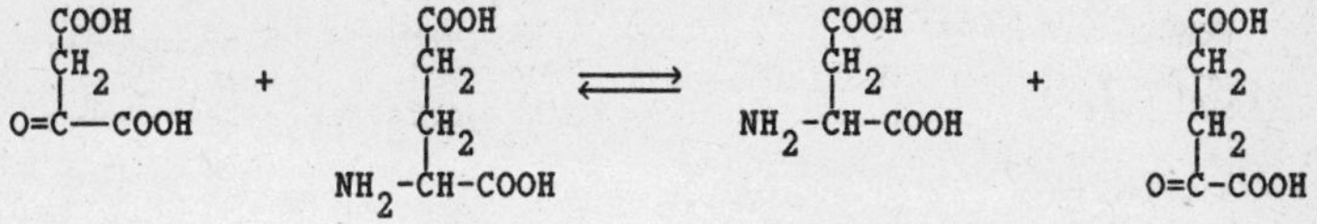

oxaloacetic acid + L-glutamic acid ⇌ L-aspartic acid + α-ketoglutaric acid

23. L-glutamic acid is involved in the majority of transaminations and is, thus, a central compound in nitrogen transfer. L-glutamic acid is also the major nitrogen source for the urea cycle.

24. The nitrogen in L-glutamine is less basic than ammonia and, thus, is less toxic to biological organisms.

25. L-aspartic acid is not a ketogenic amino acid because a dietary supply of L-aspartic acid does not give rise to ketone body formation.

26. a. L-alanine yields 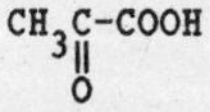

b. L-aspartic acid yields 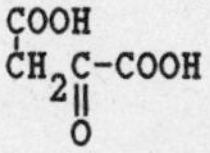

c. L-phenylalanine yields 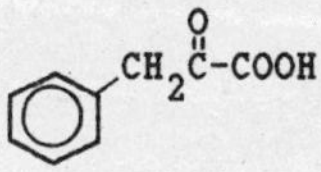

d. L-serine yields 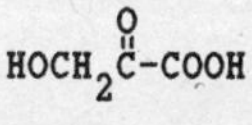

27. The structure of urea is 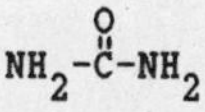

28. Three ATPs are used for every urea produced. The production of urea is important because this molecule provides a nontoxic means for excreting nitrogen.

29. The two common amino acids which contribute nitrogen to the urea cycle are:

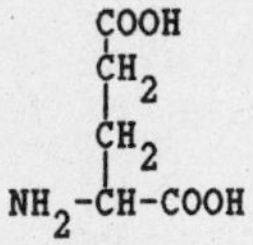

L-glutamic acid

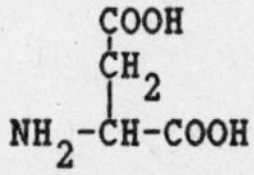

L-aspartic acid

30. The correct statements are: c, f, g, h, j, l, m, n, o, r, s, t, v. The others are not correct for these reasons:

a. Palmitic acid, as with all fatty acids, is a good source of stored metabolic energy because it contains many reduced carbons.

b. Fatty acids provide more energy than carbohydrates on a per-gram basis because fatty acids contain more carbons on a per-gram basis and these carbons are more reduced than those in carbohydrates.

d. Fatty acid oxidation can only continue if molecular oxygen is present to allow for mitochondrial electron transport and oxidative phosphorylation. Thus, fatty acid oxidation is aerobic.

e. β-oxidation directly produces no ATPs but depends on mitochondrial electron transport and oxidative phosphorylation to form ATP.

i. Lipogenesis is a process which forms lipids. Catabolism of fatty acids is the process which breaks down these compounds.

k. The acyl carrier protein is used only in fatty acid synthesis. Beta oxidation uses coenzyme A instead.

p. Nitrogen fixation occurs in lower organisms but is not carried out by higher plants.

q. A positive nitrogen balance means that more nitrogen is consumed than is excreted.

u. The acetyl group is linked to coenzyme A via a thioester bond in acetyl-CoA.

w. Urea is an especially nontoxic form of nitrogen because this compound is not a strong base.